8th Conference on Material Science and Engineering (UgalMat 2018)

IOP Conference Series: Materials Science and Engineering
Volume 485

Galati, Romania
11 - 13 October 2018

Editors:

Florentina Potecasu
Gheorghe Gurau
Mariana Busila

ISBN: 978-1-5108-8482-3
ISSN: 1757-8981

Printed from e-media with permission by:

Curran Associates, Inc.
57 Morehouse Lane
Red Hook, NY 12571

Some format issues inherent in the e-media version may also appear in this print version.

Printed by Curran Associates, Inc. (2018)

For permission requests, please contact the Institute of Physics
at the address below.

Institute of Physics
Dirac House, Temple Back
Bristol BS1 6BE UK

Phone: 44 1 17 929 7481
Fax: 44 1 17 920 0979

techtracking@iop.org

Additional copies of this publication are available from:

Curran Associates, Inc.
57 Morehouse Lane
Red Hook, NY 12571 USA
Phone: 845-758-0400
Fax: 845-758-2633
Email: curran@proceedings.com
Web: www.proceedings.com

Table of contents

Volume 485

8th Conference on Material Science and Engineering (UgalMat 2018)

11–13 October 2018, Galati, Romania

Accepted papers received: 16 January 2019
Published online: 27 March 2019

Preface

Papers

IOP Conf. Series: Materials Science and Engineering **485** (2019) 011001 doi:10.1088/1757-899X/485/1/011001

8th International Conference on Material Science & Engineering

PREFACE

F Potecasu, G Gurau*, M Busila

[1]Faculty of Engineering "Dunărea de Jos" University of Galati, Domnească Street, 47, RO-800008, Galati, Romania

gheorghe.gurau@ugal.ro

Abstract. This issue constitutes the Proceedings of the 8th International Conference on Material Science & Engineering, UGALMAT 2018. The conference, held October 11-13 2018 in Galati Romania, was organized by "Dunărea de Jos" University of Galati, Faculty of Engineering, Department of Materials Science and Engineering.

1. Preface to the UGALMAT 2018 Proceedings

UGALMAT is a traditional conference in the Romanian scientific landscape focused on advanced materials and technologies, nanomaterials and nanotechnologies. UGALMAT 2018 is in the eighth of the UGALMAT series; the first UGALMAT having been organised over sixteen years ago by the Department of Materials Science and Engineering, "Dunărea de Jos" University of Galati.

The aims of UGALMAT 2018 are to provide a platform for scientists, professors, research scholars and students, as well as industrial professionals to present their research in the filed of materials science, to exchange new ideas and to establish new relations for future collaboration.

UGALMAT 2018 was attended by over 100 delegates, including outstanding scientists, as well as young researchers from representative universities and research institutes from Romania, Europe and Asia who presented 96 scientific research works involving more than 300 authors. The scientific program consisted of 4 plenary lectures, 35 oral presentations and 57 poster presentations on the three symposia: ADVANCED MATERIALS AND TECHNOLOGIES (TMA 2018), CASTING, FROM TECHNIQUE RIGORS TO ART (ARTCAST 2018) and UGALNano8.

The present edition of the UGALMAT 2018 proceedings includes 28 papers accepted after careful peer-review processing of the submitted manuscripts. The selected papers present the newest findings related to metals and alloys, composite materials and structures, surface processes and surface characterization, mathematical modelling of material processing and material properties. Other papers address research topics in advanced materials such as thin films and submicron memory alloys.

We would like to thank the Scientific and Honorary Committee for their valuable assistance in organizing the scientific program. We would also like to thank the referees for their work which contributed to the high scientific standard of this issue of IOP Conference Series: Materials Science and Engineering.

IOP Conf. Series: Materials Science and Engineering **485** (2019) 011001 doi:10.1088/1757-899X/485/1/011001

We acknowledge financial support offered by ours sponsors: "Dunărea de Jos" University of Galati, ArcelorMittal Galaţi, Bureau Veritas România, S.C. Myosotis S.R.L. Galaţi, S.C. Company Med S.R.L. Galati, S.C. Rulmenţi S.A. Bârlad and S.C. Iprolam S.A. Bucureşti.
Finally, we would like to express our gratitude to all members of the Organizing Committee for their involvement and permanent availability to solve all the details of the conference.

Editors of UGALMAT 2018
Prof. Florentina Potecasu
Assoc.Prof. Gheorghe Gurau
Assist. Prof. Mariana Busila
"Dunărea de Jos" University of Galati, October 2018

IOP Conf. Series: Materials Science and Engineering **485** (2019) 011002 doi:10.1088/1757-899X/485/1/011002

Peer review statement

All papers published in this volume of *IOP Conference Series: Materials Science and Engineering* have been peer reviewed through processes administered by the proceedings Editors. Reviews were conducted by expert referees to the professional and scientific standards expected of a proceedings journal published by IOP Publishing.

IOP Conf. Series: Materials Science and Engineering **485** (2019) 012001 doi:10.1088/1757-899X/485/1/012001

Neural network architecture optimization by analyse the mechanical characteristics of civil engineering materials

C I Alecu[1], L Judele[2], M Movila[2], D Lepadatu[3]

[1]Senior scientific researcher at Institute of Economic and Social Research Gh. Zane - Iasi, Romanian Academy Branch
[2]Senior Lecturer at Technical University "Gh. Asachi" Iaşi, Faculty of Civil Engineering and Building Services, Romania
[3]Associate Professor at Technical University "Gh. Asachi" Iaşi, Faculty of Civil Engineering and Building Services, Romania

daniel.lepadatu@gmail.com

Abstract. Our proposal for this paper is to diversify the architecture of neural networks in order to optimize it and to obtain the best performing configurations that minimize errors of predictive mechanical properties of polymeric concrete. In this paper different architectures of artificial neural networks will be used for investigating the flexural strength of polymer concrete with fly ash and fibres. In the present study the epoxy resin was used for binding the aggregates. In the composition were introduced near the fly ash, used as filler, the cellulose fibres for improving the properties. The characteristics of these artificial neural networks architectures will be presented and analysed in order to choose the one that minimizes the prediction errors of the mechanical characteristics of polymer concrete and presents an optimal configuration that allows a high working speed that can adapt to this type of approaching the problem with a strong nonlinear character. By using this modern predictive methods, it was attempted to highlight its basic character - learning by examples specific to the human brain but much more efficient due to the mathematical models of the activation functions and the interconnection between the layers of neurons that exponentially increase their ability to adapt to strong nonlinear phenomena. Thus one can say that such a prediction helps to reduce the number of real experiences and can greatly contribute to obtaining the optimal configuration of parameters necessary to obtain a desired mechanical characteristic of the analysed concrete.

1. Introduction

The present paper proposes a modern approach to the prediction concept for phenomena with a strong non-linear character using artificial neural networks. The precision of such predictions is the ability of the user experience to find the optimal network configuration that approximates the numerical results obtained by artificial intelligence from the experimental tests. This method is used by many researchers in different domains [1-5] because of its adaptability and prediction power which, through a process similar to that of the human neuron, brings consistency, rigor and speed to phenomena of black box type that involves far more resources to achieve the same results. From vocal recognition [1] to bioengineering [2], medicine [3, 4] or civil engineering [5-7], ANN use is an increasingly common option for analyzing and predicting complex phenomena.

To illustrate this phenomenon, we considered the choice of experimental testing in the area of concern of authors and the need to show that experimental design and optimization require modern methods of investigation and analysis. To highlight what has been said the aim in the paper is to study the polymer concrete with epoxy resin, fly ash and cellulose fibers, all composition with different dosage of components. The experimental results obtained for mechanical characteristics were compared with that obtained by using ANN model.

2. Materials and methods

2.1 Materials

The experimental researches on polymer concrete were made by using the following materials: polymer (X_1), fly ash as filler (X_2), crushed aggregates and fiber type ARBOCEL. The polymer was type epoxy resin, called ROPOXID, made in Romania by POLICOLOR Bucureşti [8, 9]. The fly ash (FA) from the power plant CET Holboca Iasi was added to the fine aggregates. The principal characteristics of FA are: colour gray to black function of carbon unburned, particles sizes between 0.01 to 100 µm; the shape of particles is spherical, specific surface is between 4800-5200, the density is between 2400 and 2550 kg/m3 [2]. The aggregates were used in two sorts: 0-4 mm (X_3) and 4-8 mm (X_4), with continuous granulosity, obtained from crushed river gravel. The ARBOCEL fibers are natural cellulose fibers, produced by J. Rettenmaier and Sohne GMBH. ARBOCEL is produced from cellulose in a various qualities (fiber lengths, thicknesses, purities, etc.) The properties of ARBOCEL cellulose fibers are: mean fiber length of 10 µm, completely safe, insoluble in water and organic solvents, resistant to dilute acids and bases. The fiber was used in proportion of 3% from the mass of resin plus the hardener. For the study of polymer concrete properties nine compositions were prepared in the experimental program (Table 1) using Design of Experiment Method [10-13].

Table 1. Polymer concrete mixtures and mechanical characteristic investigated

Mix	Resin (%) X_1	FA (%) X_2	Aggregate (%) Sort I X_3	Aggregate (%) Sort II X_4	Response Y_1
1	12.4	12.8	37.4	37.4	14.69
2	12.4	6.4	43.8	37.4	14.18
3	15.6	9.6	37.4	37.4	17.09
4	15.6	6.4	40.6	37.4	14.09
5	12.4	9.6	40.6	37.4	14.60
6	16.4	7.2	38.2	38.2	17.03
7	13.2	10.4	38.2	38.2	15.35
8	13.2	7.2	41.4	38.2	13.55
9	14.0	8.0	39.0	39.0	17.57

2.2 Methods

This paper approach a field of high interest in research eco-concrete or ecological concrete with the possibilities of optimizing the additions of different recyclable wastes (ash, glass fiber for the present study), while maintaining its mechanical performances using a modern experimental planning methodology - Design of Experiment Method – DoE [10-12] specifies the investigation of new materials, supplemented by the power of analysis and optimization given by Artificial Neural Networks - ANN [1-7]. Concrete in combination with recyclable materials [8-11] such as ash and ultra-fine silica has unpredictable effects on concrete properties and therefore we have chosen to use these methods of investigation, since statistics and mathematics can capture the non-linear behaviour of this new additive material with the addition of recyclable materials. These additions can, in certain configurations of the recipe, greatly improve its mechanical behaviour doubled at the same time as the effective and sustainable take-over of the effects of waste [13-15] on society and the environment. Based on learning from the examples, neural networks are able to predict the mechanical characteristics of ecological concrete with very high precision within the field of study, and sometimes even outside, so that to achieve superior performance of these features, it only takes their experimental validation, and therefore a substantial reduction in the cost of testing for the necessary mix formulas.

2.3 Artificial neural network (ANN) architecture

The ANN modelling approach is a computer methodology with artificial neurons that attempts to be similarly with the human nervous system. ANN has ability to solve new problems by applying information learned from past experience, as human brain [7, 12, 16]. Multi Layer Perceptrons (MLP) have been applied successfully to solve some difficult and diverse problems by training them in a supervised manner with a highly popular algorithm knows as the error back-propagation algorithms. The most frequent structure for neural networks in investigations specific to engineering is MLP type. A back propagation algorithm can be used to train these multilayer feed-forward networks with differentiable transfer functions to perform function approximation, pattern association, and pattern classification. The training of ANNs by back propagation involves three stages:

1. feed-forward of the input training pattern,
2. calculation and back propagation of the associated error
3. adjustment of the weights.

This process can be used with a number of different optimization strategies [2, 7, 16]. Once the network is trained, it can be tested on a different set of data than that used for training. It is a good approach to divide the given input/output data into two parts: one part (about 70%) is used for training, whereas the other part, usually smaller, is used for testing the neural network model. The testing data set is reserved to validate the trained network. The training and testing of the networks were performed by means of the Matlab software. In this study, the performance of the ANNs were compared with respect to: mean squared error (MSE), mean absolute error (MAE), linear correlation coefficient (r). These performance measures are defined in the following equations (Eq. 1).

$$MSE = \sum_{i=0}^{N} \frac{(d_i - y_i)^2}{N}, \quad MAE = \frac{\sum_{i=0}^{N} |d_i - y_i|}{N}, \quad r = \frac{\frac{\sum_i (x_i - \bar{x})(d_i - \bar{d})}{N}}{\sqrt{\frac{\sum_i (d_i - \bar{d})^2}{N}} \sqrt{\frac{\sum_i (x_i - \bar{x})^2}{N}}} \quad (1)$$

where: N is the total of training or testing exemplars, y_i is network output for exemplar i, d_i is desired output for exemplar i, x_i is network output, $\bar{x}$ is average network output and $\bar{d}$ is average desired output.

3. Results and Discussion

According to EN 12390 (2001) [17] the mechanical characteristic of polymer concrete with cellulose fiber experimentally determined: flexural strength (N/mm^2) –Y_1 are given in Table 1. The epoxy resin, fly ash and aggregate were considered as input variables, while Y_1 were considered as the output value. Thus, each ANN presents three input variables and three output variable. In this study, the MLP (Fig. 2) was used for the modelling of the flexural properties of polymer concrete. In the present study, ANN network was trained using 6 data sets, they were than tested using 3 test data sets for verification of result. For neural modelling the following parameters have been used: the ANN is trained using the Backpropagation algorithm and the Conjugate Gradient method, maximum epochs of 20,000 and 0.001 learning rate. In our case in the process of learning, training and testing with several types of neural networks (100000) have chosen those configurations that approximates as closely as mechanical characteristic considered (Table 2).

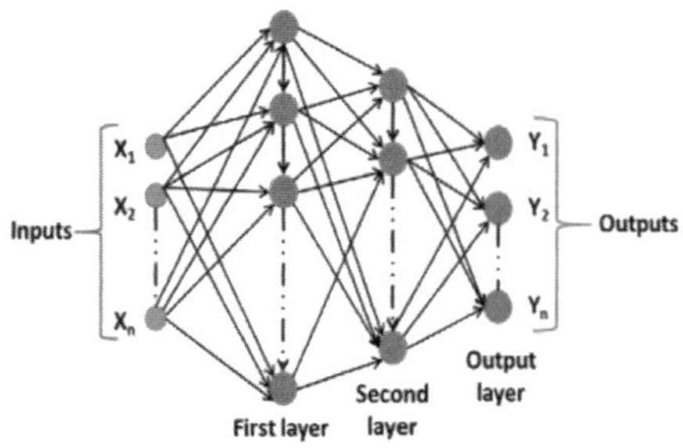

Figure 1. MLP theoretical modelling

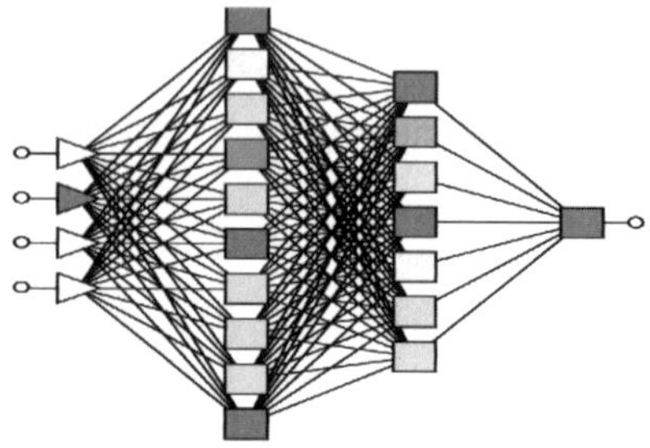

Figure 2. ANN architecture – real modelling

So was kept three network architectures marked ANN 1 to 3 which have 4 input neurons and one output neuron or two hidden layers with variable number of neurons from any neuron in the second layer of the first type ANN - 1 to 10 neurons in the first hidden layer (Fig. 2) for the other two types of ANN - 2 and ANN - 3.

Table 2. ANN architectures for polymer concrete mixtures

Type	Profile	Inputs neurons	Haiden 1	Haiden 2	Outputs neurons
ANN - 1	MLP 4:4-6-1:1	4	6	0	1
ANN - 2	MLP 4:4-10-7-1:1	4	10	7	1
ANN - 3	MLP 4:4-10-6-1:1	4	10	6	1

In table 3 you can see the determined response values using three neural networks and residual error values for each.

Table 3. ANN prediction for flexural strength – Y_1 of polymer concrete

		Network 1		Network 2		Network 3	
	EXP	**ANN-1**	**Residual**	**ANN-2**	**Residual**	**ANN-3**	**Residual**
1	14,69	13,75	-0,94	14,68	-0,01	15,10	0,41
2	14,18	14,17	-0,01	13,64	-0,54	14,22	0,04
3	17,09	17,09	0,00	17,05	-0,04	17,07	-0,02
4	14,09	14,06	-0,03	14,27	0,18	14,25	0,16
5	14,6	14,91	0,31	14,57	-0,03	14,58	-0,02
6	17,03	16,75	-0,28	17,04	0,01	17,02	-0,01
7	15,35	15,35	0,00	15,66	0,31	15,34	-0,01
8	13,55	13,57	0,02	14,10	0,55	13,46	-0,09
9	17,57	17,56	-0,01	17,57	0,00	17,58	0,01

In table 4 is presented some statistical data corresponding to the three types of neural architectures and could see that the coefficient of correlation - R is between 0.97 and 0.99 for all network architecture which shows a strong correlation of type 3 network. Standard error is a measure of degree of impedance in flexural strength data vs. mean. If we look at this (Table 4) we notice that its highest value is 0.33 for the Y1.1 network, which can also induce a greater error probability with half its decrease for the higher-precision network Y1.3. Standard deviation is a measure of variability or diversity in flexural strength data. This means that Standard Deviation keeps the same trend. While mean has the same value for all three networks, which implies a relatively small difference in processed data with almost identical scattering around the centre value.
In the figure above are the four datasets - experimental and those obtained by using artificial neural networks. Note that network no.3 provided the data closest to the experimental (Fig. 3). The residual errors are very small in value, all subunit and ANN-3 network errors are between 0.4 and 0.01 which confirms a good approximation obtained through this network (Fig. 4).

Table 4. Statistical data of three ANN architecture prediction

	Y1.1	Y1.2	Y1.3
Data Mean	15,35	15,35	15,35
Data S.D.	1,41	1,41	1,41
Error Mean	0,10	0,05	0,05
Error S.D.	0,33	0,28	0,14
Abs E. Mean	0,18	0,18	0,08
S.D. Ratio	0,23	0,20	0,10
Correlation – R	**0,97**	**0,98**	**0,99**

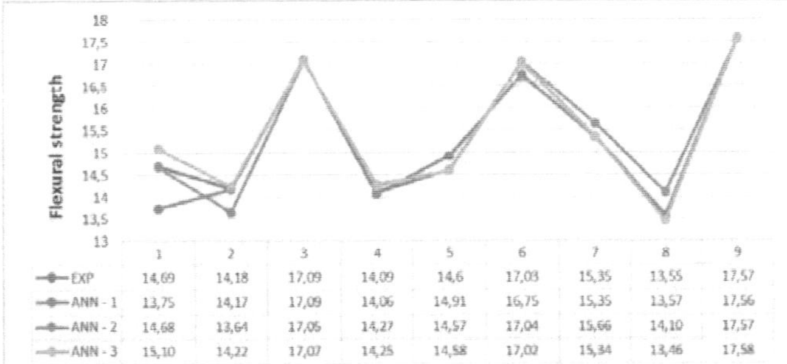

Figure 3. Experimental flexural strength versus three ANN architecture

Through this paper we brought a novelty element in experimental practice based on testing the individual tendency of some parameters or phenomena by recommending the use of two modern methods of experimental planning based on DoE [3-7, 12, 18] and then by prediction and optimization using the concept of ANN [6, 7, 16, 19, 20] but insisting on obtaining an optimal neural architecture that draws as much as possible the numerical results from experimental ones.

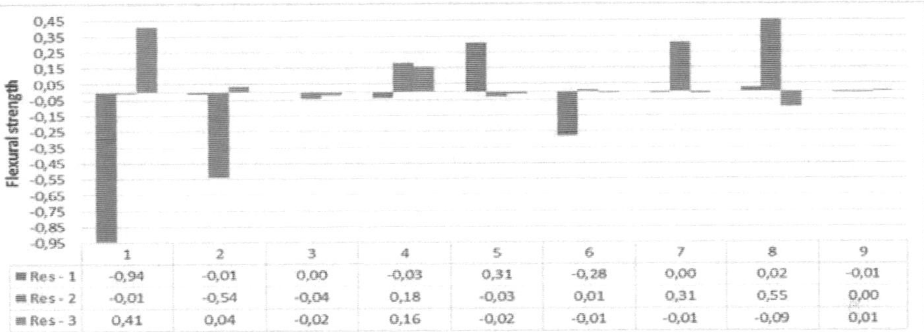

Figure 4 Residual errors for the three ANN architecture

4. Conclusions

The purpose of the study was to point out that compared to the traditional methods of predicting some mechanical characteristics of new materials that required multiple experimental testing and were very costly depending on the complexity of the prescription and the number of controlled parameters, the proposed method uses only the experimental data set existing for training. The prediction can then be done with great precision on desired configurations, leaving only validation to be done under experimental conditions. Artificial neural networks prove to be a powerful and modern interdisciplinary subject that helps solving various different engineering problems that cannot be solved by traditional modeling and statistical methods.

The power of these networks consists in their ability to adapt and learn in particular processes with a non-linear character. It was also stressed the need to use in the experimental process and the DoE the significant reduction of the required number of tests to obtain results with maximum scientific reliability using experimental matrices adapted to each study but capable of capturing the complex behaviour of chemistry in the process of reaction hardening of the polymer concrete studied. Flexural strengths of polymer concrete with fly ash and cellulose fibre were analysed using a different ANN architecture. Thus we tested several configurations of neural architectures in order to choose the one that gives us the best accuracy of the results and minimal residual errors. Were tested over 100,000 networks to find the optimal configuration that ensures great precision of predictive analysed response by an optimal neural network architecture. The results of this study indicate that the appropriate selection of training algorithm is essential for successful data modelling by ANN and it depends on the number of experimental results which are used.

IOP Conf. Series: Materials Science and Engineering **485** (2019) 012001 doi:10.1088/1757-899X/485/1/012001

References

[1] Mirsamadi S Hansen J HL 2018 Multi-domain adversarial training of neural network acoustic models for distant speech recognition *Speech Communication* In press

[2] Forro C et all 2018 Modular microstructure design to build neuronal networks of defined functional connectivity Biosensors & Bioelectronics **122** pp 75-87

[3] Adem K 2018 Exudate detection for diabetic retinopathy with circular Hough transformation and convolutional neural networks, *Expert Systems with Applications* **114** pp 289-295

[4] Lee H Kim D H Jeong S N Choi S H 2018 Detection and diagnosis of dental caries using a deep learning-based convolutional neural network algorithm *Journal of Dentistry* **77** pp 106-111

[5] Aziminezhad M Mahdikhani M Memarpour M M 2018 RSM-based modeling and optimization of self-consolidating mortar to predict acceptable ranges of rheological properties, *Construction and Building Materials* **189** pp 1200-1213

[6] Elevado KJT et all 2018 Artificial neural network (ANN) modelling of concrete mixed with waste ceramic tiles and fly ash *International Journal of Geomate* **15** (51) pp 154 -159

[7] Lepadatu D et all. 2014 Mechanical characteristic prediction on polymer concrete with silica fume using artificial neural network *Proceedings of 14th SGEM GeoConference on Nano, Bio And Green–Technologies* **2** pp 235-242

[8] Arpitha G.R Yogesha B 2017 An Overview on Mechanical Property Evaluation of Natural Fiber Reinforced Polymers *Materials Today:Proceedings* **4:2** (A) pp 2755-2760

[9] Barbuta M Diaconescu R Harja M 2012 Using neural network for prediction of proprieties of polymer concrete with fly ash *Journal of Material Civil Engeneering* ASCE **24** pp 523-528

[10] Montgomery D C 2013 *Design and analysis of experiments*, 8th ed Wiley New York

[11] Dean A Voss D Draguljić D 2017 *Design and analysis of experiments* Second edition Springer International Publishing AG

[12] Bliuc I Lepadatu D et all 2017 Assessment of thermal bridges effect on energy performance and condensation risk in buildings using DoE and RSM methods *European Journal of Environmental and Civil Engineering*, **21** (12) pp 1466-1484

[13] Cucos I 2014 Research on the impact of technical, economic and environmental impact of energy management systems for integrated waste *The Yearbook of the Gh. Zane Institute of Economic Research*, Romanian Academy Publishing House **23** (1)

[14] Cucos I 2015 Systems management methods of waste biomass energy administration in mountain areas, *Journal of Montanology* **III** (3) pp147-156

[15] Cucos I 2014, Research on the impact of technical, economic and environmental impact of energy management systems for integrated waste, *The Yearbook of the Gh.Zane Institute of Economic Research* **23** (1) pp 63-72

[16] Khan M I 2012 Predicting properties of High Performance Concrete containing composite cementations materials using Artificial Neural Networks *Automation in Construction* **22** pp 516–524

[17] European Normative EN 12390 2001 *Testing hardened concrete* European Committee for Standardization Brussels Belgium

[18] Barbuta M et all 2015 Statistical Analysis of the Tensile Strength of Coal Fly Ash Concrete with Fibers Using Central Composite Design *Advances in Materials Science and Engineering* Article ID 486232 p 7

[19] Nunes da Silva I et all 2017 *Artificial Neural Networks. A Practical Course.* Springer International Publishing Switzerland

[20] Lepadatu, D. et al. 2018, Fly Ash Concrete with Fibers: Comparison of Tensile Strength Using Neural Network and Design of Experiments Methods. *Environmental Engineering & Management Journal* **17** (6) pp1321–1328.

Acknowledgements. This paper was realized with the support of project EFECON – ECO-INNOVATIVE PRODUCTS AND TECHNOLOGIES FOR ENERGY EFFICIENCY IN CONSTRUCTION, POC/71/1/4 - Knowledge Transfer Partnership, Cod MySMIS: 105524, ID: P_40_295, Project co-financed by the European Regional Development Fund.

IOP Conf. Series: Materials Science and Engineering **485** (2019) 012002 doi:10.1088/1757-899X/485/1/012002

Advanced systems for heat treatment controlling of electrical furnaces

I. Antonescu[1], C. I. Alecu[2], D. Lepadatu[3]

[1]Senior Lecturer at Technical University "Gheorghe Asachi" of Iasi, Faculty of Civil Engineering and Building Services, Romania
[2]Senior scientific researcher at Institute of Economic and Social Research Gheorghe Zane - Iasi, Romanian Academy Branch
[3]Associate Professor at Technical University "Gheorghe Asachi" of Iasi, Faculty of Civil Engineering and Building Services, Romania

lepadatu@gmail.com

Abstract. This paper presents a modern electric heat treatment which has implemented an electric furnace control system through thermoregulatory with PID control laws. An important aspect is the development of a mathematical model for controlling electrical heating systems for heat treatments based on mathematical modelling to predict the mechanical properties and structure of the play resulting from heat treatment. Validation of the experimental research of heat treatment designed by the mathematical model for predicting mechanical properties and structure of the finished part and model for controlling electrical heating system based on PID adjustment algorithms consists in structural analysis to determine the chemical composition, structural analysis by light microscopy, electron microscopy SEM and EDX quantitative analysis, determination of hardness alloys and physical and mechanical of the turbine blades component of a plasma torch on hydrogen made of heat-treated steel.

1. Introduction

The way to achieve a modern heating system for thermal treatments equipped with equipment for automatic control of the thermal processes in order to improve the mechanical and structural properties of the heat treated parts, which is determined by the need of obtaining low manufacturing costs and a level higher quality [1-6]. The main problem with the management of the heat treatment furnaces is to control the temperature in the furnace so that it evolves after the profile imposed by the technologist [1, 2, 4, 7, 8]. Thermal treatment processes can be accurately modeled with transfer functions for dead-time first order elements. PID algorithms are sometimes successfully implemented in the control of such machines but dead time has unfavorable adjusting effects, the best adjustment can be made using a one-step predictive algorithm. PID-type regulation having a simple structure and being easy to assign is one of the first control strategies used and has the widest applicability in industrial process control, the main feature of PID-type regulation is that it is very little dependent on accurate modeling of the process [7, 9-11].

2. Material and method

Technological multivariable system consists in an electric furnace heat treatment with two different temperature zones, maximum temperature of 1200 °C with total power of 1.6 kW / zones with digital temperature control (figure 1).

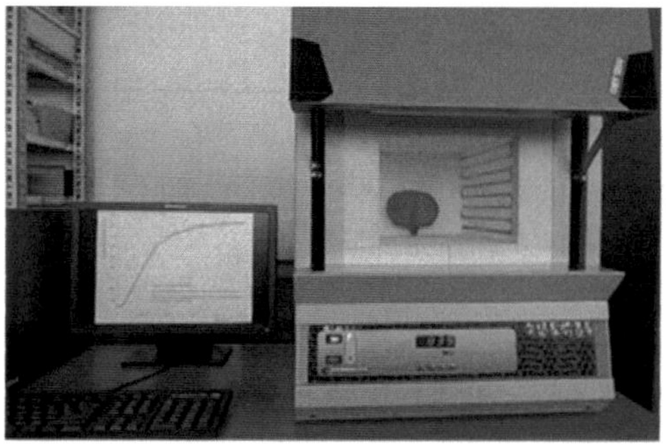

a)

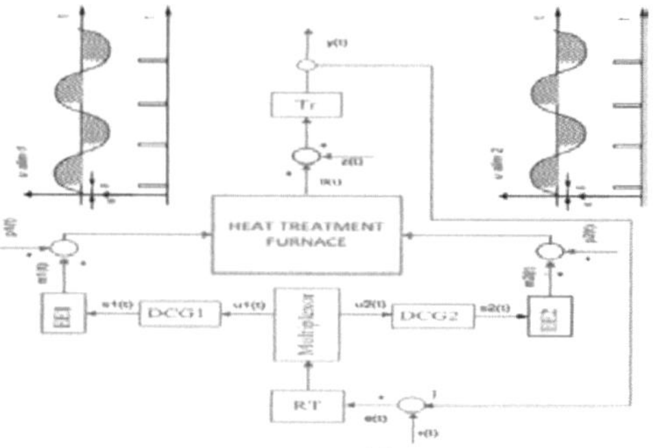

b)

Figure 1. The multivariable heat treatment system with thermoregulatory controlled PID
a) heat treatment system b) block diagram for furnace control

The mathematical model for designing the command system for heat treatment furnace involves the following steps:

- mathematical characterization of the behavior of the controlled heat treatment furnace and acting on its exterior sizes depending on operating mode, setting goals adjustment determined by the type of furnace process management;
- choice of method design involves optimizing parameter values agree, establish criteria for the selection and award regulators and determining optimal control algorithms;
- identification experimental model driven process, identification is performed online in safe working condition without piece inside the furnace, as input data becomes available through measurement;
- simulating the control structure of the heat treatment furnace using the Simulink package;
- test design and analysis of algorithms, possibility of achieving implementation, the optimal choice of equipment that ensures a precise implementation of the algorithm of management;
- mathematical modelling to predict the microstructure and mechanical properties of steel depending on the brand, size and number of semi batch, mechanical properties and quality of the piece prior to heating for optimal heat treatment diagram setting;
- finite element analysis of the evolution in time of the temperature and heat flow of the piece subjected to heat treatment conducted in the furnace with thermostatic controlled PID;
- implementing the heat treatment furnace control structures and tracking simulated time evolution of track parameters;
- validation solution by analysing the performance of the entire control system implemented allows the furnace if necessary redesign or adjustment parameters for the operating agreement.

Thermal treatment processes are modeled according to the control algorithms used, for the PID type control it is often used a one-time deceleration transfer function (f.d.t.) model with dead time and for predictive regulation on one step ARX model [1, 2, 7, 10, 11].

Both models are obtained experimentally by applying test signals to the input element of the furnace temperature probe assembly and by properly processing the output signals.

The transfer function is obtained for the processes described by the models:

$$G_f(s) = \frac{k_f}{(1 + sT_{f1})(1 + sT_{f2})} e^{-sL} \tag{1}$$

The grant procedure is based on simplifying process poles with regulator zeros, resulting in an open-circuit transfer function dependent only on the dead time L.

Knowing that G_d (s) = G_f (s) GR (s), by applying the Haalman method for the process transfer function (1) a PID regulator can be given [5, 6, 8]:

$$\frac{k_f}{(1 + sT_{f1})(1 + sT_{f2})} e^{-sL} k_R \frac{s^2 T_d T_i + sT_i + 1}{sT_i} = \frac{2}{3Ls} e^{-sL} \tag{2}$$

The simplification of process poles immediately leads to finding regulator time constants:

$$T_i = T_{f1} + T_{f2}, T_d = \frac{T_{f1}T_{f2}}{T_{f1} + T_{f2}} \tag{3}$$

and then to the calculation of the proportionality factor with the formula:

$$k_R = \frac{2\left(T_{f_1} + T_{f_2}\right)}{3Lk_f} \tag{4}$$

The mathematical model was obtained by parametric identification using Matlab-Simulink software platform with following diagram (figure 3).

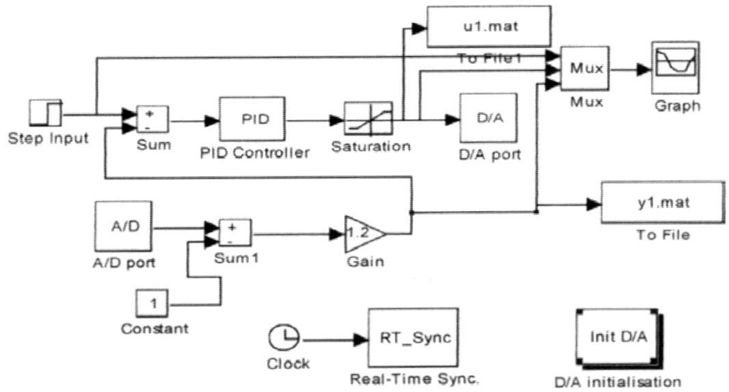

Figure 2. Real-time simulink diagram for implementing a PID regulator [6, 8].

Temperature control in most heat treatment processes is accomplished using PID regulators, considering an estimated off-line of the order I with dead time of form (1).

3. Results and discussions

Experimental validation of the results was made by applying a heat treatment to normalize a turbine blades component of a plasma torch on hydrogen made of heat-treated steel C45-RO1.0503, normalizing heat treatment has the effect of changing their structure and tenacity. The technological parameters were determined with expert software based on mathematical model for predicting mechanical properties and structural piece heat treated. Technological characteristics for normalizing heat treatment: Ac_1=713,20 [^{0}C], Ac_3=787,24 [^{0}C], Ms=331,72 [^{0}C], heating temperature = 841 [^{0}C], heating time = 5040 [sec], maintaining the temperature = 1048 [sec], cooling conditions air [7, 11, 12].

Figures 3 and 4 shows the experimental results, taken from data acquisition board installed in the electric furnace control system for the treatment of normalization, heating diagram calculated with classical algorithms and the actual temperature of the furnace resulting in following predictive PID control algorithm shown by thermocouples placed in the furnace vault.

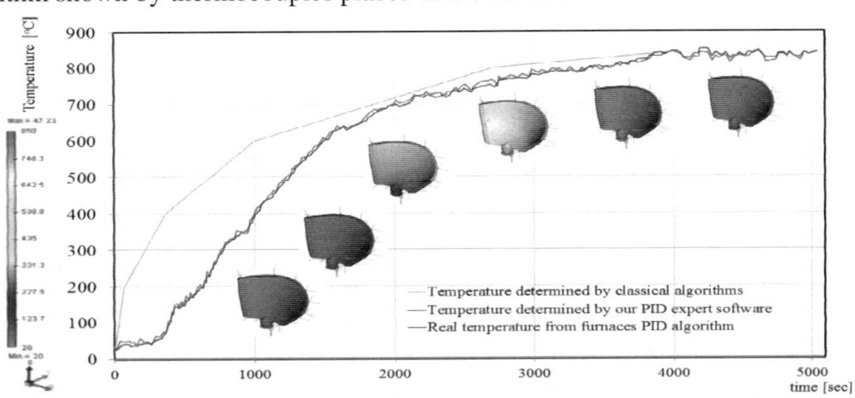

Figure 3. Normalizing heat treatment applied to piece in furnace with PID control system.

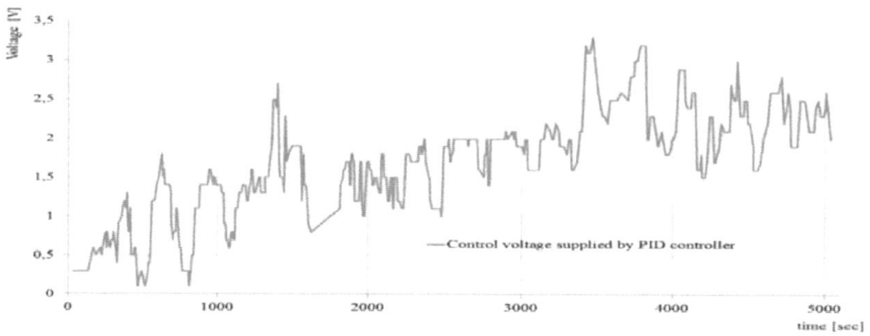

Figure 4. Treatment of normalization - control voltage provided by PID controller

The studies presented below relate to finite element analysis of the evolution in time of the temperature and heat flux normalization piece subjected to heat treatment conducted in the furnace with thermoregulatory controlled PID (figure 5).

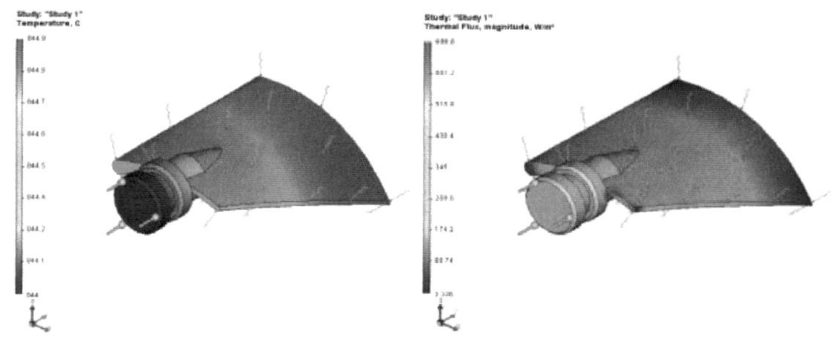

a) *b)*

Figure 5. Temperature and heat flux magnitude normalization piece in furnace PID system
a) temperature, b) heat flux magnitude

From the results obtained by adjusting the electric furnace where it is found that traditional regulation is respected diagram heat treatment required with an accuracy of ± 10 ^{0}C acceptable for some types of heat treatment and grades of steel (alloy steel and low alloy) this type of control in turn has an adverse effect on the heating elements resulting in premature wear and a decrease in heating performance. When adjusting the PID controller tuning accuracy is ± 3 ^{0}C good for most conventional heat treatment diagrams, this setting determines not strain heating elements extending their operating results are influenced by the choice of the operating point and disturbance but is observed to improve performance when using conventional control relay.

In figure 6 are shown images of secondary electrons obtained by electronic microscopy Scanning Electron Microscopy (SEM) Vega Tescan LMH II, working - 46 High using detector type LFD (Large Field Detector), acceleration voltage electron beam used was 30 kV, and working distance was 15 mm.

EDAX (Energy Dispersive X- Ray Analysis) analysis was performed on the sample surface layer to highlight its chemical composition is observed the presence of chemical elements Fe, C, Mn, S, Si and O_2 in different proportions corresponding to the chemical composition of steel C45-RO1.0503.
If the heat treatment classical distribution of chemical elements in the sample material is uneven, pearlite is found in large islands scattered, surrounded by large areas consisting of ferrite.

In samples of heat treated blades in furnace control system based on PID algorithm shows that iron is observed following elemental analysis is found uniformly distributed in the base material as can be seen in the distribution map. Carbon is present in the iron alloy, uniformly distributed, but is observed uniformly dispersed islands alloys thereof chemical compounds with Mn, S and Si.

The manganese is in the form of large islands uniformly dispersed in the mass of base material.

IOP Conf. Series: Materials Science and Engineering **485** (2019) 012002 doi:10.1088/1757-899X/485/1/012002

Sulfur is a small scale and appears in different alloys with Fe, Mn, C, uniformly dispersed in the base material. Silicon is insignificant and appears only as an alloying element to the base material.
If the heat treatment temperature sensitive heat treatment furnace realized ordered thermostatic control systems with PID structure material consists of fine pearlite crystalline grains uniformly distributed in the ferrite matrix, the mechanical properties are superior parts have a high resistance to mechanical wear and high toughness.

Determination of chemical composition by EDAX spectrometry (figure 6) allowed a comparative analysis of the distribution of chemical elements in samples of material palettes microcell reveals the fact that if the heat-treatment furnace control system based on PID algorithm is more uniform dispersion of chemical elements relative to the treated samples heat in microwave classic control system that allows the conversion of a higher percentage of austenite perlite dispersed to obtain uniform fine structure that leads to an improvement of the mechanical properties of the part.

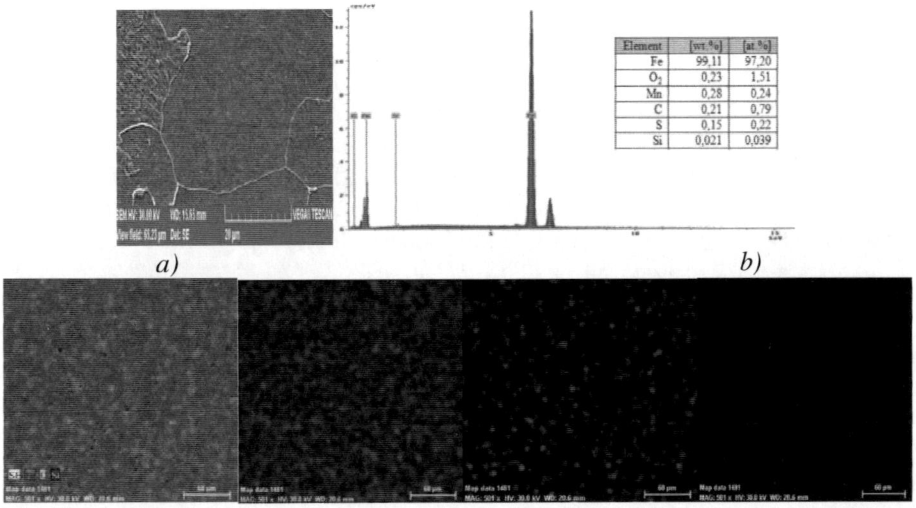

Element	[wt %]	[at %]
Fe	99,11	97,20
O_2	0,23	1,51
Mn	0,28	0,24
C	0,21	0,79
S	0,15	0,22
Si	0,021	0,039

a) b)

c)

Figure 6. SEM and EDAX for turbine blades component of a plasma torch made of heat-treated steel C45-RO1.0503 with control system based on PID algorithm
a) SEM image, b) EDAX elemental chemical analysis, c) the distribution map of chemical elements all elements (Fe, C, Si) map, Fe map, C map and Si map

Microstructure blade material heat treated in heat treatment furnace is controlled with conventional systems composed of ferrite single phase crystalline grains which have flat and pearlite grains which are a mixture of two phases and presented in embossed steel microstructure is typical of overheating, identifies separation of the ferrite around the precipitates, the structure is unsuitable from the point of view of mechanical properties.

Material microstructure of heat-treated blade heat treatment furnace with thermostatic controlled temperature sensitive PID consists of ferrite and pearlite evenly distributed, perlite transformation of austenite made islands shaped ferrite embedded in the table, this determines superior mechanical properties. From the analysis of samples by electron microscopy SEM shows that if the heat treatment of pearlite grains are coarse classic uniformly distributed in the ferrite matrix therefore blade material's mechanical properties are reduced, the piece has a good running low, low durability and low mechanical strength. If the heat treatment temperature sensitive heat treatment furnace realized ordered thermostatic control systems with PID structure material consists of fine pearlite crystalline grains uniformly distributed in the ferrite matrix, the mechanical properties are superior parts have a high resistance to mechanical wear and high toughness.

4. Conclusions

The central objective of the research is to design and implement a management system for electric furnaces advanced mathematical modelling management system furnaces for heat-sensitive treatments based on algorithms evolve in order to increase the mechanical properties and structural parts heat treated and reducing the percentage of reboot due to classical technologies. Heat treatment processes can be modeled accurately with transfer functions for elements with stroke-order retarders, PID algorithms are implemented successfully in the control of such equipment but time is detrimental to the setting retarder. The main directions of continuation of the scientific researches are the development of mathematical models for the design and control of electrical heating systems for thermal treatments based on the prediction of mechanical properties and the structure of the workpiece resulting from thermal treatment by the diversification of databases in the field of alloyed steels.

References

[1] Avram P Imbrea M S Istrate B Strugaru S I Cucos I Axinte C Munteanu C 2014 Wear Resistance and XRD Analyses of CMoCuNiCrSiBO Coatings Obtained by Thermal Deposition on OLC45 Substrate *Applied Mechanics and Materials* Trans Tech Publications **659** pp 10-15

[2] Avram, P Imbrea M S Istrate B Strugaru S I Cucos I Axinte C Munteanu C 2014 Wear Resistance and XRD Analyses of CNiCrSiBO Coatings Obtained by Thermal Deposition on OLC45 Substrate *Applied Mechanics and Materials* Trans Tech Publications **659** pp16-21

[3] Carlone P Palazzo G S 2011 Development and Validation of a Thermo-Mechanical Finite Element Model of the Steel Hardening Process Including Solid-Solid Phase Changes *International Applied Mechanics* **46** pp 955-971

[4] Ferguson B Freborg A Li Z 2007 Residual Stress and Heat Treatment-Process Design for Bending Fatigue Strength Improvement of Carburized Aerospace Gears, *Proceeding of 5th International Conference on Hardening and Control Distortion IFHTSE* Press Berlin pp 95-104

[5] Judele L Lepadatu D Boboc A Rosu A R Cucos I 2017 Cold waste asphalt recycling using fly ash, cationic emulsions and natural aggregated added GeoConference SGEM 2017 SGEM2017 27 - 29 November *Vienna GREEN Conference Proceedings* **17** pp 43-50

[6] Prantil V C Callabresi M L Lathrop J F Ramaswamy G S Lusk M T 2003 Simulating Distortion and Residual Stresses in Carburized Thin Strips ASME *Journal of Engineering Materials and Technology* **125** (2) pp 116-124

[7] Cucos I Munteanu C 2015 The experimental researches of advanced systems for managing thermosensitive heat treatments *Journal of Optoelectronics and Advanced Materials* **17** (3-4) pp 409-420

[8] Trzaska J 2013 Calculation of the steel hardness after continuous cooling *Archives of Materials Science and Engineering* **61** (2) pp 87-92

[9] Brooks B E Beckermann C 2013 Prediction of Heat Treatment Distortion of Cast Steel C-Rings in Proceedings of the 61st SFSA Technical and Operating Conference Steel Founders Society of America, Chicago IL pp 1-30

[10] Cucos I 2014 Research on the impact of technical, economic and environmental impact of energy management systems for integrated waste *The Yearbook of Gheorghe Zane Institute of Economic Research* Romanian Academy Publishing Hous **23** (1) 63-72

[11] Cucos I Munteanu C Avram P 2014 The Advanced System for Conducting the Electric Furnaces for Heat Treatments *Applied Mechanics and Materials* Trans Tech Publications **659** pp 359-364

[12] Cucos I Munteanu C Mirea C 2015 The electric furnaces for heat treatment with PID predictiv system *Applied Mechanics and Materials* Trans Tech Publications **809-810** pp 778-783

Acknowledgments

This paper was realized with the support of project EFECON – Eco-innovative products and technologies for energy efficiency in construction, POC/71/1/4 - Knowledge Transfer Partnership, Cod MySMIS: 105524, ID: P_40_295, Project co-financed by the European Regional Development Fund.

IOP Conf. Series: Materials Science and Engineering **485** (2019) 012003 doi:10.1088/1757-899X/485/1/012003

Studies and researches regarding the obtaining of multifunctional nanocomposites coatings

S Boiciuc, V Muşat

„Dunarea de Jos" University of Galati

simonaboiciuc@yahoo.com

Abstract. The paper presents the preparation of ZnO - SiO_2 nanocomposite coatings by successive layers deposition using sol-gel method, dip-coating technique. The influence of the drawing speed and number of layers on the film morphology, optical properties and corrosion resistance were monitored. It has been found that the increase in drawing speed and the number of layers results in an increased film thickness and a modified film morphology by spherical restructuring. The films deposited on a glass substrate have a transparency ranging from 45-90% in the VIS-NIR range depending on the film thickness. In the case of non-transparent substrates, it has been found that these films diminish the reflectivity due to the surface morphology and the presence of ZnO nanoparticles that absorb some of the light radiation. The corrosion test has showed that all of the deposited films provide superior protection to the substrates.

1. Introduction

The field of thin films has grown spectacularly in recent years due to their use in many branches of science and technology. Their most important applications are in microelectronics, optoelectronics, telecommunications, computer industry, aerospace industry, mechanical parts manufacturing, machine parts and cutting tools.

Functional / multifunctional coatings are generally micronized or thin films in the order of nanometers which, in addition to a specific functionality such as electrical conductivity, optical reflexivity, anti-fouling, antibacterial action, etc., also provides additional protection (to corrosion, abrasion, erosion) and / or decorative functions.

Particular attention is now paid to achieving the smallest possible value-added coatings that have the least impact on the environment, able to apply easily, to adapt to the morphology of the surface and to ensure / increase safety and durability [1, 2, 3].

The sol-gel method is one of the most efficient chemical methods for obtaining thin nanocomposite films with SiO_2 matrices and represents an alternative to vacuum deposition methods.

In principle, the sol-gel method consists in forming a concentrated colloidal soil of oxides or metal hydroxides and converting it to a semirigid gel by one or more methods. The deposition of thin films by the sol-gel method is carried out in several stages: soil preparation, soil deposition on the substrate (with evaporation of the solvent) as a wet gel film, drying and stabilization of the deposited layer.

The dip-coating technique consists in immersing the substrate into the solution containing precursors (soil) and then taking it (drawing it) at a well-defined constant rate, to well-controlled temperature and atmospheric conditions [1-6].

Coatings and thin films obtained by this method are thermally or chemically stable, can transmit, absorb, or reflect radiation at a given wavelength, resist abrasion, erosion, or may constitute protective barriers (under alkaline and acidic environments) [1-3].

Nanocomposite coatings can be made on different substrates, such as steel, aluminum, copper or glass, depending on the application.

The paper presents a number of results obtained from the characterization of thinned nanocomposite films of ZnO-SiO_2 deposited on different substrates. The influence of the drawing speed and the number of layers on the film morphology, optical properties and corrosion resistance were monitored.

2. Experimental Conditions

2.1. *Making films*

Operational nanoparticles of ZnO (≤ 50nm) and a SiO_2 matrix were used for deposition purpose.

The steel and copper substrates (50x20 mm size) coated with thin films of ZnO-SiO_2 were polished and then subjected to ultrasonic cleaning for 10 minutes in acetone.

Aluminum substrates (50x20 mm size) were subjected to grinding, polishing, ultrasonic cleaning in acetone and alkaline pickling.

Glass substrates (20x20 mm size) were cleaned with ethanol and then dried.

The 2nd, 3rd and 4th layers of ZnO-SiO_2 were deposited on steel, copper, aluminum and glass substrates using dip-coating technique and different drawing speeds. After each deposited layer, the gel film was subjected to heat treatment by heating in a furnace (FN 055 Nüve type) at 65 ° C for 45 minutes followed by a heating at 190 ° C for 15 minutes. Cooling was done later with the oven.

The deposition regimes that were used are shown in Table 1.

2.2. *Film characterization*

2.2.1. *Morphology of the deposited films*

The microstructure of the deposited layers was obtained by optical microscopy using a Neophot 2 microscope.

2.2.2. *Optical properties of the deposited films*

The optical properties of the thin films were analyzed by a Perkin Elmer double Lambda 35 UV-VIS beam using the UV-WinLab software. The range of wavelengths in which this spectrophotometer is working is 190-1100 nm. The experiments were performed at room temperature.

2.2.3. *Corrosion behavior of the deposited films*

The tests were carried out in a corrosion chamber equipped with an automatic temperature measuring and control system, coupled with an evacuation system for the solution or condense which resulted during the tests. Samples were placed on plastic bars. The corrosion acceleration test in saline mist was performed according to SRISO 9227.

To conduct the test, the required solution was prepared at 35 ± 2° C from a quantity of sodium chloride so as to obtain a concentration of 50 ± 5 g / l. The pH of the saline solution was adjusted so that the pH of the solution collected in the spraying chamber ranged between 6.5 and 7.2. This was measured with a pH meter HI 991001, produced by "Hanna Instruments", also provided with a temperature indicator.

The corrosion test was carried out over a period of 96 hours at 35 ° C. Initially the samples were degreased with alcohol weighed by analytical balance with an accuracy of 0.1 mg.

At the end of the test, the samples were removed from the chamber, the corrosion products were removed by washing with water and then dried. The analytical balance with an accuracy of 0.01 mg was subsequently used.

Table.1. The deposition regimes of $ZnO-SiO_2$ films

Type of sublayer	Sample symbol	Drawing speed [cm/min]	Number of layers	Thickness of the deposited film [μm]
Steel	2O2	2	2	1.176
	2O3	2	3	1.839
	2O4	2	4	2.502
	10O2	10	2	1.313
	10O3	10	3	2.014
	10O4	10	4	2.715
Copper	2C2	2	2	0.606
	2C3	2	3	1.269
	2C4	2	4	1.932
	10C2	10	2	0.758
	10C3	10	3	1.459
	10C4	10	4	2.16
Aluminum	2A2	2	2	0.735
	2A3	2	3	1.398
	2A4	2	4	2.061
	10A2	10	2	0.798
	10A3	10	3	1.499
	10A4	10	4	2.2
Glass	2S2	2	2	0.372
	2S3	2	3	1.035
	2S4	2	4	1.698
	10S2	10	2	0.88
	10S3	10	3	1.581
	10S4	10	4	2.282

3. Experimental Results

3.1. Morphology of the deposited films

As a result of the microscopic analysis, it has been observed that the obtained films are generally free of cracks, homogeneous and adherent as shown in Fig. 1.

Thus the obtained films exhibit high compaction because aggregation, gelling and drying (evaporation) occur almost simultaneously during the extremely short deposition on the substrate; the atmospheric vapor pressure controls evaporation of the solvent and the transformation of the soil into gel with the formation of a transparent film; the subsequent heat treatment produces gel densification and crystal formation (crystalline).

As also shown in Table 1, the thicknesses of the obtained layer varied depending on the substrate and deposition conditions as follows: 1.1-2.7 μm for the films deposited on steel, 0.7-2.2 μm for the ones deposited on aluminum, 0.6-2.1μm for copper and 0.3-2.2μm for those deposited on glass.

For the films with a thickness of less than 1 μm, it has been found that surface imperfections can cause some inhomogeneities. In the case of the films with a thickness greater than 1.5 μm, a film restructuring was noticed due to the spherical restructuring of the structures, most likely caused by the increase in superficial tension as seen in Fig. 2.

When increasing the drawing speed, correlated with the same number of layers, the thickness of the deposited film increased too.

IOP Conf. Series: Materials Science and Engineering **485** (2019) 012003 doi:10.1088/1757-899X/485/1/012003

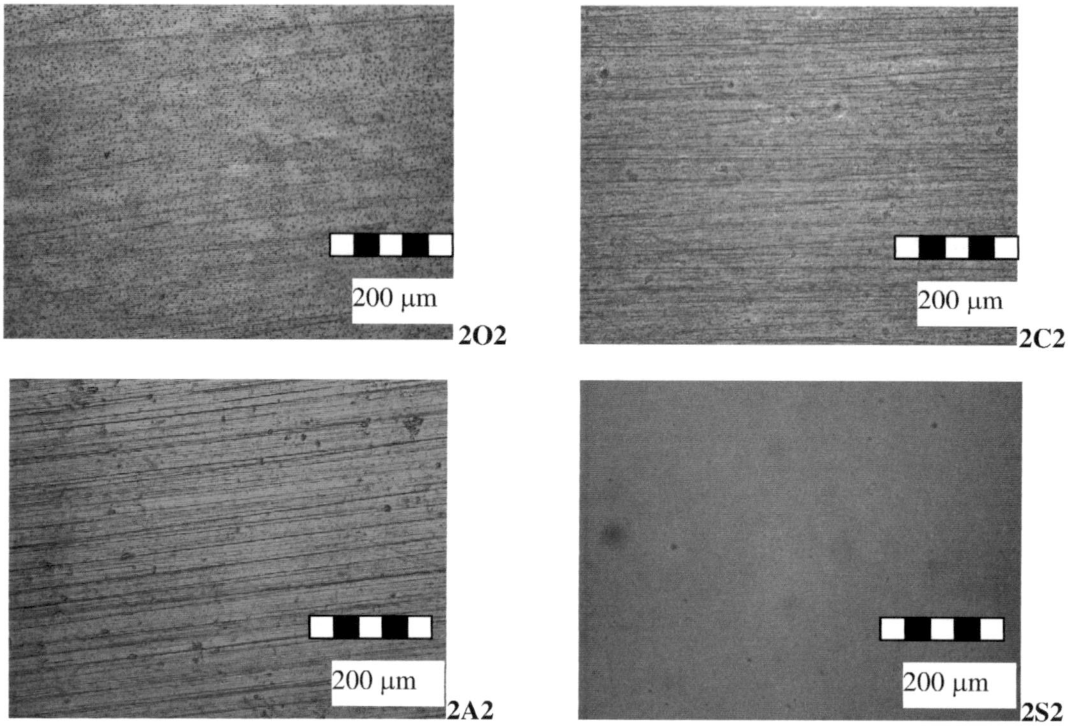

Figure 1. Micrographs of film deposited at 2 cm / min rate on different substrates

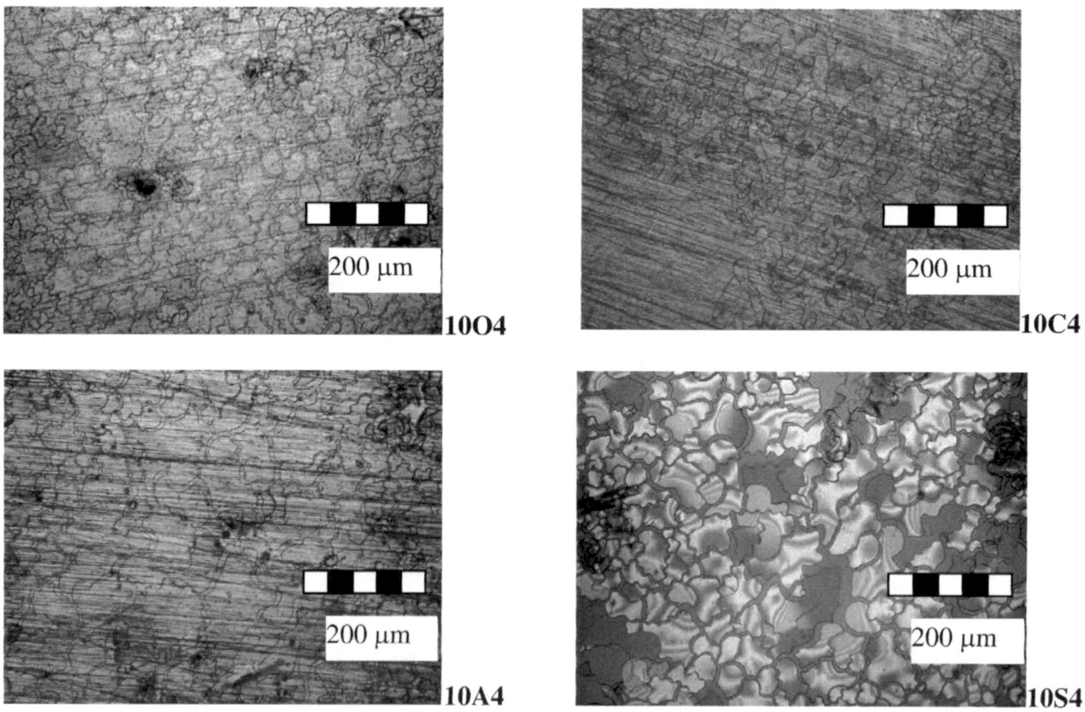

Figure 2. Micrographs of film deposited at 10 cm / min rate on different substrates

3.2. Optical properties of the deposited films

It has been found that films with a small thickness of approx. 0.3-1.6 μm deposited on the glass substrates exhibit 75-90% transparency in the VIS-NIR range, and with those of larger thickness of 1.5-2.2 μm this decreases down to about 45-85%, as can also be seen from Fig.3.

In the case of non-transparent substrates of steel, copper and aluminum, it has been observed that these films decrease reflectance due to the surface morphology and the presence of ZnO nanoparticles in the film composition that absorb some of the light radiation as highlighted by the reflectance spectra of Figs. 4 and 5. Decreasing surface roughness leads to improved reflectivity properties, which is particularly noticeable for aluminum and copper sublayers.

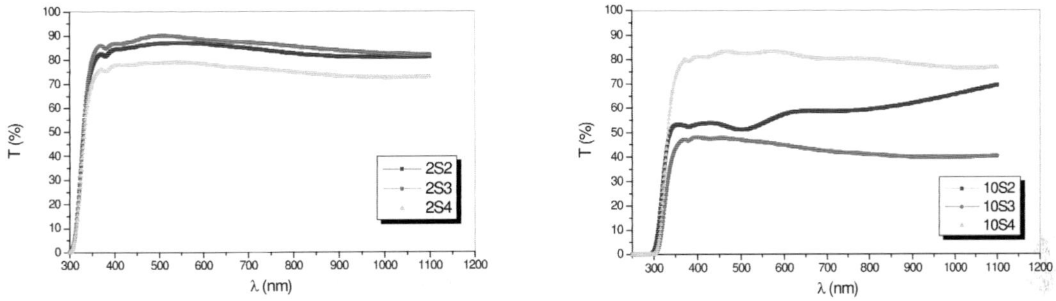

Figure 3. Transmission spectra of films deposited on glass substrates

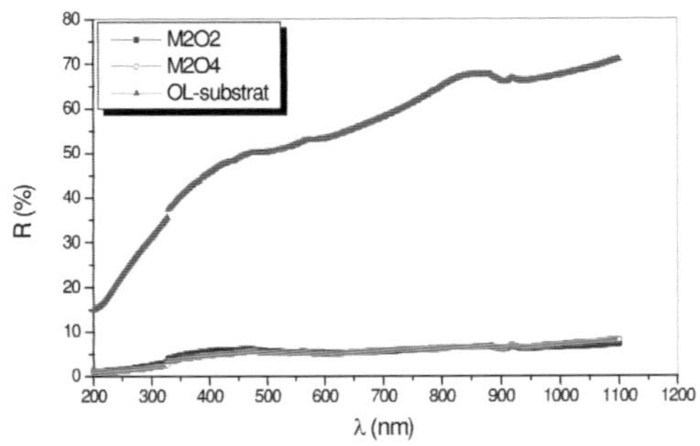

a.

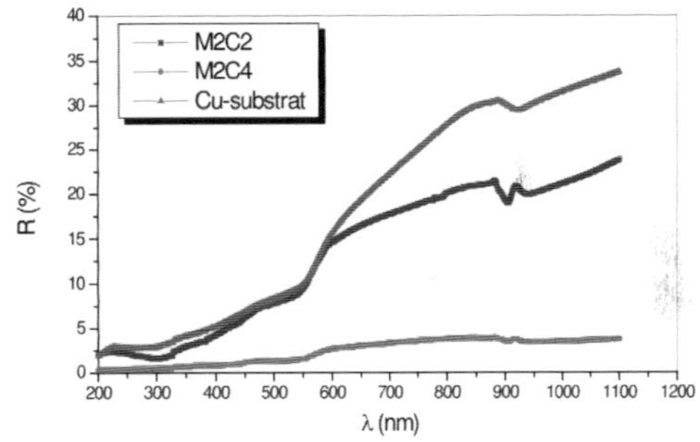

b.

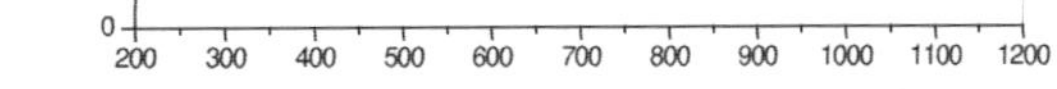

c.

Figure 4. Reflection spectra for film deposited at 2 cm / min rate on different substrates:
a - steel b - copper and c - aluminum

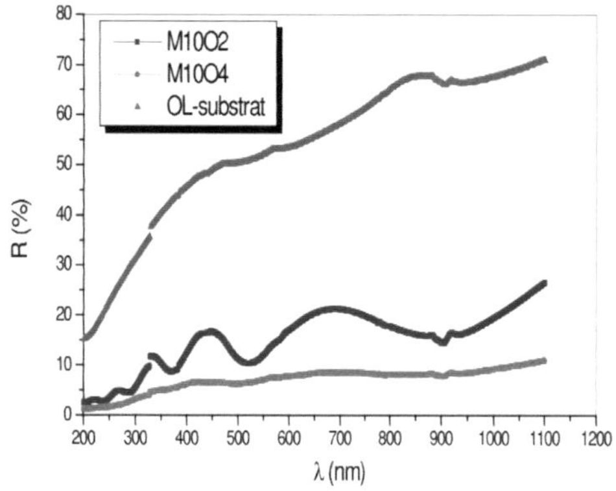

a.

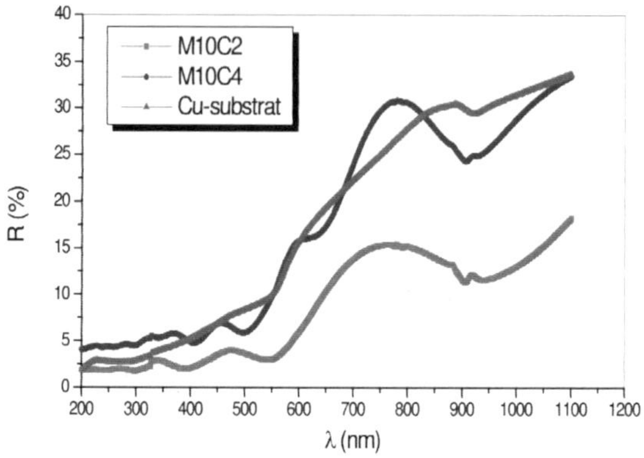

b.

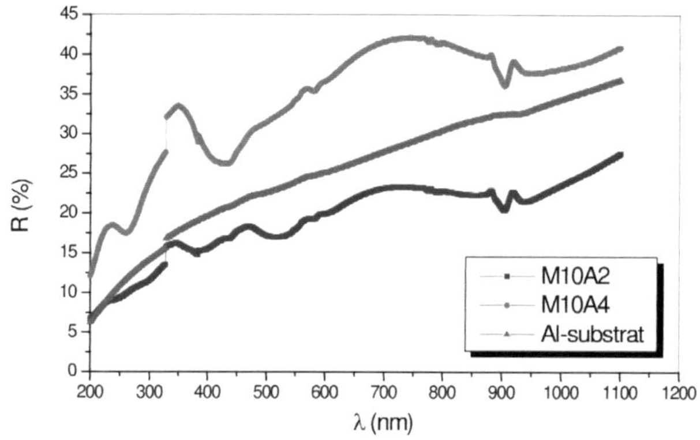

c.

Figure 5. Reflection spectra for film deposited at 10 cm / min rate on different substrates: a - steel b - copper, c - aluminum

3.3. Corrosion behavior of deposited films

Following the corrosion test in salt mist, after 96 hours, it has been found that the deposited films provide a corrosion protection superior to the substrates. The best corrosion behavior has been that of the films deposited on aluminum substrate where corrosion rate was between 0.1-14 mg/m^2h, followed by those deposited on a copper substrate where corrosion rate has ranged between 0.5-35 mg/m^2h, followed by those deposited on a steel substrate with a corrosion rate between 118-392 mg/m^2h.

Corrosion rates for non-transparent substrates were 692 mg/m^2h for steel, 125 mg/m^2h for copper and 73 mg/m^2h for aluminum.

However, it has been found that higher thickness films feature lower corrosion rate than the thinner ones where corrosion was more obvious. Thus, in the first case we can consider that there is a localized corrosion while in the second case the corrosion is generalized. These aspects can also be seen in figure 6.

IOP Conf. Series: Materials Science and Engineering **485** (2019) 012003 doi:10.1088/1757-899X/485/1/012003

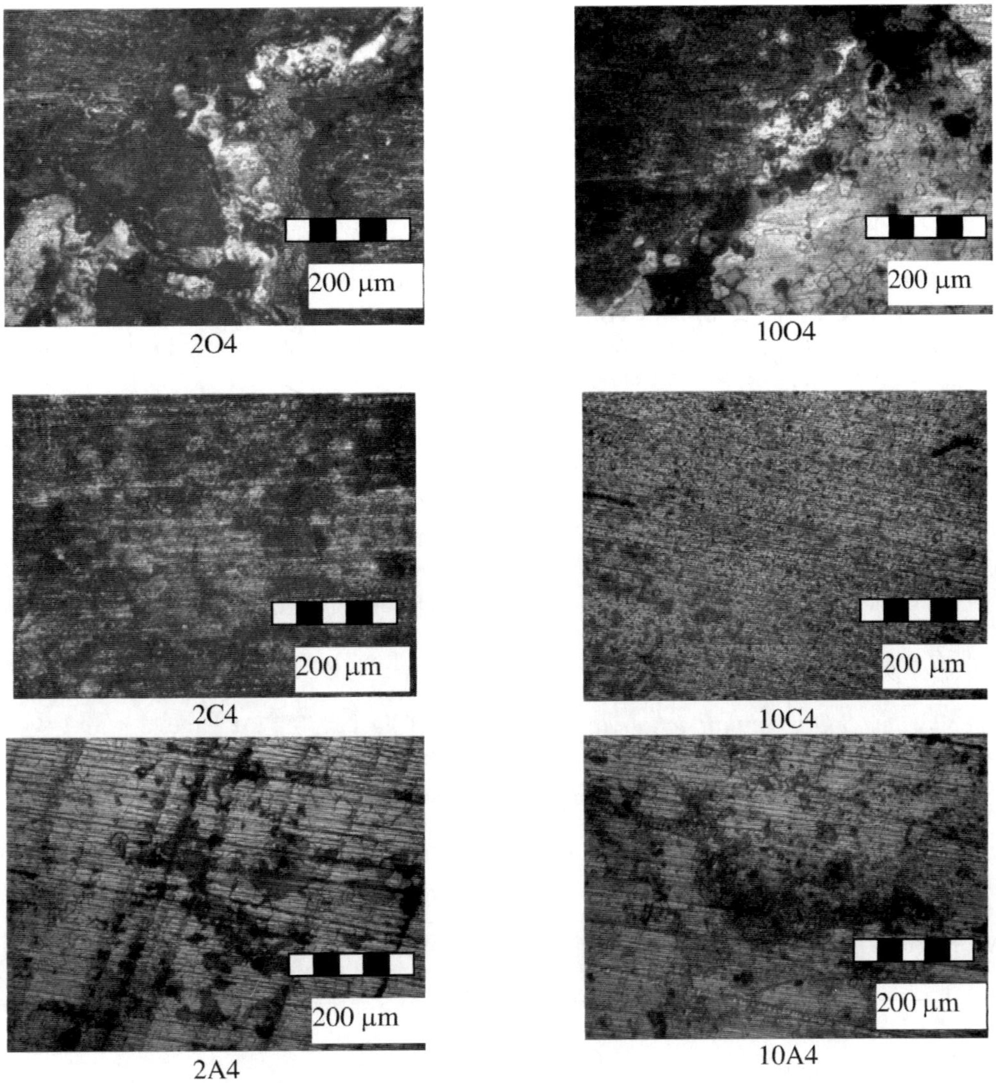

Figure 6. Micrographs of films subject to corrosion

4. Conclusions

As a result of characterizing ZnO-SiO_2 films deposited on different substrates, the following conclusions can be drawn:

a. The technique adopted is a low-cost, affordable method for obtaining oxide films, including conductive and transparent films, with performance similar to films obtained by physical methods;

b. The deposits made have resulted in homogeneous, compact films adhering to the substrate with thicknesses between 0.6-2.7μm;

c. The microstructural analysis of the deposited layers revealed that the films generally do not show cracks, but there are morphological differences depending on the type of the support and the thickness of the film; with increased layer thickness there is a decrease in roughness and a reorganization of the film by spherical restructuring due to the increase of the surface tensions;

d. Increasing the drawing speed, as compared to the same number of layers, the thickness of the deposited film is increased;
e. Low-thickness films (0.3 – 1.6 μm) deposited on glass sublayer have a transparency of 75-90% in the VIS-NIR range, and those with a higher thickness (1.5 – 2.2 μm) it decreases to about 45-85%; In the case of non-transparent sublayers of steel, copper and aluminum, it is noticeable that these films decrease reflectivity due to the surface morphology and the presence of ZnO nanoparticles in the film composition that absorb some of the light radiation;
f. The corrosion test revealed the following: all deposited films, regardless of their thickness, provide superior corrosion protection to the substrates used; thicker layers exhibited a lower corrosion rate than thinner ones; the best corrosion behavior is found with films deposited on aluminum sublayers;
g. The ZnO-SiO_2 films deposited on steel and aluminum sublayers can be used for solar thermal devices, solar collectors and glass-coated ones for optoelectronic devices, transparent conductors, solar cells, sensors.

References

[1] Muşat V, Acoperiri multifuncţionale şi protecţia împotriva coroziunii, *Revista de Coroziune şi Protecţie Anticorozivă*, 2009, Vol IV, Nr 2, page 4.
[2] Muşat V, Acoperiri de protecţie obţinute prin metoda sol-gel pentru benzi metalice, *Revista de Coroziune şi Protecţie Anticorozivă*, 2010.
[3] Muşat V, *Filme Subţiri Multifuncţionale*, Editura CERMI Iaşi (2005) 20, pages 33- 59.
[4] Panaite V, Boiciuc S, Muşat V - ZnO Nanoparticles–Epoxy Resin Hybrid Nanocomposite with Anticorrosive and Antifouling Properties as Coatings for Naval Steel, *Revista de chimie*, No. 2, 2015.
[5] Davies G., s.a, Functional Materials. Future and Directions, 2005, University of Birmingam
[6] Brincker C.J. şi Scherer G.W., Sol-gel The Physices and Chemistry of Sol-gel Processing, *Academic Press INC*, Harcoaut Brance Javonovick Publisher, 1990, Boston san Diego New Work London, ISBN 0-12-134970-5.

IOP Conf. Series: Materials Science and Engineering **485** (2019) 012004 doi:10.1088/1757-899X/485/1/012004

Thermomechanical processing effects on the structure and properties of Fe-based SMAs. I. Evolution of phase structure

V Bulbuc[1], B Pricop[2*], Popa[2], E Mihalache[2], B Özkal[3], L G Bujoreanu[2]

[1] Faculty of Materials Science and Engineering, "Gheorghe Asachi" Technical University of Iasi, Blvd. Dimitrie Mangeron, 61 A, 700050 Iasi, Romania
[2] The National Railway Company "CFR" SA, Regional Branch of Passenger Rail Iasi, Str. Piața Gării 1, 700090 Iași, Romania
[3] Particulate Materials Laboratory, Metallurgical and Materials Engineering Department, Istanbul Technical University, 34469 Maslak, Istanbul, Turkey

*bogdan.pricop@tuiasi.ro

Abstract: The paper analyses the effects of heat treatment (HT) and mechanical alloying (MA) on the evolution of phase structure of an Fe-Mn-Si-Cr-Ni Shape Memory Alloy (SMA) obtained by powder metallurgy (PM). Using various amounts of mechanically alloyed (MA'ed) powders, namely 0 and 40 vol. %, two groups of specimens with chemical composition Fe-14Mn-6Si-9Cr-5Ni (mass.%) were produced by PM. After blending, pressing, sintering and hot-rolling, each set of samples was subjected to five different heat treatment temperatures 973, 1073, and 1373 K, respectively. Tensile pre-straining tests were performed on all samples, with increasing pre-straining degrees up to 4.5 %, in order to stress-induce ε (hcp) martensitic phase in the material. Structural analysis was implemented by means of X-ray diffraction and SEM observations. XRD and SEM investigations revealed the formation of thermally induced α'-bcc martensite, besides ε-hcp, as a particularity of this powder metallurgy, mechanically alloyed, shape memory alloy.

1. Introduction

Fe-Mn-Si-based shape memory alloys (SMAs) were used for practical applications [1] since 1991due to a number of advantages, i.e., low cost, good mechanical processability and high corrosion resistance [2]. Heating-induced reversion to γ-fcc austenite of ε-hcp stress-induced martensite, between critical temperatures A_s (for the start) and A_f (for the finish), represents the mechanism of SME occurrence [3]. At low Mn % or at high deformation degrees, besides ε-hcp, α'-bcc martensite can be additionally induced by cooling or deformation, which diminishes the crystallographic reversibility of $\gamma \rightarrow \varepsilon$ transformation [4]. The formation of stress-induced α'-bcc martensite, in a three-phase structure, has been responsible for the decrease in SME [5].

The great majority of SMAs applications used the alloys obtained through ingot metallurgy (IM, comprising melting-alloying-casting) which, especially in the case of quintenary Fe-Mn-Si-Cr-Ni SMAs, has several technological drawbacks, such as compositional segregation, difficult incorporation of Si into melt, demanganization on melting and heat treatment, time consuming chemical composition homogenization, cracking enhancement due to cooling contraction during solidification and quenching, tempering embrittlement, etc. [6].

A viable alternative for replacing IM materials could be powder metallurgy coupled with mechanical alloying. PM MA'd routine has been used for making shape memory alloys from other systems, such as Ti – Ni [7], Cu – Zn – Al [8] or Cu – Al – Ni [9]. Some of the present authors have showed that the amount of ε hcp-martensite in sintered samples could exceed 31%, if as-blended powders were used, after a final rolling pass performed at room temperature (RT) [10]. The formation of thermally induced martensite was enhanced, at PM specimens, by both MA and the increase of heat treatment (HT) temperature [11], while an optimized combination of MA'ed powder fraction and technological parameters of hot rolling (HR) and HT enabled an increase of shape recovery degree [12].

The present paper investigates the effects of heat treatment (HT) and mechanical alloying (MA) on the evolution of phase structure of an Fe-Mn-Si-Cr-Ni Shape Memory Alloy (SMA) obtained by powder metallurgy (PM).

2. Experimental details

The specimens, with nominal chemical composition 66Fe-14Mn-6Si-9Cr-5Ni (mass %), were pressed and sintered under argon atmosphere and designated as: (i) 0_MA, from as-blended elemental powders and (ii) 40_MA comprising 40 vol. % MA'ed powders, obtained after 4 hrs. high-energy ball milling under Ar atmosphere in order to inhibit the chemical reactivity of oxygen-sensitive alloying elements. In order to further increase of compactness six consecutive hot rolling passes at 1373K were performed, until samples' thickness decreased from 4 to 1 mm. Low-temperature annealing, i.e. 923 – 1073 K with 30 minutes maintaining, was also applied after hot rolling by Drucker et al. [13] but in our case the specimens were porous and the risk of chemical altering of the superficial layer was augmented. For this reason, we reduced holding time to 5 minutes that would be enough for stress relieving, after hot rolling. Before further experiments, rectangular, "dog-bone" (gauge dimensions 1×4×20 mm) were cut by spark-erosion wire cutting and carefully grinded under water flow, in order to remove the outer layers damaged by oxidizing and demanganization.

Tensile pre-straining tests were performed on an INSTRON testing machine at RT, with increasing pre-straining degrees (up to 4.5 %) during a static loading-unloading with a crosshead displacement rate of 3x10-3mm s^{-1}[14].

The gauges of "dog bone" specimens, both in initial and pre-strained conditions were cut, embedded into cold mounting resin and metallographically prepared for performing XRD patterns and SEM micrographs. The former were recorded, on 2θ = 40-100°, using a Brucker ASX D8 Advance diffractometer with Cu Kα radiation and the latter on a SEM—VEGA II LSH TESCAN microscope. Crystallographic databases were used for the identification of, α'-bcc, γ-fcc and ε-hcp phases.

3. Experimental results and discussion

Structural analysis was further focused on the specimens that were HT'ed at 1273 K (1000^{0}C) due to their prominent effect, as compared to the other samples. Their corresponding XRD patterns are shown in figure 1.

XRD patterns recorded on the hot rolled heat treated specimens in unstrained and pre-strained conditions, revealed the presence of the three phases, α'-bcc, γ-fcc and ε-hcp, identified by means of the non-overlapping diffraction maxima (110)α', (200)α', (101)ε, (103)ε, (200)γ and (222)γ.

It is noticeable that the increase of pre-straining degree, in the case of 0_MA samples, enhanced the formation of martensite plates due to the increase in intensity of the main diffraction maxima of the ε-martensite phase (002). Another thing to notice is that, if in initial state the 0_MA samples presented a high α' martensite intensity peak (110) with increasing the pre-straining degree its intensity decreased with almost 80%. This kind of behavior is atypical because in theory α' martensite forms at the intersections of ε-martensite and is favorited by large deformation degrees. In this case the main diffraction maxima (110) decreased with increasing deformation degree.

The behavior of 40_MA samples is as expected, α' main intensity peak increased with deformation degree while a new plane (200) can be observed at 2θ = 63°. Due to increasing pre-straining degree a large quantity of α' martensite was induced in this types of samples which will be detrimental to shape memory effect.

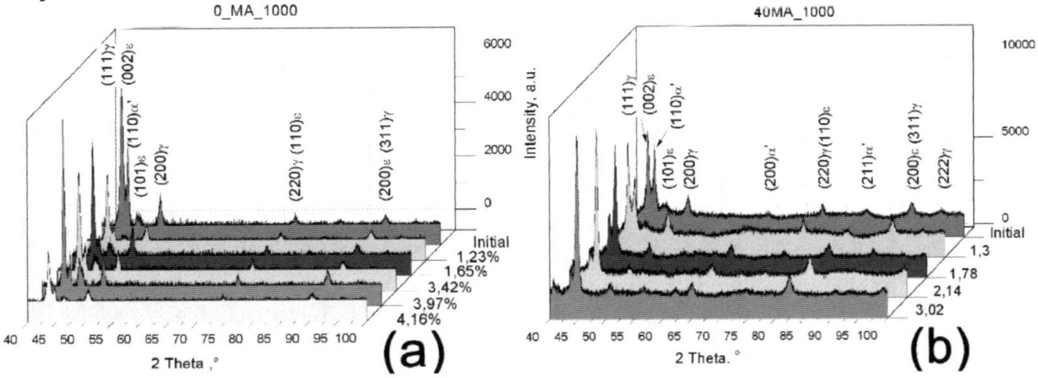

Figure 1. XRD patterns of the specimens heat treated at 1000°C, (a) 0 MA and (b) 40 MA

SEM micrographs from figure 2 reveal that ε-hcp martensite displays a typical "triangular" morphology, with narrow plates that completely cross austenite grains, from one border to the other, while α'-bcc martensite forms shorter bands with lenticular shape and deeper surface relief, which do not cross austenite grains.

At this point, one could argue that the introduction of α'-bcc martensite in an amount observable by SEM is unlikely in this kind of alloys. This observation was valid in the case of undeformed Fe-12.60 Mn-6.00 Si-9.27 Cr-4.74 % Ni-0.27 Al-0.13 Mo-0.05 C which contained 2.7 % α'(bcc) martensite [15] and in the case of Fe-14.2 Mn-5.1 Si-9.4 Cr-5.1 Ni (mass.%) that was nitrogen quenched and annealed at 923 K. Yet, in the latter alloy, increasing annealing temperature to 1223 K caused a marked rise in the amount of α'(bcc) martensite, visible on XRD pattern but not sustained by optical micrographs [16]. This tendency of α'(bcc) martensite, to increase with increasing annealing temperature, was also previously observed at the alloys under study, where the amount of this type of martensite ranged between 20-90 % [17]. It is noticeable that, with increasing MA'ed powder fraction, martensite plates became finer. This refinement should normally enhance shape recovery but a complete analysis of this effect in the pre-stressed specimens is beyond the scope of this article.

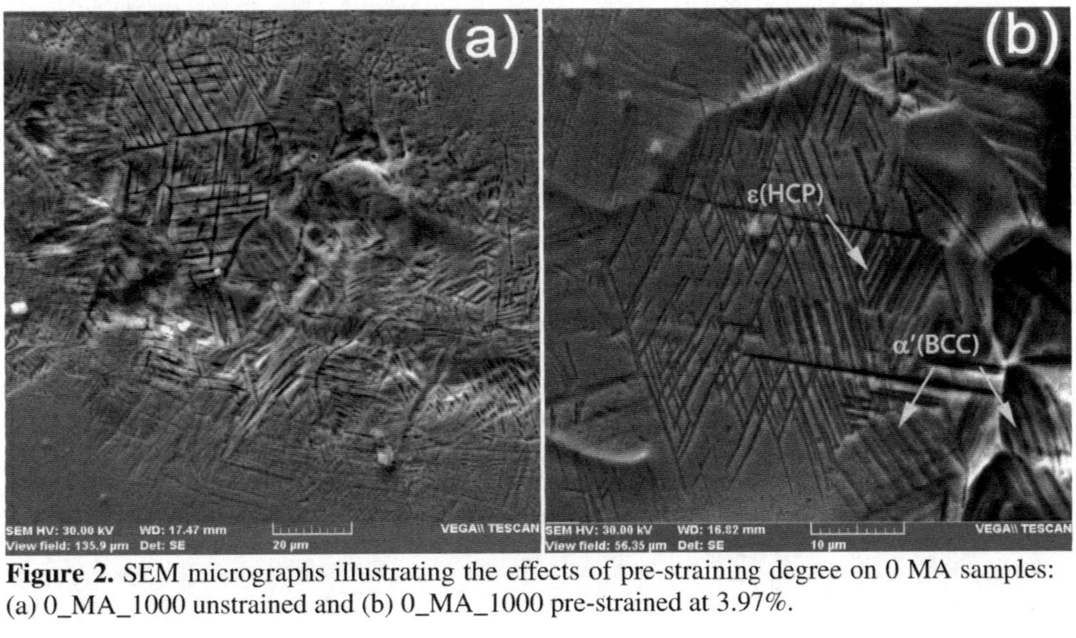

Figure 2. SEM micrographs illustrating the effects of pre-straining degree on 0 MA samples: (a) 0_MA_1000 unstrained and (b) 0_MA_1000 pre-strained at 3.97%.

Finally, the effects of MA fraction on the microstructure of PM-MA Fe-Mn-Si-Cr-Ni specimens, were presented in figure 3. Figure 3(b) shows the microstructure of specimen 40_MA_1000 which was pre-strained with 3.02 %. Although the ε-martensite quantity, which can be observed in the XRD patterns from figure 1(b), decreased the SEM micrograph showed a more complex martensite structure as compared to 0_MA.

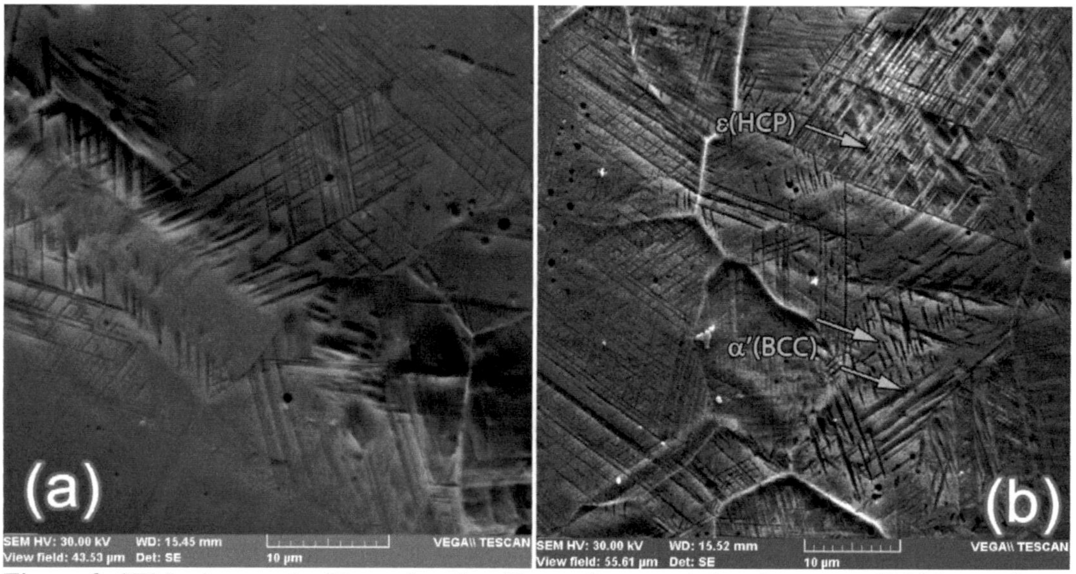

Figure 3. SEM micrographs illustrating the effects of pre-straining degree on 40 MA samples: (a) 40_MA_1000 unstrained and (b) 40_MA_1000 pre-strained at 3.02%.

4. Summary and conclusions

- XRD and SEM revealed the formation of thermally induced α'-bcc martensite, besides ε-hcp, as a particularity of these PM-MA'ed alloys.
- Martensite plate variants became more diversified with the increase of MA fraction and more complexly intersected.
- Introducing MA fraction lead to a decrease in the γ-fcc amount which transformed to α'-bcc and ε-hcp stress-induced martensites.

Acknowledgements: This research work was supported by UEFISCDI by means of Project PN-III-P4-ID-PCE-2016-0468, Contract 76/ 2017.

References

[1] Moriya Y, Kimura H, Ishizaki S, Hashizume S, Suzuki S, Suzuki H and Sampei T 1991 Properties of Fe-Cr-Ni-Mn-Si (-Co) shape memory alloys *Le J. Phys. IV* **01** C4-433-C4-437

[2] Maki T and others 1999 Ferrous shape memory alloys *Shape Mem. Mater.* 117

[3] Kajiwara S 1999 Characteristic features of shape memory effect and related transformation behavior in Fe-based alloys *Mater. Sci. Eng. A* **273** 67–88

[4] Pricop B, Söyler U, Özkal B, Lohan N M, Paraschiv A L, Suru M G and Bujoreanu L G 2013 *Influence of Mechanical Alloying on the Behavior of Fe-Mn-Si-Cr-Ni Shape Memory Alloys Made by Powder Metallurgy* vol 738–739

[5] Spiridon I P, Pricop B, Suru M G, Paraschiv A L, Lohan N M and Bujoreanu L-G 2013 The influence of heat treatment atmosphere and maintaining period on the homogeneity degree of a Fe-Mn-Si-Cr-Ni shape memory alloy obtained through powder metallurgy *J. Optoelectron. Adv. Mater.* **15**

[6] Söyler A U, Özkal B and Bujoreanu L G 2011 Investigation of Mechanical Alloying Process Parameters on Fe-Mn-Si Based System *Suppl. Proc.* 577–83

[7] Jabur A S, Al-Haidary J T and Al-Hasani E S 2013 Characterization of Ni-Ti shape memory alloys prepared by powder metallurgy *J. Alloys Compd.* **578** 136–42
[8] Wong C R, Whang S H and others 1993 Fabrication and damping capacity of Cu-Zn-Al matrix composites processed by powder metallurgy route *Mater. Sci. Eng. A* **165** 35–43
[9] Ibarra A, Juan J S, Bocanegra E H and Nóa M L 2006 Thermo-mechanical characterization of Cu-Al-Ni shape memory alloys elaborated by powder metallurgy *Mater. Sci. Eng. A* **438–440** 782–6
[10] Ariapour A, Yakubtsov I and Perovic D D 1999 Effect of nitrogen on shape memory effect of a Fe--Mn-based alloy *Mater. Sci. Eng. A* **262** 39–49
[11] Sawaguchi T, Bujoreanu L-G, Kikuchi T, Ogawa K and Yin F 2008 Effects of Nb and C in solution and in NbC form on the transformation-related internal friction of Fe-17Mn (mass%) alloys *ISIJ Int.* **48** 99–106
[12] Verbeken K, Van Caenegem N and Raabe D 2009 Identification of ε martensite in a Fe-based shape memory alloy by means of EBSD *Micron* **40** 151–6
[13] Druker A, Baruj A and Malarría J 2010 Effect of rolling conditions on the structure and shape memory properties of Fe-Mn-Si alloys *Mater. Charact.* **61** 603–12
[14] Suru M-G, Lohan N-M, Mihalache E, Pricop B, Mocanu M and Bujoreanu L-G 2016 AFM evaluation of pre-straining degree effects on the dimensions of stress induced martensite plates in Fe-Mn-Si based SMAs *J. Test. Eval.* **45** 419–27
[15] Verbeken K, Van Caenegem N and Verhaege M 2008 Quantification of the amount of ε martensite in a Fe-Mn-Si-Cr-Ni shape memory alloy by means of electron backscatter diffraction *Mater. Sci. Eng. A* **481** 471–5
[16] Akhondzadeh A, Zangeneh-Madar K and Abbasi S M 2008 Influence of annealing temperature on the shape memory effect of Fe-14Mn-5Si-9Cr-5Ni alloy after training treatment *Mater. Sci. Eng. A* **489** 267–72
[17] Mocanu M, Mihalache E, Pricop B, Borza F, Grigoraş M, Comaneci R I, Özkal B and Bujoreanu L G 2018 The Influence of α′(bcc) Martensite on the Dynamic and Magnetic Response of Powder Metallurgy FeMnSiCrNi Shape Memory Alloys *Proceedings of the International Conference on Martensitic Transformations: Chicago* pp 99–108

IOP Conf. Series: Materials Science and Engineering **485** (2019) 012005 doi:10.1088/1757-899X/485/1/012005

Antibody functionalized magnetic nanoparticles for circulating tumor cells detection and capture using magnetophoresis

T A Burinaru[1,4], M Volmer[3], M Avram[1], V Țucureanu[1,3], A Avram[1], B Țîncu[1,2], C Mărculescu[1], A Matei[1], R Marinescu[1,2], M Militaru[4]

1. National Institute for Research and Development in Microtechnologies – IMT Bucharest, Romania
2. University Politehnica of Bucharest, Romania
3. Transilvania University of Brasov, Romania
4. University of Agronomic Sciences and Veterinary Medicine of Bucharest, Romania

tiberiu.burinaru@imt.ro

Abstract. Circulating tumor cells (CTCs) are cells present in the blood stream during the metastasis process. They can originate from primary or secondary tumors. Circulating tumor cells can be used for early diagnosis or they can be used for prognosis evaluation and even treatment efficiency evaluation. Circulating tumor cells can be captured based on specific antigens found on their surface that differ from those of normal blood cells, they can be captured using specific electrical signatures using dielectrophoresis and they can also be captured using induced magnetic properties and magnetophoresis. In this paper we describe a method for synthesizing and functionalizing superparamagnetic nanoparticles. The nanoparticles will be covered with polyethylenglicol (PEG) molecules to reduce agglomeration and non-specific cell adhesion or blood proteins fouling. The PEG covered magnetic nanoparticles will be functionalized with anti-EpCAM antibodies that are going to make the nanoparticles specifically bind to CTCs present in the blood sample. The samples will be further processed in a microfluidic device that will separate the targeted cells through magnetophoresis.

1. Introduction

In recent years nanotechnology has sky rocketed in the research and industrial fields with chemical, environmental and medical applications. Nanotechnology is the engineering and manufacturing of technologies or materials at nanoscale. One of the materials of great interest is represented by magnetic nanoparticles which gained a lot of attention for different medical applications. They are used for magnetofection, cell labelling, cell separation, cell targeting and controlled drug delivery and release, and localized magnetic hyperthermia for cancer treatment [1]. There are two types of superparamagnetic nanoparticles commonly used in medical applications: maghemite (γFe_2O_3) and magnetite (Fe_3O_4). Also they can be further devised in superparamagnetic iron oxide particles (SPIOs) with a particle mean diameter >50 nm, ultrasmall superparamagnetic iron oxide (USPIOs) nanoparticles with a mean diameter between 10 50 nm and very small superparamagnetic iron oxide (VSPIOs) with a mean diameter <10 nm.[2] Their application is dictated by properties like the size of the iron oxide crystals, their coating, surface charge and the hydrodynamic size of the particles.[3]

Most of their applications are in magnetic resonance imaging (MRI), since they are being used as contrast agents [4-7]. Xie et al. used PEG, PEG/PEI and PEG/PEI/Tween80 modified superparamagnetic nanoparticles to perform *in vivo* mouse brain MRI. The used nanoparticles demonstrated vascular imaging effects in different lobes of the brain after a 24 hours intravenous injection.[4] Sanjai et al. created superparamagnetic iron oxide nanoparticles encapsulated in chitosan-triphosphate by using an ionotropic gelation method. They demonstrated that their chitosan covered nanoparticles can be used as tissue-specific MRI contrast agent. [8] Unterweger et al. exhibited that their developed dextran coated SPION nanoparticles ($SPION^{Dex}$) exceptionally safe and perfect candidates for further clinical trials [7].

Because superparamagnetic nanoparticles respond to magnetic fields they are suitable for targeted drug and gene delivery. If stimulated by high-frequency magnetic fields they can also be used for heat-triggered drug release and hyperthermia treatment.[9-12] Apart magnetic hyperthermia there are two other emerging methods for cancer treatment: photodynamic therapy (PDT) and photothermal therapy (PTT) [13, 14]. Magnetic particles have also been used for cell identification and separation techniques [15-20].

Based on their application, iron oxide nanoparticles are synthesized with controlled shape, size and composition. The manufacturing methods can be divided into top-down or bottom-up technologies. Top-down methods refer to obtaining nanoparticles by reducing the size from the bulk materials. Bottom-up technologies refer to creating magnetic nanoparticles from molecular reagents to the desired nano-size. The preferred method is the bottom-up approach, since it gives you a good control over their size and shape. The simplest and most efficient way of producing magnetic nanoparticles is by using the method of iron salts co-precipitation in aqueous solution. The mean diameter of the obtained particles is smaller than 50 nm. However, because of the broad size distribution of particles obtain this way, the nanoparticles require further size sorting to reduce the polydispersity index.[21] The size distribution can be better controlled in the presence of surfactants like alkanolamines,[22] (cetyl trimethylammonium bromide (CTAB), polyvinyl pyrrolidone (PVP) and sodium cholate [23]. Solvents like diethyleneglycol (DEG) can play the role of surfactants. It has chelating properties and it gives you the possibility to work at high temperatures [24].

In this paper we describe a method for synthesizing and functionalizing superparamagnetic nanoparticles for the detection and separation of circulating tumor cells (CTCs). The nanoparticles can be used in a microfluidic device that will separate the targeted cells through magnetophoresis. The nanoparticles will be covered with polyethylenglicol (PEG) molecules to reduce agglomeration and non-specific cell adhesion or blood proteins fouling. In order to make the superparamagnetic nanoparticles attach to the targeted cells, the PEG covered nanoparticles will be further functionalized with anti-EpCAM antibodies that will bind only to the cells with epithelial origin.

2. Experimental

2.1. Materials used for magnetic nanoparticles synthetisation

We synthesized the maghemite superparamagnetic nanoparticles by using the following chemical reagents: $FeCl_3{\bullet}6H_2O$ (236489 Sigma-Aldrich), $FeSO_4{\bullet}7H_2O$ (215422 Sigma-Aldrich), NH_4OH-28% v/v (320145, Sigma-Aldrich), HCl (320331, Sigma-Aldrich), poly(ethylene glycol) diamine (Mn6000) (752444, Sigma-Aldrich) and EDC (39391, Sigma-Aldrich). We functionalized the nanoparticles with anti-EpCAM mouse monoclonal antibodies (BSH-7402-100, Nordic BioSite).

2.2. Superparamagnetic nanoparticles synthesis method

The γ-Fe_2O_3 nanoparticles were synthesized by dissolving $FeCl_3{\bullet}6H_2O$ and $FeSO_4{\bullet}7H_2O$ (Sigma-Aldrich) into DI water (2:1 molar ratio). While stirring the solution, small droplets of NH_4OH were added with a pipette. The precipitate is collected with the help of a magnet and rinsed in DI water.

2.3. PEG covered superparamagnetic nanoparticles synthesis and antibody functionalization method

In order to increase biocompatibility and to reduce agglomeration and non-specific cell adhesion or blood proteins fouling and to facilitate directed antibody functionalization, the superparamagnetic nanoparticles were covered in poly(ethylene glycol) diamine. We did that by mixing 0.675g $FeCl_3{\bullet}6H_2O$, 0.350g $FeSO_4{\bullet}7H_2O$ and 6.0g PEG-diamine in 4 ml (DI water). The reaction was alkalinized with 800μL NH_4OH and then brought to acid pH (4.0) by using HCl (6M). The formed precipitate was separated with a magnet and washed in DI water. Antibodies were crosslinked to the amino groups of the PEG-diamine by using carbodiimides (EDC) crosslinking chemistry.

2.4. Sample characterisation

The bond between PEG molecules and maghemite nanoparticles was analysed by Fourier Transform Infrared (FTIR) spectrometry at room temperature using a Bruker Tensor 27 spectrometer, in the wavenumber 4000-400 cm^{-1} by averaging 64 scans and with a resolution of 4 cm^{-1} and KBr pellet technique. The magnetic characterization of our maghemite nanoparticles and functionalized maghemite nanoparticles with PEG-diamine was made by using the VSM unit from 7T Mini Cryogen Free Measurement System.

3. Results and discussion

3.1. Sample phase analysis

Infrared molecular absorption spectroscopy is an analytical technique used for the qualitative analysis of materials based on absorption spectra produced by vibrating atoms present in different molecular groups. Chemical processes and different chemical structures of chemical compounds can be studied with the help of Fourier Transform Infrared (FTIR) spectrometry. Thus, we obtained FTIR spectra of PEG, maghemite and PEG coated maghemite (Fig.1). PEG spectrum is characterized by spectral bands which can be attributed mainly to the stretching and deformation vibration modes of C-H bonds (in the spectral region 3000-2700 cm^{-1}, 1400-1250 cm^{-1} and 950-840 cm^{-1}). It can also be characterized by the vibration modes of the C-O bonds (1200-950 cm^{-1}) of the ether group. Maghemite spectrum shows spectral bands at 584 cm^{-1} and 444 cm^{-1} that can be attributed to the stretching vibration mode of the Fe-O bond. The spectral bands that are showing from the 4000-800 cm^{-1} range, are attributed to the solvent used during the washing process. It can be seen that the maghemite covered in PEG spectrum is composed of spectral band belonging to organic molecules and inorganic molecules of the sample. A shift can be observed in the spectral bands for both organic and inorganic components. The shift in the spectral band the 1100-1000 cm^{-1} domain and in the 630-400 cm^{-1} domain clearly indicates that PEG-diamine has bonded to the maghemite nanoparticles through Fe-O-C bonds. We can also observe a change in the 1400-1200 cm^{-1} region of the C-H bond vibration mode that shifted to 2950-2700 cm^{-1} region after PEG binding.

The magnetic characterization of our maghemite nanoparticles and functionalized maghemite nanoparticles with Polyethylene glycol PEG-diamine was made by using the VSM unit from 7T Mini Cryogen Free Measurement System. The blocking temperature is determined by Zero Field Cooling-Field Cooling (ZFC-FC) magnetization measurements. From ZFC-FC measurements, Fig. 2(a), made on our maghemite nanoparticles, 10 nm in diameter, we obtained T_B=252 K. For temperatures larger than T_B a superparamagnetic behaviour is expected that is desirable for biomedical applications.

The room temperature magnetization curve for our maghemite MNPs, Fig. 2(b), is well fitted by the Langevin function (1) indicating the superparamagnetic behavior of the nanoparticles.[25]

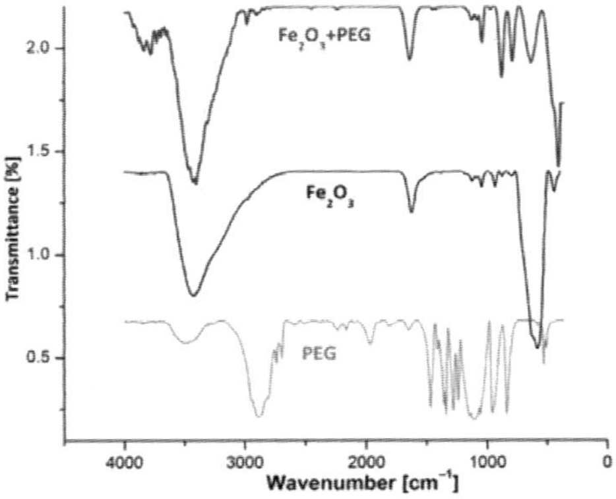

Figure 1. FTIR spectra of PEG, maghemite and maghemite + PEG

$$M(H) \approx M_S(1/\tanh(\alpha) - 1/\alpha), \quad \alpha = \frac{\mu_0 H \cdot m_S}{K_B T}$$

where $m_s = \pi d^3 M_S/6$ and M_S = 480 emu/cm^3=480 kA/m (for maghemite).

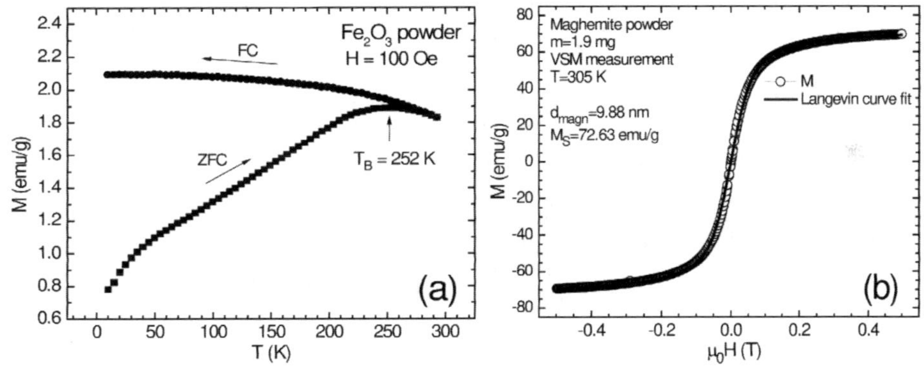

Figure 2. (a) The ZFC-FC magnetization curves and (b) the magnetization curves obtained by VSM for maghemite nanoparticles; the particle diameter is determined from the fitting curve.

From the fitting curve we found an average value of the particle diameter, d_{magn}=9.88 nm. In many applications, like detection of magnetically labelled molecules of interest, the functionalised MNPs are found in aqueous solutions. Fig. 3(a) shows the magnetization curve at room temperature for 30 μl of aqueous solution containing maghemite nanoparticles functionalised with PEG-diamine. By comparing the magnetization curves presented in Figs. 2(b) and 3(a), we can see how large is the diamagnetic contribution of water and PEG molecules at fields higher than 0.1 T. Fig. 3(b) presents the magnetisation curves for the same solution when diamagnetic correction is applied. This behaviour shows that diamagnetic contribution of water and PEG molecules can lower the field produced by the functionalized MNPs when large magnetic fields are used for magnetophoresis and detection experiments. To lower the diamagnetic contribution, the maximum value of the magnetizing field must be carefully tuned. From Fig. 3(a) it comes that optimal value of the applied field in order to get a maximum response from the functionalized MNPs is 0.1 T.

Fig. 4(a) presents the high field magnetisation curve, as measured, for 1.27 mg powder of functionalized maghemite nanoparticles. The high field magnetization curve, allows us to estimate the diamagnetic contribution of the PEG-diamine which has a linear dependence with the applied field as seen in Fig. 4(a). To obtain 1.27 mg of powder, we used 107.7 mg of aqueous solution which was dried at 56 °C. A higher temperature would destroy the PEG molecule. After applying the diamagnetic correction and fitting the magnetization data from Fig. 4(b) with the Langevin function, we obtained the "magnetic diameter" of the functionalized maghemite nanoparticles as d_{magn}=11.48 nm.

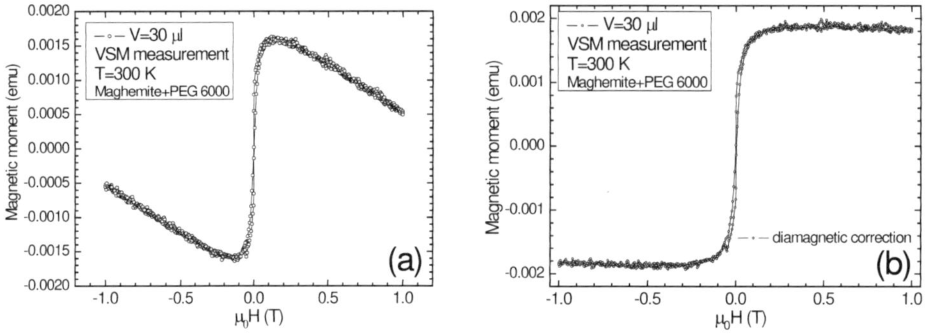

Figure 3. (a) The magnetisation curve obtained by VSM at room temperature for 30 μl of aqueous solution containing maghemite nanoparticles functionalised with PEG-diamine and (b) the same curve when diamagnetic correction is applied.

The PEG-diamine covered nanoparticles have been functionalized with anit-EpCAM antibodies using carbodiimides (EDC) crosslinking chemistry, in order to make the nanoparticles attach to CTCs of epithelial origin. To enhance cell attachment, the antibodies have to be linked with the crystallisable fragment (Fc) to the nanoparticles so that the antigen binding region (Fab), especially the paratope will be facing outwards, so that the availability of antigen binding sites will be maximized. EDC will react with the carboxylic acid (-COOH) that exists at the end of the Fc region of the antibody activating it, thus linking to the NH_2 molecules present on the magnetic nanoparticles. The quantity of used PEG-diamine, EDC, antibodies and the antibody functionalization process needs further optimization.

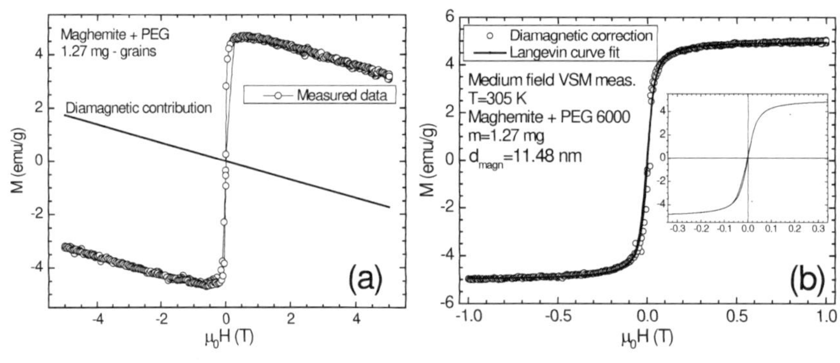

Figure 4. (a) Magnetization curve obtained by VSM for maghemite nanoparticles functionalised with PEG-diamine as powder and (b) the extracted curve after diamagnetic correction; the inset shows the low field magnetization curve.

4. Conclusions

The γ-Fe_2O_3 superparamagnetic nanoparticles coated with PEG-diamine have been synthesized and characterized by FTIR spectrometry and magnetically characterized by using the VSM unit from 7T Mini Cryogen Free Measurement System. Phase analysis was performed by FTIR spectrometry, showing spectral bands at 584 cm^{-1} and 444 cm^{-1} which can be attributed to the Fe-O bond. After PEG-diamine bonding, FTIR spectra demonstrated the formation of Fe-O-C bonds. Magnetic characterization showed that the diameter of the nanoparticles was d_{magn}=9.88 nm. The diamagnetic contribution of water and PEG molecules can lower the field produced by the functionalized MNPs when large magnetic fields are used for magnetophoresis and detection experiments. The "magnetic diameter" obtained with the Langevin function of the PEG-diamine coated superparamagnetic nanoparticles was d_{magn}=11.48 nm. Future work is focused on optimizing the antibody functionalization process, which will improve CTCs detection.

Acknowledgements

This work was supported by a grant of the Ministry of National Education And Scientific Research, RDI Program for Space Technology and Advanced Research—Star, project number 639/2017 and by UEFISCDI in the partnership framework: PN-III-P1-1.2-PCCDI-2017-0214 (project no. 3PCCDI/2018).

References

[1] Li X, Wei J, Aifantis K E, Fan Y, Feng Q, Cui F-Z and Watari F 2016 *J. Biomed. Mater. Res. Part A* **104** 1285–96
[2] Corot C, Robert P, Idée J-M and Port M 2006 *Adv. Drug Deliv. Rev.* **58** 1471–504
[3] Hedayatnasab Z, Abnisa F and Daud W M A W 2017 *Mater. Des.* **123** 174–96
[4] Xie S, Zhang B, Wang L, Wang J, Li X, Yang G and Gao F 2015 *Appl. Surf. Sci.* **326** 32–8
[5] Wang J, Zhang B, Wang L, Wang M and Gao F 2015 *Mater. Sci. Eng. C* **48** 416–23
[6] Estelrich J, Sánchez-Martín M J and Busquets M A 2015 *Int. J. Nanomedicine* **10** 1727–41
[7] Unterweger H, Dézsi L, Matuszak J, Janko C, Poettler M, Jordan J, Bäuerle T, Szebeni J, Fey T, Boccaccini A R, Alexiou C and Cicha I 2018 *Int. J. Nanomedicine* **13** 1899–915
[8] Sanjai C, Kothan S, Gonil P, Saesoo S and Sajomsang W 2014 *Carbohydr. Polym.* **104** 231–7
[9] Mohammed L, Gomaa H G, Ragab D and Zhu J 2017 *Particuology* **30** 1–14
[10] Czugala M, Mykhaylyk O, Böhler P, Onderka J, Stork B, Wesselborg S, Kruse F E, Plank C, Singer B B and Fuchsluger T A 2016 *Nanomedicine* **11** 1787–800
[11] Aires A et al. 2016 *Nanotechnology* **27**
[12] Das R, Alonso J, Nemati Porshokouh Z, Kalappattil V, Torres D, Phan M-H, Garaio E, García J Á, Sanchez Llamazares J L and Srikanth H 2016 *J. Phys. Chem. C* **120** 10086–93
[13] Wu M and Huang S 2017 *Mol. Clin. Oncol.* **7** 738–46
[14] Shen S, Wang S, Zheng R, Zhu X, Jiang X, Fu D and Yang W 2015 *Biomaterials* **39** 67–74
[15] Andree K C, van Dalum G and Terstappen L W M M 2016 *Mol. Oncol.* **10** 395–407
[16] Karabacak N M et al. 2014 *Nat. Protoc.* **9** 694–710
[17] Gajaila G, Gajaila I, Cotor G and Ioniţă L 2015 *Journal of Biotechnology* vol **208** S23
[18] Gajaila G, Gajaila I, Dobrea M and Cotor G 2009 *Scientific works, C Series, Veterinary Medicine*
[19] Gajaila G, COTOR G and Gajaila I 2014 *Bull. Univ. Agric. Sci. Vet. Med. Cluj-Napoca. Vet. Med.* **71** 59–61
[20] Gajaila G, Gajaila I, Cotor G and Dobrea M 2010 *Bull. Univ. Agric. Sci. Vet. Med. Cluj-Napoca. Vet. Med.* **67**
[21] Stanicki D, Elst L Vander, Muller R N and Laurent S 2015 *Curr. Opin. Chem. Eng.* **8** 7–14
[22] Pereira C et al. 2012 *Chem. Mater.* **24** 1496–504
[23] Filippousi M, Angelakeris M, Katsikini M, Paloura E, Efthimiopoulos I, Wang Y, Zamboulis D and Van Tendeloo G 2014 *J. Phys. Chem. C* **118** 16209–17
[24] Stanicki D, Boutry S, Laurent S, Wacheul L, Nicolas E, Crombez D, Vander Elst L, Lafontaine D L J and Muller R N 2014 *J. Mater. Chem. B* **2** 387–97
[25] Brzeska M, Panhorst M, Kamp P B, Schotter J, Reiss G, Pühler A, Becker A and Brückl H 2004 *J. Biotechnol.* **112** 25–33

IOP Conf. Series: Materials Science and Engineering **485** (2019) 012006 doi:10.1088/1757-899X/485/1/012006

A study on specific heat of nanoparticle enhanced fluids

E I Chereches[1], M Chereches[1] and A A Minea[1]

[1]Department of Technologies and Equipment for Materials Processing, Technical University "Gh. Asachi" from Iasi, ROMANIA

aminea@tuiasi.ro

Abstract. A real necessity in the field of materials engineering and beyond is the development of new smart materials that effectively combine state-of-the-art technology with competitive costs to meet current technical requirements and challenges. In our days, fluids used in engineering occupy a significant place, and for their application in the industry, special thermophysical, thermal and chemical properties are required. Nanofluids are new fluids that contains nanometric particles, called nanoparticles. Traditional fluids used for nanofluids are water, oil and ethylene glycol, and nanoparticles are typically made of metal oxides, carbides or carbon nanotubes. Ionanofluids, on the other hand, are new heat transfer fluids based on ionic liquids, with superior thermal properties compared to the conventional fluid. Ionanofluids are recently developed by adding nanomaterial particles to an already novel fluid, as the ionic liquid. From the available literature, it is easy to understand that a new heat transfer fluid should have good thermal and flow properties and the iononofluids thermophysical properties can be improved by juggling with the amount and type of nanoparticles added in the ionic liquid. The goal of ionanofluids development is to obtain the highest possible thermal properties at the lowest possible concentrations, thus this paper aims to experimentally study the specific heat of these new nanoparticle enhanced fluids.

1. Introduction

It is well known that regular fluids used in the heat transfer applications are much less conductive to heat than some metals, and nanofluids, even at extremely low concentrations, considerably improve heat transfer coefficient of a specific application. The term "nanofluid" was proposed by Choi [1] in 1995, wishing to develop a new class of heat transfer fluids. Nanofluids are fluids that exhibit superior thermal properties compared to conventional liquids. Ionanofluids are an innovative research area and due to the growing need to identify non-pollutant heat transfer environments while reducing energy consumption. Ionanofluids are actually another category of nanofluids recently developed by the addition of nanoparticles in ionic liquids. The use of ionic liquids opens up new possibilities for obtaining new types of fluids prepared using an environmentally friendly, recyclable process and applicable in the industry. However, there is a limited number of industrial processes that use ionic liquids due to their high cost

[2]. In order to minimize the price of ionic liquids, an efficient recycling of ionic liquids is essential for their use in the industry.

Nevertheless, the concept of "Ionanofluids" was firstly proposed by Nieto de Castro and co-workers [2] in Portugal in 2009. The main properties that have been measured for ionanofluids are density, heat capacity, viscosity and thermal conductivity [3]. Incorporation of the nanoparticles into the base fluid leads to a change in the thermophysical properties, such as thermal conductivity, viscosity, density and specific heat, which highly affect convective heat transfer [4-13]. The nanoparticle concentration, purity level, nanomaterial shape and size are some of the main factors that significantly modify the thermophysical properties of these new fluids. Nonetheless, the preparation of ionanofluids is the key step in the use of nanoparticles to improve the specific heat of the fluids as well as the quality of the instruments used.

Most ionanofluid research studies focus on improving the thermal conductivity and viscosity of ionic liquids, largely by addition of multi-walled carbon nanotubes, but also graphene or Al_2O_3, and it is therefore necessary to intensify research on the thermophysical properties of ionanofluids. Applications of ionanofluids in several fields of science and technology are based on the unique properties of ionic liquids, fused to those of nanosystems. Addition of nanoparticles may lead to increases in thermal conductivity of an ionic liquid by over 30% and thermal capacity by up to 10% [8, 10, 11]. Moreover, ionanofluids have also shown a great potential to be used as new heat transfer fluids in many important applications, such as solar energy technology [7-9].

In this study we will consider the specific heat as the base point while we developed an experimental study on specific heat variation for several nanoparticles enhanced ionic liquids, named shortly as ionanofluids. As far as the authors are aware, these new fluids were not studied before and we intent to obtain an overview about their possible behavior as heat transfer fluids.

2. Materials and methods

Specific heat is one of the fundamental properties of ionanofluids and plays an important role in the heat transfer of ionanofluids, which has been investigated through a series of studies.

The first model is based on the volume concentration of the nanoparticles and was presented by Pak and Cho [4], taking the idea of the liquid-particle mixture.

$$C_{P_{nf}} = \phi(C_p)_p + (1-\phi)(C_p)_{bf} \tag{1}$$

The second model is based on the heat equilibrium mechanism and this was presented by Xuan and Roetzel [5].

$$C_{P_{nf}} = \frac{(1-\phi)(\rho C_p)_{bf} + \phi(\rho C_p)_p}{(1-\phi)\rho_{bf} + \phi\rho_p} \tag{2}$$

where:

C	heat capacity, KJ/Kg K
ϕ	concentration
p	particle
b	bulk
f	fluid
ρ	density, g/m^3

IOP Conf. Series: Materials Science and Engineering **485** (2019) 012006 doi:10.1088/1757-899X/485/1/012006

Also, some experimental results on thermal conductivity, specific heat and viscosity, were performed by Titan [6]. These investigations have noted an improvement in the thermophysical properties studied for ionanofluids compared to basic ionic liquids. Titan [6] studied the specific heat of the ionic liquid [C_4mmim][NTf_2] and ionanofluids with different concentrations of 0.5, 1.0 and 2.5 wt% Al_2O_3. Specific heat was measured in the temperature range of 25-345°C (at a range of 10°C).

2.1. Methodology

The ionic liquid were manufactured by American chemicals company (Sigma-Aldrich) and water was produced by a Milli-Q 185 Plus system (Millipore Ltd., Watford, United Kingdom) with a resistivity of 18.2 MW·cm. Several physical and chemical properties of the ionic liquid (1-ethyl-3-methylimidazolium methanesulfonate) used in this study are shown in table 1.

Table 1. Physical and chemical properties of the 1-ethyl-3-methylimidazolium methanesulfonate.

CAS Number	145022-45-3
Molecular formula	$C_7H_{14}N_2O_3S$
Molecular Weight	206.26 g/mol
Water Content	0.029 wt%
Flash Point (C)	286°C
Refractive index	1.4970
Water Solubility	Soluble in water
Impurities	≤0.5%

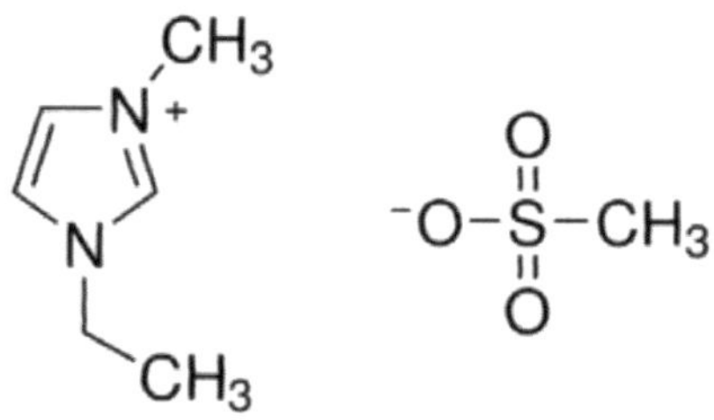

Figure 1. Structure of investigated ionic liquids.

We started preparing fluids based on [C_2mim][CH_3SO_3] and H_2O at various concentrations (0.25, 0.50, 0.75 for H_2O) and in table 2 the molar concentrations are presented.
The water adding in the ionic liquid was subjected to ultrasonic vibrations for about one minute to ensure that a uniform dispersion was obtained.

Table 2. Quantities in molar percentages for ionic liquid and water.

Molar fraction	0.25	0.50	0.75
M_{H2O} [g]	0.09201	0.26781	0.79197
$m_{[C2mim][CH3SO3]}$ [g]	3.15948	3.06531	3.02581

IOP Conf. Series: Materials Science and Engineering **485** (2019) 012006 doi:10.1088/1757-899X/485/1/012006

Specific heat, c_p, was measured as a function of the temperature, T, by using differential scanning calorimeter equipment, Q2000, TA Instruments, New Castel, USA. Specific heat was measured in the temperature range 10-60°C.

3. Results and discussion

To obtain a morphological characterization, the transmission electron microscope technique was employed. The equipment is a JEOL JEM 1010 microscope, operating at an acceleration voltage of 40 kV to 100kV, and equipped with a 2k x 2k AMT CCD camera for digital image acquisition. The images obtained are shown in figure 2, and one can see that the fluids are composed of quasi spherical particles of Al_2O_3.

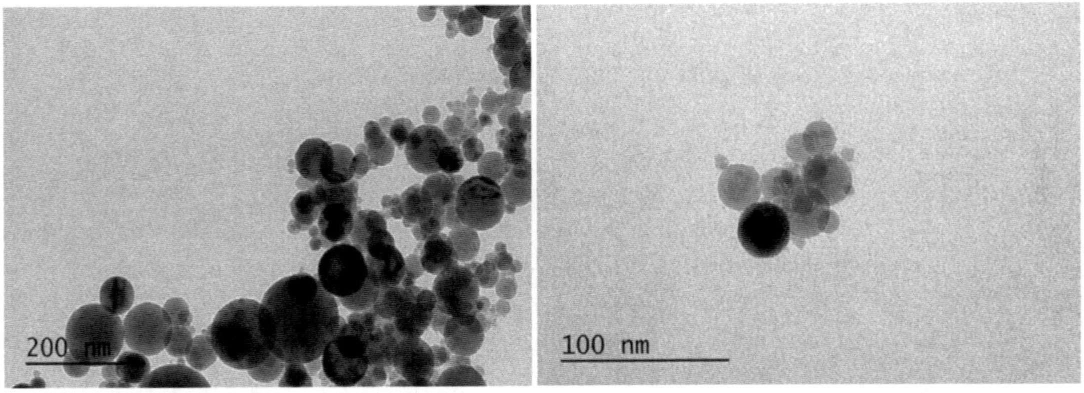

Figure 2. Image of the Al_2O_3 with transmission electron microscopy (TEM).

The results of the heat capacity measurements for pure ionic liquid and water are shown in figure 3. Unfortunately, datasets have not been previously reported for similar systems, to date, which limits the comparison that can be made. Moreover, as can be seen the heat capacity of all data increases almost linearly with temperature, as expected.

In this study we measured the specific heat of the ionic liquid $[C_2mim][CH_3SO_3]$ and the ionic liquid with different percentages of water, namely 0.25, 0.50 and 0.75. From the specific heat experimental results, it can be seen that the specific heat of ionanofluids is much higher than pure ionic liquid over the measured temperature range (which ranges from 10-60°C). The prior knowledge of the specific heat capacity as a function of temperature is crucial, from an engineering perspective.

Plus, we calculated the specific heat of alumina based ionanofluids by adding alumina nanoparticles in percentages of 1, 2 and 3 % vol. The corresponding results can be seen in figure 4, which was obtained using Eq. (1).

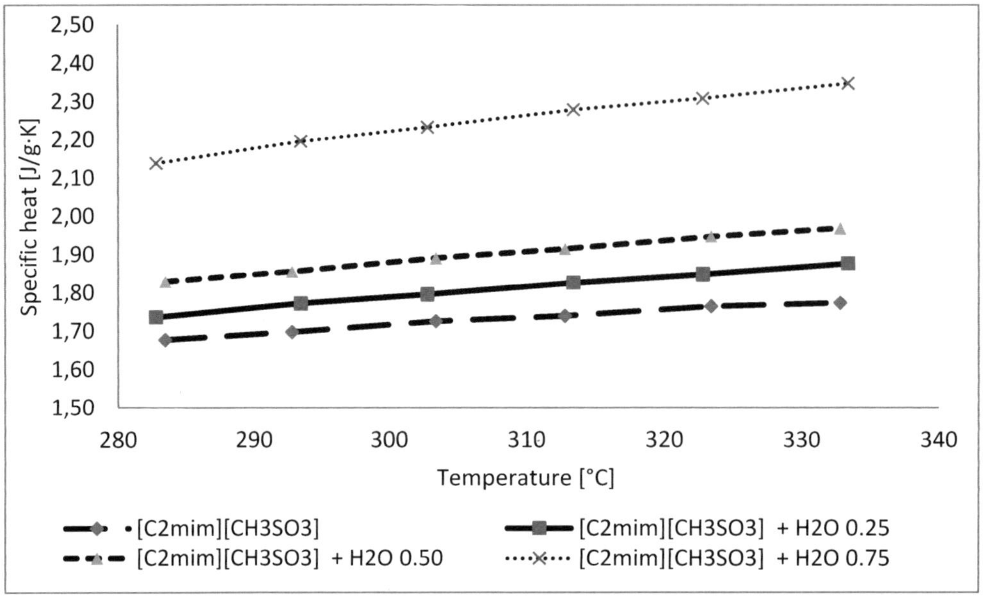

Figure 3. Specific heat of pure ionic liquid [C_2mim][CH_3SO_3] and the ionic liquid with different percentages of water, namely 0.25, 0.50 and 0.75 depending on the temperature.

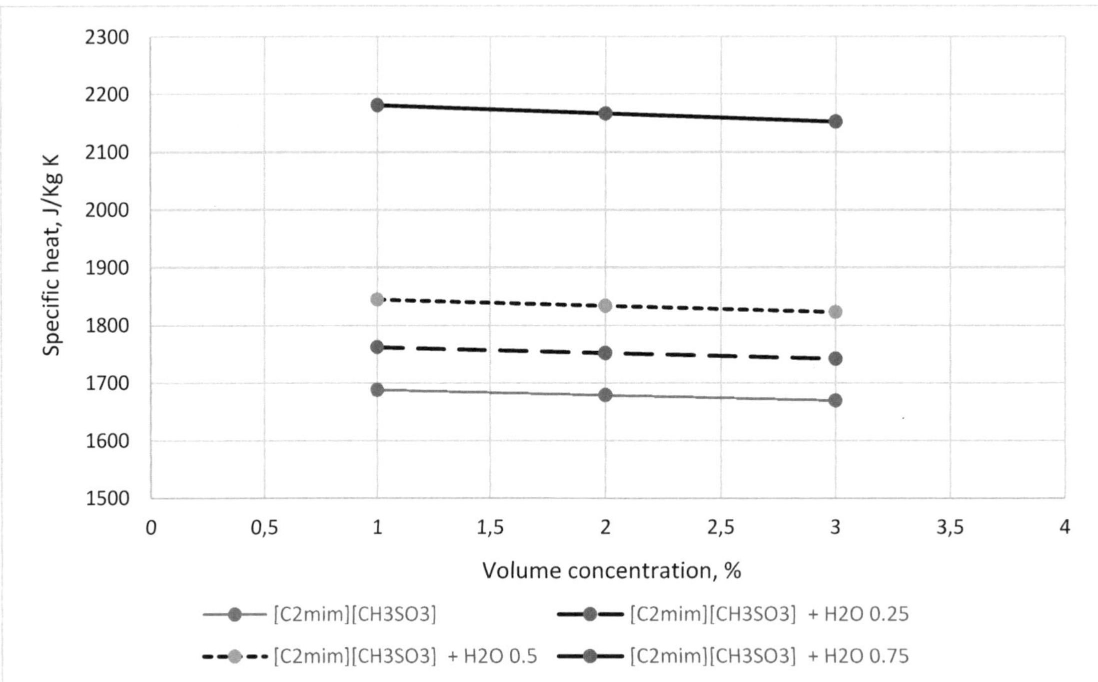

Figure 4. Variation of specific heat while adding nanoparticles to manufactured mixtures

One can notice from Figure 4 that adding nanoparticles to base liquids, the specific heat is slightly decreasing, phenomenon noticed for all nanoparticles enhanced fluids.

IOP Conf. Series: Materials Science and Engineering **485** (2019) 012006 doi:10.1088/1757-899X/485/1/012006

Moreover, to make a comprehensive comparison of the two equations used in the literature (see eq. (1) and (2)) we decided to plot the data for all ionanofluids, as can be seen in figure 5.

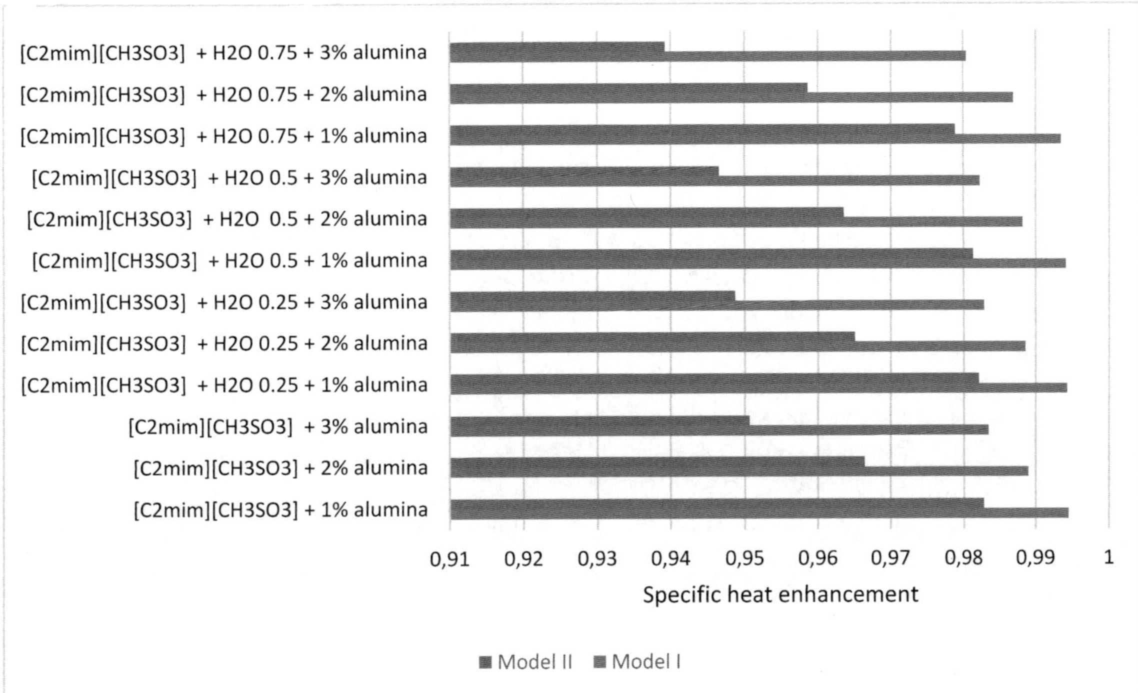

Figure 5. Specific heat variation of ionanofluids.

If it looks to figure 5 it can notice that Model II (corresponding to eq. (2)) is under predicting the specific heat results as compared to Model I.

4. Conclusion

Ionanofluids are a new class of nanoparticle enhanced fluids. Ionanofluids can be obtained by dispersing nanoparticles into base ionic liquids or ionic liquids mixtures. In this paper some experimental and analytical results are depicted in order to outline the changes in specific heat while mixing ionic liquids with water and adding alumina nanoparticles.

The results show that adding water to ionic liquids the specific heat is augmented and the improvement goes to almost 80% for $[C_2mim][CH_3SO_3] + H_2O$ 0.75.

On the other hand, adding nanoparticles to prepared base liquids decrease sligfhtly the specific heat.

As a general conclusion, ionanofluids studies needs to be intensified in order to completely describe their properties and their advantages over classical heat transfer fluids.

References

[1] Murshed S and Nieto de Castro C 2014 Nanofluids: synthesis, properties and applications *NOVA Science Publishers, Inc., New York, C.A.*

[2] Ribeiro A P C, Lourenço M. J V, Nieto de Castro C and Hardacre C 2008 Thermal conductivity of "Bucky Gels" *In: Proceedings of Conference on Molten Salts and Ionic Liquids (EUCHEM2008)*, Copenhagen, Denmark, pp 24-29

IOP Conf. Series: Materials Science and Engineering **485** (2019) 012006 doi:10.1088/1757-899X/485/1/012006

[3] Nieto de Castro C 2010 Thermophysical properties of ionic liquids: do we know how to measure them accurately? *Journal of Molecular Liquids* **156** pp 10-17

[4] Pak B C and Cho Y I 1998 Hydrodynamic and heat transfer study of dispersed fluids with submicron metallic oxide particles *Exp Heat Transf Int J* **11(2)** pp 151–70

[5] Xuan Y and Roetzel W 2000 Conceptions for heat transfer correlation of nanofluids *Int J Heat Mass Transf* **43(19)** pp 3701–3707

[6] Titan C. P, Morshed A K M M, Fox E and Khan A J 2017 Enhanced thermophysical properties of NEILs as heat transfer fluids for solar thermal applications *Applied Thermal Engineering* **110** pp 1–9

[7] Minea A A and Murshed S 2018 A review on development of ionic liquid based nanofluids and their heat transfer behavior *Renewable and Sustainable Energy Reviews* **91** pp 584-599

[8] Minea A A and El-Maghlany W M 2018 Influence of hybrid nanofluids on the performance of parabolic trough collectors in solar thermal systems: recent findings and numerical comparison *Renewable Energy* **120** pp 350-364

[9] Chereches E I, Viswanatha Sharma K and Minea A A 2018 A numerical approach in describing ionanofluids behavior in laminar and turbulent flow *Continuum Mechanics and Thermodynamics* pp 1-10

[10] Minea A A and El-Maghlany Wael M 2017 Natural convection heat transfer utilizing ionic nanofluids with temperature-dependent thermophysical properties *Chemical Engineering Science* **174** pp. 13-24

[11] Minea A A, Moldoveanu M G, and Dodun O 2017 Thermal conductivity enhancement by adding nanoparticles to ionic liquids *Solid State Phenomena* **261** pp 121-126

[12] Dinu (Chereches) E I, Chereches M and Minea A A 2017 Nanoparticle enhanced ionic liquids thermophysical properties variation and the benefits of their development *First international conference on energy systems engineering ICESE 2017* (2-4 november, Karabuk, Turkey)

[13] Chereches E I and Chereches M 2018 *Conference of the Doctoral School-TUIASI* (23-24 May, Iasi, Romania)

IOP Conf. Series: Materials Science and Engineering **485** (2019) 012007 doi:10.1088/1757-899X/485/1/012007

Study on the changes induced by the Pressure-Assisted Thermal Processing (PATP) in polymer films used as packaging by the meat industry

V Filimon[1], D Borda[2], G Gurău[3], S Butan[4], P Alexe[2], M Stoica[1]

[1]Cross-Border Faculty, "Dunarea de Jos" University of Galati, Domneasca Street, 47, RO-800008, Galati, Romania
[2]Faculty of Food Science and Engineering, "Dunarea de Jos" University of Galati, Domneasca Street, 47, RO-800008, Galati, Romania
[3]Faculty of Engineering, "Dunarea de Jos" University of Galati, Domneasca Street, 47, RO-800008, Galati, Romania
[4]Faculty of Sciences and Environment, "Dunarea de Jos" University of Galati, Domneasca Street, 47, RO-800008, Galati, Romania

maricica.stoica@ugal.ro

Abstract. The purpose of this study was to assess the changes induced by pressure-assisted thermal processing (PATP) treatment in two multilayer polymer films used as packaging material by the meat industry. The changes induced by the treatment at 600 MPa, 70 °C for 10 minutes on the structural, mechanical and thermal properties of two multilayer polymer films (Pev-70 coextruded high barrier multilayer film with the structure PA / EVOH / PA / PE and Pob-60 combination of biaxially oriented polyamide with a co-extruded barrier film of the polyethylene / EVOH / m-polyethylene structure) were studied. Numerous changes induced by the PATP treatment on the structural, mechanical and thermal properties of the two multi-layered polymer films used as packaging materials in the meat industry have been identified. The choice of a packaging material according to the PATP treatment requires a better understanding of all the changes induced by this treatment. So, this study has preliminarily identified a part of those changes that occurs and could irreversible affect the quality of foods.

1. Introduction

Pressure assisted thermal processing (PATP) is a combined technique designed to prolong food shelf-life. Very limited information is available on impact of combined pressure-heat treatment on packaging material during PATP treatment [1-6].

The integrity maintenance of flexible polymeric films used for packaging during PATP treatment has a particular importance for the food safety and food shelf-life [2-4, 6]. Prior to PATP treatment, the packaged products should be preheated to a target temperature. The preheating efficiency depends on the polymeric film used. The results have shown that multilayer films require a shorter preheating time than thin polymeric materials. In addition, it has been observed that multilayer polymer films represent a viable option for the packaging materials used in the PATP treatment process in terms of the strength and the overall appearance [7]. Lately, various kind of packaging materials are thermally processed ensuring pre-packed food quality during the shelf-life [8]. The main goal of this study is to select the most suitable packaging material able to withstand the PATP treatment.

2. Materials and methods

To achieve this objective, the analysis of structural, thermal and mechanical properties induced by PATP (pressure-assisted thermal processing) were performed on two multilayer polymeric materials: Pev-70 (coextruded high barrier multilayer film with the structure PA/EVOH/PA/PE) and Pob-60 (combination of biaxially oriented polyamide with a coextruded barrier film of the structure polyethylene/EVOH/m-polyethylene), used in the meat industry.

These materials were used to obtain the bags for pork packed under vacuum-conditions. The bags containing pork were pressurized at 600 MPa for 10 min at 70 °C (PATP treatment) and at 600 MPa for 10 min at 20 °C (HPP treatment) in a Resato HPP equipment with water-glycol mixture as a pressure transmission medium. The high pressure processing treatment (HPP) at 600 MPa for 10 min at 20 °C, was applied to evaluate only for the effect of high pressure on the system, and it was considered an equivalent to the pasteurization treatment. The PATP treatment applied is equivalent to the thermal sterilization. The conditions chosen in this case are above the limits currently applied by the industry but in line with the further expected developments while the purpose of this study was to identify the limits imposed by PATP for the polymeric films used as packaging material. The structural and thermal behaviour of polymeric materials were evaluated using SEM (Scanning Electron Microscopy) and DSC (Differential Scanning Calorimetry), respectively. For the SEM analysis, the packaging films were analyzed using Quanta 200 microscope at various magnification levels and a low voltage (20 kV). Mechanical tests were performed to determine Young's modulus of elasticity, breaking strength and elongation, using an universal test machine, Testometric M350-5AT. The mechanical parameters were calculated according to ASTM D882-02 (ASTM 2002). Differential scanning calorimetry was used to evaluate the effect of PATP on the thermal behaviour of polymeric materials. Also, the changes of thermodynamic parameters can be attributed to the thermal damage that occurs during preheating [7].

The data were statistical analyzed using MINITAB 16 software (Tukey analysis and single-factor analysis). These experimental data were obtained from mechanical tests.

3. Results and discussion

From the micrographs of all cross sections of the treated films it can be seen that processing at 600 MPa, 70 °C, during 10 minutes induces structural defects (Figure 1).

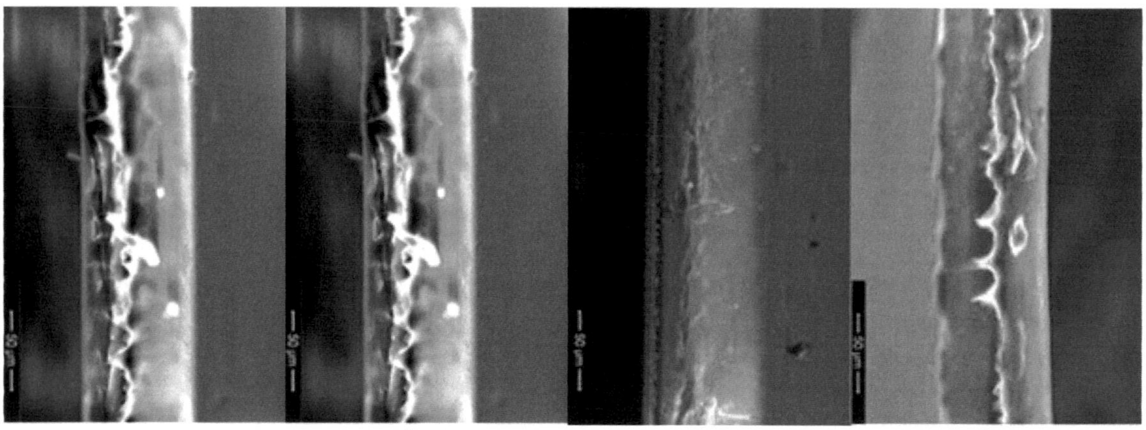

a b c d

Figure 1. Micrographs of all cross sections of the treated films
a) Untreated Pob-60, b) PATP Pob-60, c) Untreated Pev-60, d) PATP Pev-70

Compared to the untreated film, the boundary limit between the layers of the Pob-60 treated PATP film is not clear, even if the layers are parallel. Obviously, there are some areas of low separation between the layers. A clear film-air interface can be observed in the PATP-60 film that occurred most probably

IOP Conf. Series: Materials Science and Engineering **485** (2019) 012007 doi:10.1088/1757-899X/485/1/012007

during decompression. In case of the Pev-70 treated PATP film, the layers are not well defined, compared to the untreated film. Unclear film layers could give less vapor permeability and/or weaker mechanical properties, confirmed by tear resistance analysis.

In order to obtain more in-depth results, the treated HPP (PM) and PATP (PTM) films from Pob-60 and Pev-70 packaging materials were mechanically tested. In addition, comparative studies of the bands obtained from films treated at 70 °C (T1) were performed. The influence of the treatment on the value of three parameters of multilayered polymer films (Young's modulus of elasticity, breaking stress and elongation) was analyzed.

The effect of the treatments applied on the breaking elongation (%) of the studied polymeric films is shown in Figure 2. Statistically, the treatment effect on the breaking elongation has a significant influence ($p < 0.05$), but it does not depend on the type of the polymer film studied, all the samples being influenced equally by the applied treatment (Figure 2).

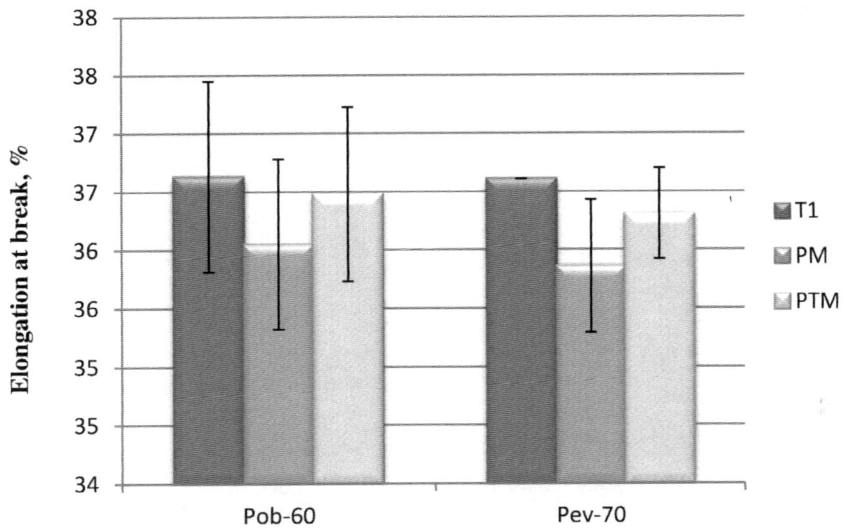

Figure 2. The influence of the treatment on the value of the breaking elongation of multilayer polymer films

The breaking elongation of the Pob-60 films has a very small reduction (Figure 2), regardless of the treatment applied (HPP or PATP), this behavior revealing the formation of a rigid structure [9], confirmed by the SEM analysis. The high resistance of these films can indicate that no breaks occurred between the polymer layers, and did not hindered the film elongation. In the case of the Pev-70 film, the breaking elongation is practically constant (Figure 2). Actually, the applied treatment has induced changes on the breaking elongation of the treated samples, being 2 % for the sample treated with HPP (PM) and 0.8 % for the PATP treated sample (PTM) and proving good mobility of the polymer chain.

The elasticity modulus is significantly influenced ($p < 0.05$) by the type of the polymer film used as packaging material in case of the HPP and PATP treatments. As shown by the graph, the elasticity modulus of Pob-60 and Pev-70 polymer films is influenced by high pressure treatments in the same way (Figure 3). This behavior is similar with the one reported for polymeric structure containing PE and EVOH of these multilayer films [7]. The strongest influence of HPP and PATP treatments on the elasticity modulus can be seen in the Pob-60 film in case of the PM sample. The elasticity of the Pob-60 film decreases by 64 % for the treated PM sample and 48 % for the PTM sample, compared to the blank sample (T1), which showed the highest elastic modulus value (1247.15 MPa). The rigid behavior of this highly elastic polymeric film can be explained by the intra- and interstructural changes induced by high-pressure treatment. The overall tendency of the elasticity modulus for the analyzed films is to decrease more under the influence of HPP treatment than under the influence of PATP.

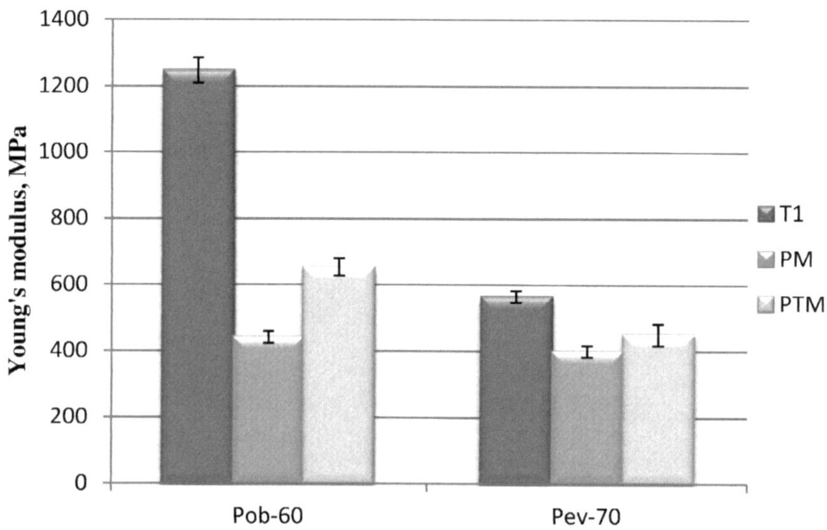

Figure 3. The influence of the treatment on the elasticity modulus of multilayer polymer films

The influence of the treatment on the tensile strength of the multilayer polymeric films is significant ($p < 0.05$), depending on the type of polymeric film used for packaging (Figure 4).

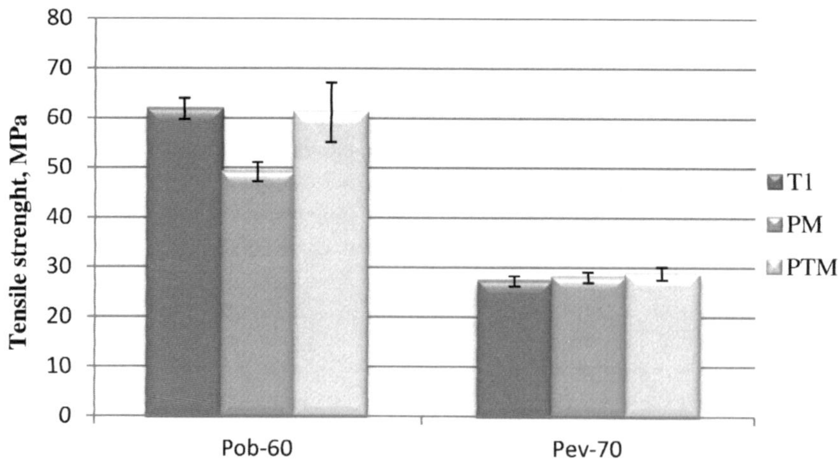

Figure 4. The influence of the treatment on the tensile strength value of multilayer polymer films

IOP Conf. Series: Materials Science and Engineering **485** (2019) 012007 doi:10.1088/1757-899X/485/1/012007

In the case of the Pob-60 polymer film, the effect of PATP treatment has a small influence on this parameter, which is characterized by a similar value as the blank sample (T1) and PTM sample. The behaviour of the Pev-70 film shows that the treatment does not cause strong changes in the breaking stress. Thus, the voltage value exhibits a small increase depending on the temperature of the applied treatment, this behavior being observed also by other researchers [10, 11]. This increase demonstrates a slightly rigidity occurrence in the polymeric film structure when a specific treatment is applied. However, the breaking stress value of the treated samples is almost constant compared to the control film (T1). In industry, a variation of up to 25 % is allowed for this parameter [10]. The two polymeric films studied are fitted for this restriction, which demonstrates a low influence of the applied treatments on the breaking stress of these packaging materials.
The melting temperature (Tm) and melting enthalpy (ΔHm) for each polymer were determined from endothermic curves (melting). The results were presented on the thermograms (Figure 5). The Pob-60 multilayer polymer film has PA, PE, EVOH and m-PE polymers in its structure. Figure 5 shows the endothermic curves for the unprocessed Pob-60 multilayer polymer film and the PATP processed Pob-60 multilayer polymeric film, respectively. There is no significant difference between the thermal melting behavior of the processed PATP and unprocessed Pob-60 film, the value of the thermodynamic parameter Tm (melting temperature) remaining almost constant. By integration of the curves, important data on the thermal behavior of this film were obtained.

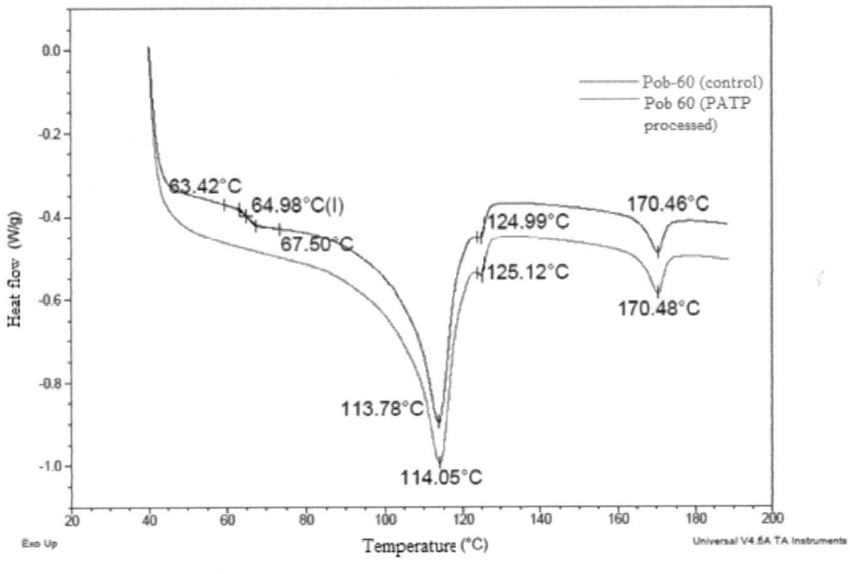

Figure 5. Endothermic DSC curves for Pob-60 multilayer polymer film before and after PATP processing

From the red endothermic DSC curve, no heat related event associated with the vitreous transition was detected for the untreated Pob-60 film. Three distinct melting peaks were recorded. The major (strong) and sharp endothermic peak centered at 114.05 °C can be attributed to the PE polymer, whose melting temperature (Tm) is in the range of 105-115 °C. The second peak with a small area was centered at 125.12 °C corresponding to the melting m-PE. The third peak with an intermediate concentration was recorded between the first two peaks at 170.48°C, indicating the melting temperature of EVOH, whose Tm is in the range of 160-190 °C. The weak peak, centered at 125.12 °C, could also be attributed to the crystallization of PE polymer [12]. From the blue endothermic curve of the Pob-60 multilayer polymer film, four heat events were recorded: a glass transition at 64.98 °C and three endothermic signals. The first signal was shown at 113.78 °C, signifying the PE melting point, the second peak, much weaker, at

124.99 °C, indicating melting m-PE and the third one at 170.46 °C corresponding to the melting point of EVOH.

The following table (Table 1) shows the influence of PATP treatment on the ΔHm (melting enthalpy) value of Pob-60 multilayer polymer film.

Table 1. The influence of PATP treatment on the ΔHm (melting enthalpy) value of Pob-60 multilayer polymer film

Analysed film	ΔHm (J/g) PE	ΔHm (J/g) EVOH
Pob-60 (control)	35.19	3.26
Pob-60 (PATP treated)	43.38	2.53

In case of the Pev-70 film, a heat event associated with the vitreous transition was identified at 70.29 °C (Figure 6). The same Tg can also be seen on the endothermic curve of the PATP-treated Pev-70 film, its value being very close to that of Tg for the control sample (68.33 °C).

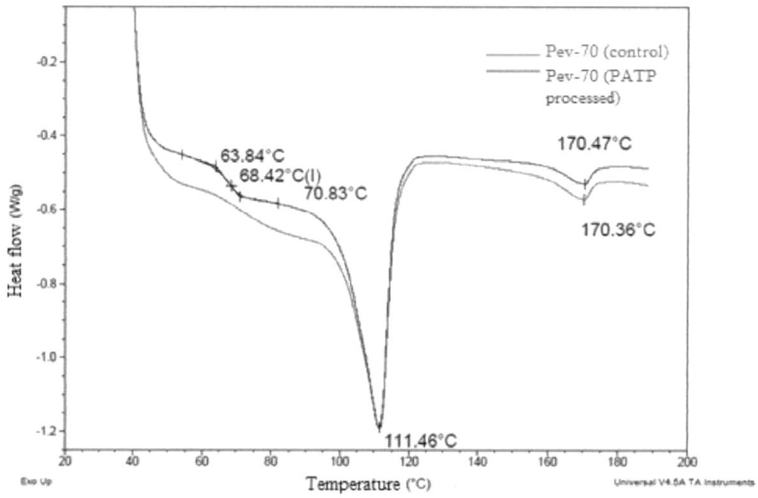

Figure 6. Endothermic DSC curves for the Pev-70 multilayer polymer film before and after PATP processing

Both endothermic curves of the Pev-70 multilayer polymer films exhibit two distinct melting peaks. The major (strong) and sharp endotherm peak, centered at 111.46 °C, has the same value for both the blank and PATP treated samples (Figure 6). This value can be attributed to the melting point of PE. The second peak with a lower area, centered at 170.36 °C for the control sample and 170.47 °C for the PATP treated sample, indicates the melting temperature of EVOH.

Table 2 shows the influence of PATP treatment on the value of the ΔHm parameter for the Pev-70 multilayer polymer film.

Table 2. The influence of PATP treatment on the value of the ΔHm parameter for the Pev-70 multilayer polymer film

Analysed film	ΔHm (J/g) PE	ΔHm (J/g) EVOH
Pev-70 (control)	33.02	2.75
Pev-70 (PATP treated)	44.97	2.49

By integration, the value of ΔHm for the polymers identified on the endothermic curve was obtained. The value of ΔHm for PE increases by 27 % in the case of the Pev-70 sample – treated PATP compared to the control sample. This increase of the melting enthalpy in both films after treatments can be explained due to an improvement of the crystallinity of the PE components. So, in this way, an increase of the crystallinity of the PATP polymer film was recorded. Higher values of the melting enthalpy lead to higher values for the breakage resistance and elongation percentage [13].

4. Conclusions

Various structural changes of the two studied materials were found after the PATP treatment. Pob-60 material displayed no delamination or opaque areas and thus the foil-air interface remains clear. The results of DSC analysis showed an increase of the crystallinity of multilayer polymers, which implies the improvement of barrier and mechanical properties. Minor changes were identified in the mechanical properties of both multilayer polymeric materials. The mechanical parameters of the Pob-60 material showed the smallest changes. Thus, from mechanical characteristics standpoint, the Pob-60 multilayer material represents a better option for meat based products packaging, demonstrating a better ability to withstand PATP treatments, and a better stability compared to Pev-70 polymeric material.
In conclusion, PATP treatment induce structural, thermal and mechanical properties changes in the multilayer polymer films used as packaging materials applied in meat industry. The choice of the packaging materials adequate for PATP treatments requires a deep knowledge of the changes that occurs during PATP and the transformations that takes place during food processing and storage.

References

[1] Ayvaz H, Schirmer S, Parulekar Y, Balasubramaniam V M, Somerville J A and Daryaei H 2012 Influence of selected packaging materials on some quality aspects of pressure assisted thermally processed carrots during storage *LWT - Food Science and Technology* **46** 2 pp 437-447
[2] Filimon V, Borda D, Alexe P. Stoica M 2016 Study of PATP impact on food packaging materials. *Revista de Materiale plastice* **53** 1 pp 48-51
[3] Filimon V, Stoica M, Alexe P, Borda D 2013a Microstructural changes of some multilayer polymer films applied in patp food treatment *Journal of Agroalimentary Processes and Technologies* **19** pp 79-82
[4] Filimon V, Stoica M, Borda D, Alexe P 2013b Evaluation of PATP impact on aesthetic qualities of food packaging materials *Global Research Analysis* **2** pp 59-60
[5] Mensitieri G, Scherillo G and Iannace S 2013 Flexible packaging structures for high pressure treatments *Innovative Food Science and Emerging Technologies* **17** pp 12–21
[6] Stoica M and Borda D 2017 *Flexible Packaging Structures for High-Pressure Thermal Processing (HPTP)* (In Reference Module in Food Science) ed G Robertson (Melbourne – Victoria: Elsevier) chapter online DOI 10.1016/B978-0-08-100596-5.21415-7
[7] Koutchma T, Song Y, Setikaite I, Juliano P, Barbosa-Canovas G V, Dunne C P, Patazca E 2010 Packaging evaluation for high-pressure high-temperature sterilization of shelf-stable foods *Journal of Food Process Engineering* **33** 6 pp 1097-1114
[8] Liu D, Zhang Y 2017 *Journal of Food Processing and Preservation*, **41** 5
[9] Hernández-Muñoz P, Villalobos R and Chiralt A 2004 Effect of cross-linking using aldehydes on properties of glutenin-rich films *Food Hydrocolloids* **18** 3 pp 403-411
[10] Galotto M J, Ulloa P A, Hernández D, Fernández-Martín F, Gavara R and Guarda A 2008 Mechanical and thermal behaviour of flexible food packaging polymeric films materials under high pressure/temperature treatments *Packaging Technology and Science* **21** 5 pp 297-308
[11] Guillard V, Mauricio-Iglesias M and Gontard N 2010 Effect of novel food processing methods on packaging: Structure, composition, and migration properties *Critical reviews in food science and nutrition*, **50** 10 pp 969-988
[12] Fuentes-Audén C, Martínez-Boza F J, Navarro F J, Partal P and Gallegos C 2007 Formulation of new synthetic binders: Thermo-mechanical properties of recycled polymer/oil blends *Polymer testing*

IOP Conf. Series: Materials Science and Engineering **485** (2019) 012007 doi:10.1088/1757-899X/485/1/012007

26 3 pp 323-332
[13] Chiou B S, Avena-Bustillos R J, Bechtel P J, Imam S H, Glenn G M and Orts W J 2009 Effects of drying temperature on barrier and mechanical properties of cold-water fish gelatin films *Journal of food engineering* **95** 2 pp 327-331

Acknowledgments
The work was supported by project SOP HRD – TOP ACADEMIC 107/1.5/S/76822. The authors wish to thank POR RE-Spia (SMIS 11377) project for the infrastructure provided in this study.

IOP Conf. Series: Materials Science and Engineering **485** (2019) 012008 doi:10.1088/1757-899X/485/1/012008

Fatigue analysis of the connecting rod in internal combustion engines

M Frătiţa[1], F Popescu[1], K Uzuneanu[1], V Mereuţă[2], I Ion[1]

[1]Department of Thermal Systems and Environmental Protection, "Dunărea de Jos" University of Galaţi, RO
[2] Mechanical Department, "Dunărea de Jos" University of Galaţi, RO

kristina.uzuneanu@ugal.ro

Abstract. The connecting rod along with the crankshaft turn the translation movement of the piston into rotation. The alternating forces acting on the connecting rod caused micro-fissures, resulting in materials breaking at much lower forces than the maximum allowable stress. This work presents the finite element analysis to determine the fatigue strength of the connecting rod. The results of the study show the dependence of fatigue resistance on the configuration of the connecting rod processing surface for the two types of the steel being analysed.

1. Introduction

Internal combustion engines convert chemical energy into mechanical energy in the form of reciprocating motion of the piston. Crankshaft and connecting rod convert reciprocating motion into rotary motion. Connecting rod (CR) is one of the important driving parts of light vehicle engines. The connecting rod experiences various forces of both piston acceleration and deceleration from the stroking motion, friction-generated loads and the load caused by the cylinder pressure during the combustion stroke. If not strong enough, CR fatigue failure would occur, further leading to component fracture and engine failure [1].

To get the most competitive prices, car manufacturers choose software specifically for the design, analysis and execution of the parts required for the assemblies. Thus, sizing and checking the components shall be made for a certain number of cycles and for a specific operating period. Analysis programs have been developed to predict over which period of time the first damaging micro-fissures begin to occur.

In the fatigue processes the surface effects are very important. In most cases the degradation, as a result of the local plastic deformations, takes place at the surface layer [2], where the resistance to deformation is modified at the same time with the cyclic fatigue loading based on a hardening/and/ or softening mechanism. In his studies, Seiler revealed that the processed surfaces had a better fatigue resistance [3]. The work is intended to study the occurrence of the phenomenon encountered in the literature as the fatigue of the material. This phenomenon occurs as a result of repeated cycles to which the connecting rod is subjected. It can be subjected to 12 cycles per second up to 66 cycles per second in the specific case of cars, and in the case of small engines (e.g. motorcycle, boat) it can reach 135 cycles per second.

2. Materials and Methods

2.1. Materials

For simulation purpose, two types of steel AISI 4340 and AISI 1050 are used [4]. The properties of these materials are presented in Table 1, as provided by Autodesk Simulation database.

Table 1. Steel physical properties

Materials	Mass Density	Yield Strength	Ultimate Tensile Strength	Young's Modulus	Poisson's Ratio	Shear Modulus
Steel AISI 4340 409 QT	7.85 g/cm^3	1371 MPa	1467 MPa	207 GPa	0.33	77.8195 GPa
Steel AISI 1050	7.85 g/cm^3	206.842 MPa	517.104 MPa	199.947 GPa	0.3	76.9027 GPa

2.2. Calculation method

To determine the theoretical parameters, the physical parameters of a 2.0 litters TDI engine, of 102 kW, with a bore × stroke of 81 × 95.5 mm and a connecting rod length of 144 mm have been adopted.

The forces acting on the connecting rod are dependent on the stroke and crankshaft position. They are calculated according to [5] as presented below [6]. The gas force that acts on the piston was calculated as [7]:

$$F_g = \frac{\pi D^2}{4}(p_c - p) \quad [N] \tag{1}$$

where: D - cylinder bore, in mm;
p_c - cylinder gas pressure, in MPa;
p - crankcase pressure, in MPa.

The inertial force was expressed as:

$$F_i = -m_i a_i = 10^{-3} m_i r\omega^2(cos\theta + \mathrm{r}\mathrm{l}^{-1} cos2\theta) \quad [N] \tag{2}$$

where: mi - reciprocating mass, which is the sum of the piston mass and a percentage of the connecting rod mass, in kg;
a_i - instantaneous acceleration of the piston, in m/s^2;
r - crank radius, in mm;
ω - crank angular velocity, in rad/s;
θ - crank rotational angle from top dead center, in °;
l - connecting rod length, in mm.

The resultant force acting on the cylinder axis was calculated as:

$$F = F_g + F_i \tag{3}$$

Conventionally, it is assumed that the resultant force acts upon the piston pin axis. The dependence of this force as function of the crankshaft angle is presented in Figure 1.

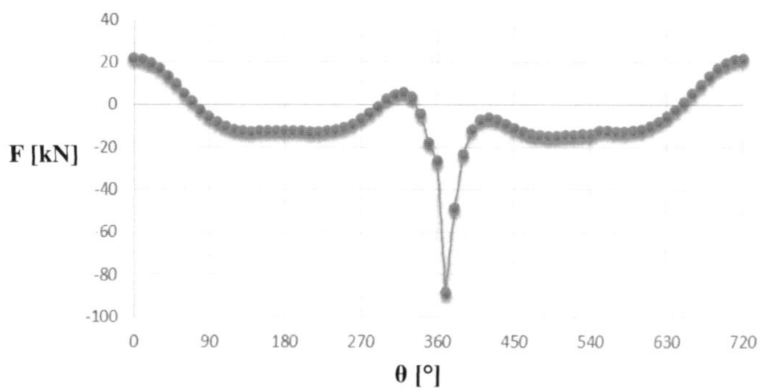

Figure 1. Resultant force function vs. crankshaft angle

IOP Conf. Series: Materials Science and Engineering **485** (2019) 012008 doi:10.1088/1757-899X/485/1/012008

2.3. Experimental determinations
Experimental determinations were focused on the influence of the degree of surface processing on the behaviour of a structural steel subjected to variable loads. The experimental results show that the roughness parameters of the surface layer change with the variable load. Thus, the microgeometry is considered as surface defect that can characterize its integrity.
Based on the experiments conducted regarding the behaviour to fatigue considering roughness as surface defect, it has been established the influence of the factors that determine the evolution of the resistance to fatigue, under different surface processing conditions.
The experimental results were presented in a number of previously published papers [2, 8, and 9].

3. FEM Analysis
The 3D model was created in the Autodesk Inventor Part Design module. With the same program the static analysis of the finite element was performed, resulting a safety factor of 1.65.
The 3D model has been exported to Autodesk Simulation Mechanical program. The simulation of alternating cycles to obtain fatigue resistance requires the selection of a coefficient that varies depending on the surface area of the connecting rod. It has several types of pre-defined processing surface, like: polished, ground, machined, rolled, forged. In most cases, a connecting rod has an unprocessed surface resulting from the forging process. The model that needs to meet more special working conditions also has the exterior surfaces mechanically processed.

3.1. Mesh generating
The automatic generation method with tetrahedral elements was used for meshing. The solid model of the H-beam connecting rod has been meshed into 11260 elements and 20090 nodes.

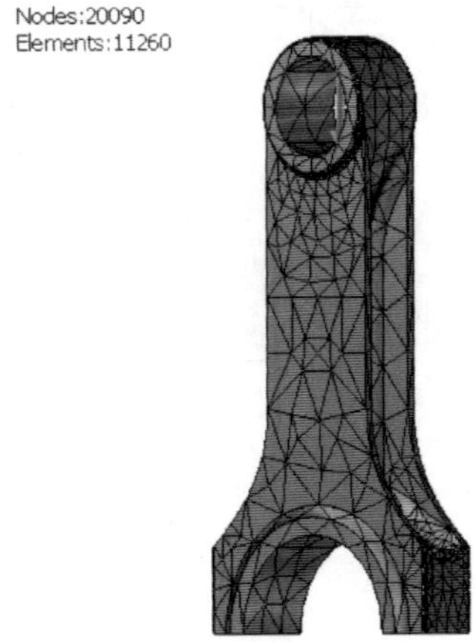

Figure 2. H-beam mesh

3.2. Boundary conditions
Constraints were applied according to the loadings in the connecting rod. Thus, the connecting rod is mainly subjected to compression and bending – because of the gas pressure inside the cylinder – as well as to tensile load – because of inertia (inertia increases when rotation speed and mass of oscillating parts increase). Inertia load is ignored when calculating the compressive load. In this assumption, calculations are more accurate since inertia load acts in opposite direction to the gas pressure load when the expansion in power stroke starts.

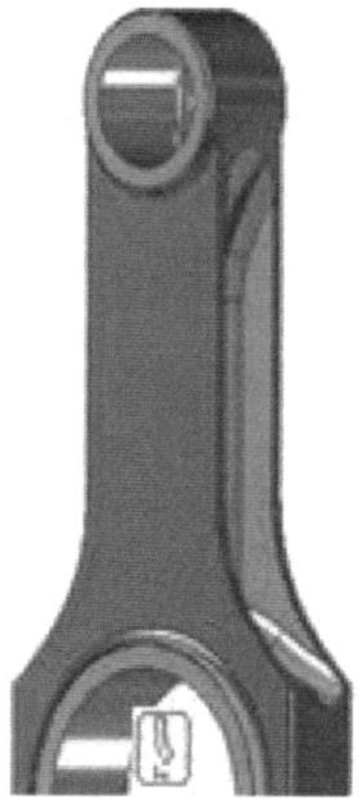

Figure 3. The constraints of the connecting rod

Based on the above aspects, the big end was considered fixed. The force calculated with formula (3) was applied to the connecting rod in vertical direction according to [10].

4. Results and Discussion

The simulation was performed in the two scenarios that are experienced in the real case. For the forged surface, the program defines Surface Factor coefficients of 0.429. If the connecting rod is processed after the forging process, the surface factor becomes 0.762. Autodesk Simulation calculates the safety factor or the cycles to failure. The calculation can be stopped if the safety factor has been reached or it can be done until the connecting rod is destroyed. Also, it may be imposed a certain "lifetime" to reach optimizations. Figure 4 illustrates the areas predisposed to fatigue. The critical areas are represented by the embedded (the threading of the end bolts) and the small end of the connecting rod.

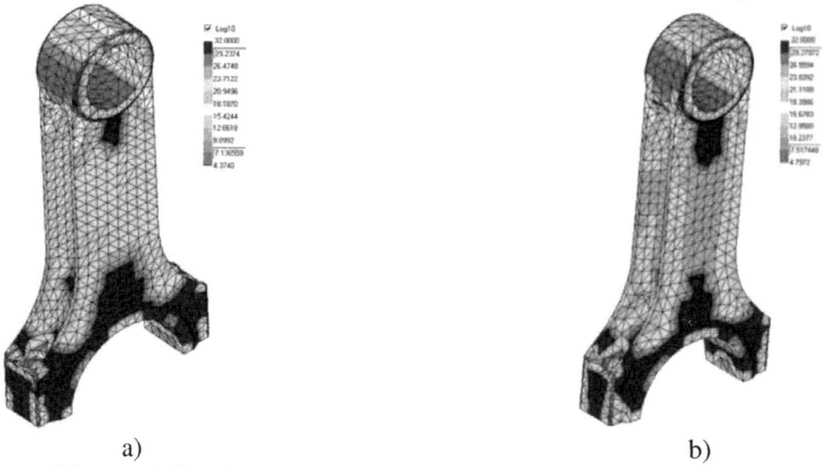

a) b)

Figure 4. Graphic results: a) surface forged, b) surface machined

The failure is predicted after 62685 cycles at the maximum power of the vehicle. These would be equivalent to an operating time of 31.34 minutes at 4000 rpm at full throttle. The evolution of the micro-fissures, occurrence and propagation are represented in Figure 5.

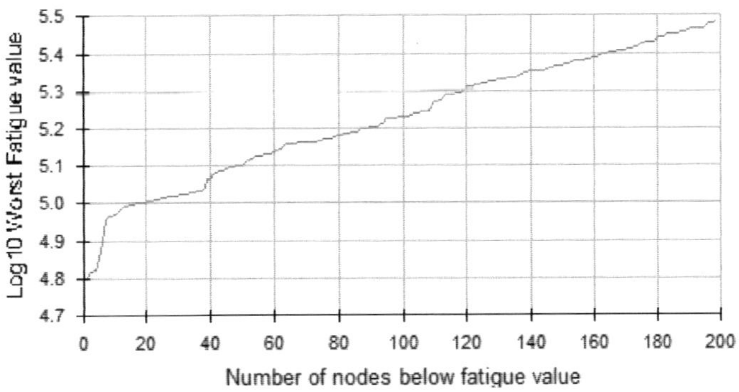

Figure 5. The evolution of micro-fissures for surface forged

If the surface of the connecting rod is mechanically machined, its roughness decreases, the surface factor increases to the value of 0.762, typical for a surface area processed by machining.

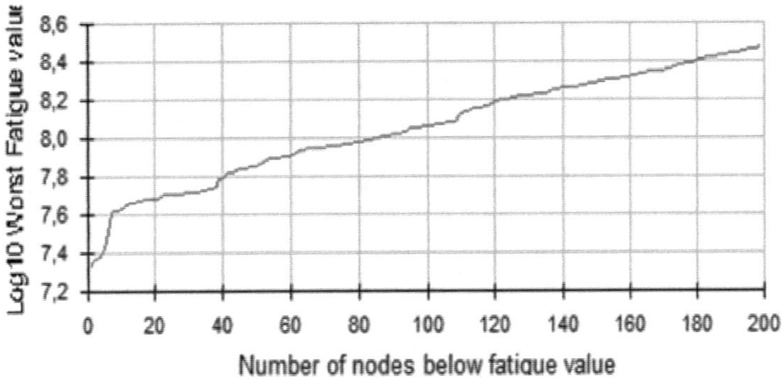

Figure 6. The evolution of micro-fissures for surface machined

In this simulation the failure is predicted after 2.16 x 107 cycles at the maximum power of the vehicle. These would be equivalent to an operating time of 180 hours at 4000 rpm at full throttle. The analysis was repeated for the AISI 1050 material, as shown in Table 2.

Table 2. Predicted number of cycles to failure

Material	Surface forged	Surface machined
AISI 4340	62.685 cycle	21.630.950 cycle
AISI 1050	15.766 cycle	171.684 cycle

Considering the external feature of the internal combustion engine, it was possible to determine the operating time of the engine until the shaped connecting rod was destroyed.

After simulation it can be noticed that the processing of all surfaces is delaying the appearance of the micro-fissures and its propagation, the small end being the most sensitive area. Regardless of the material used, the reliability of the elements subjected to alternating cycles can be increased by mechanically processing the surface, as shown in Figure 7.

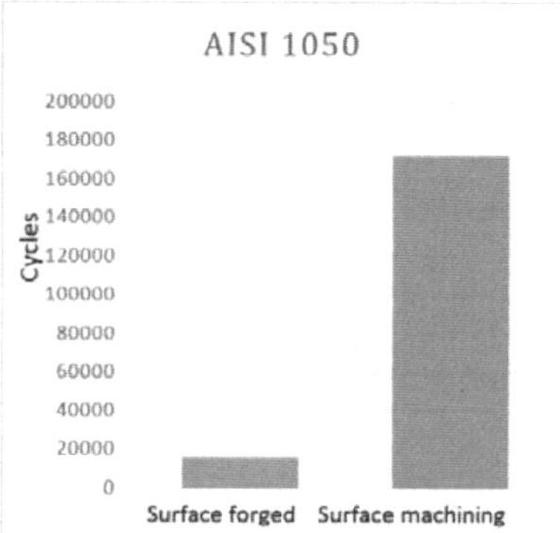

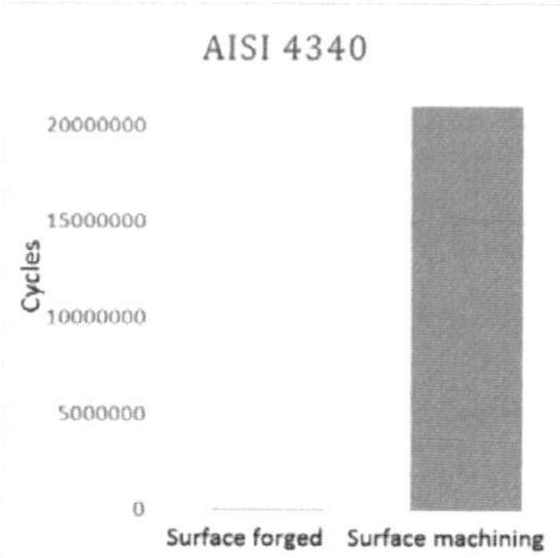

Figure 7. Working time of the connecting rod

5. Conclusions

The alternating forces to which the connecting rod is subjected, determine the aging of the material by the appearance of local plastic deformations favouring the occurrence of micro-fissures. The choice of material has a predominant role in determining the fatigue resistance of the connecting rod.
Surface quality of a connecting rod, resulting from processing, plays an important role in the degradation process by fatigue. The study shows that the surface mechanical processing can increase the life of the connecting rod for the two materials concerned.
The results obtained, following the FEM analysis, are in agreement with the experimental results, regarding the influence of the surface roughness on the fatigue strength and can be used to optimize the connecting rod manufacturing without testing prototypes series.

References

[1] Zheng Bet al 2013 *Stress and Fatigue of Connecting Rod in Light Vehicle Engine*, The Open Mechanical Engineering Journalpp 14-17
[2] Mereuţă V and Florea B 2012 *Influence of Roughness Evolution As a Surface Defect on Fatigue Crack Initiation Process*, Metalurgia International Vol. XVII, ISSN 1582-2214, pp. 5-10
[3] Seiler W et al1998 *Détermination du profil de contraintes résiduelles dans les gorges galetées de vilebrequins*
[4] Puran S Debashis P, Ran *Singh Fatigue and Structural Analysis of Connecting Rod's Material Due to (C.I) Using FEA* International Journal of Automotive Engineering and Technologies pp 245- 253
[5] Rakoşi E et al2004*Ghid de proiectare a motoarelor cu ardere internă pentru automobile* (Iaşi: Politehnium)
[6] Frăţiţa M et al 2018 *About I-beam versus H-beam connecting rod design using Inventor Autodesk 2018* ACME 2018 Section 6
[7] Chikalthankar S B and Nandedkar V M2012*Fatigue Numerical Analysis for Connecting Rod*International Journal of Engineering Research and Applications (IJERA)ISSN: 2248-9622 pp 628-632
[8] Mereuta V and Ciortan S 2016 *The Influence of Superficial Layer Parameters on Fatigue Strength*, Bulletin of the Transilvania University of Brasov, Proceedings of the IXth International Conference on Product Design, Robotics, Advanced Mechanical & Mechatronic Systems and Innovation- PRASIC Nov. 10-11, ISSN 2065-2119, pp. 169-175
[9] Mereuţă V and Palaghian L 2012 *Variable Loads and Surface Roughness Evolution*, The 3rd International Conference on Diagnosis and Prediction in Mechanical Engineering Systems DIPRE 12, 31may-1june, ISBN 978-606-8008-33-2, pp1-5
[10] Thakare N U et al 2015 *Finite element analysis of connecting rod using* Ansys International Journal of Advances in Science Engineering and Technology3 2pp 82-86.

IOP Conf. Series: Materials Science and Engineering **485** (2019) 012009 doi:10.1088/1757-899X/485/1/012009

Achieving ultra-fine grained microalloyed steel by severe plastic process

G Gurau[1], C Gurau[1*], L Gurau[2]

[1]Faculty of Engineering"Dunărea de Jos" University of Galati, Domnească Street, 47, RO-800008, Galati, Romania
[2]ArcelorMittal Europe Flat Products - Business Division East, Dabrowa Gornicza - Poland

*carmela.gurau@ugal.ro

Abstract. Latterly, it is of interest to enhance combination of strength and toughness of steel for advanced products, therefor the objective of this research is to improve the mechanical properties of high strength low - or ultra - low alloy (HSLA) steels through grain refinement. The ultra-fine grained S420NL steel were produced by high speed high pressure torsion (HSHPT) at room temperature. This severe plastic deformation method was chosen because it is one of the most feasible and rapid ways for nanostructuring various metallic materials down to ultrafine grain range. The procedure key parameters of deformation are described in detail in the paper. The influence of severe plastic deformation applied in cast state steel on microstructure changes and deformation behaviour was investigated. Optical microscopy, SEM and energy dispersive x-ray spectroscopy (EDX) analysis carried out on the samples revealed that severe plastic deformation led to the subsequent formation of an ultrafine-grained microstructure. Sample hardness increased rapidly with progress of torsion deformation degree.

1. Introduction

Since the discovery the concept of nanocrystalline materials (H.Gleiter 1989) [1] that offer excellent mechanical properties, good corrosion resistance as well as good weldability and so on, severe plastic deformation (SPD) has been developed as attractive method to refines the structure grains metallic materials to the ultrafine scale or even under 100 nm. Ultrafine grained or nanograined materials produced by HPT demonstrated extremely high strength and high hardness [2-6]. Recently, bulk ultrafine-grained (< 1 μm) tempered martensitic 9–12% chromium steels steel was processed by high pressure torsion (HPT) severe plastic deformation (SPD) method [7] to exhibit a faster minimum creep rate and higher ductility in comparison with coarse-grained alloy ductility. Withal high strength low-carbon steels for small modular reactors (SMRs) was designed by HPT [8]. Our steps proceed very much in the same way of refines the structure of low-carbon steel based on severe plastic deformation by high speed high pressure torsion (HSHPT) [9]. HSHPT consists in simultaneous induce of shear strain, pressure and friction sample-anvils by high speed rotation of superior piston. This severe plastic deformation method was chosen because it is one of the most feasible and rapid ways for nanostructuring various metallic materials down to ultrafine grain range. The contribution of this paper is to provide the effect of different degree of deformation on severely plastic deformed S420NL steel.

2. Experimental

It was carrying out a process of severe deformation by HSHPT. High speed high pressure torsion is a very efficient method which refines the structure of metallic materials to the ultrafine (< 1 µm) or even the nanometer (< 100 nm) scale [9-12]. In this study a ferritic steel with traces of pearlite grains was in use and its composition is presented in Table 1. From the middle of the steel rings with outer diameter of 20 mm, inner diameter of 10 with a thickness of 2÷3 mm were machined from transversal direction of the sheet. These rings were processed in tronconic elements by HSHPT using an initial pressure of 25barrs, nominal pressure between 0.06GPa to 1.15GPa, a rotational speed of 900r/min. The degree of deformation imposed onto the HSHPT-rings was calculated with: $\varepsilon = \ln(hi/hf)$. The logarithmic strain levels achieved were from 0.1 up to 2.04. The thinnest truncated cone shell exhibited a thickness of 0.55mm. More details to the used HSHPT-setup can be found elsewhere [10].

Table 1. Chemical composition.

C	Mn	Si	S	P	Al	N	Cu	Ni	Cr	Mo	Nb	Ti	V
[%]	[%]	[%]	[%]	[%]	[%]	[%]	[%]	[%]	[%]	[%]	[%]	[%]	[%]
0.16	1.60	0.50	0.005	0.015	0.04	0.015	0.140	0.200	0.070			0.010	0.120

To determine the influence of HSHPT severe plastic deformation a series of experimental tests were conducted to evaluate the structure and mechanical properties. A detail analysis involving macroscopic feature and microstructure analysis of the microalloyed steel was also performed by optical microscopy (OM), scanning electron microscopy (SEM) and energy dispersive spectroscopy (EDS). The examination was carried out using an OLYMPUS BX51 microscope and a ZEISS microscope, respectively. Prior to microstructural examination the samples were prepared by standard metallography techniques. The hardness measurements were carried out on a Vickers hardness tester with a load of 100 gf and a abide time of 15 s. For each data point 6 indents were made at equivalent geometrical positions and averaged.

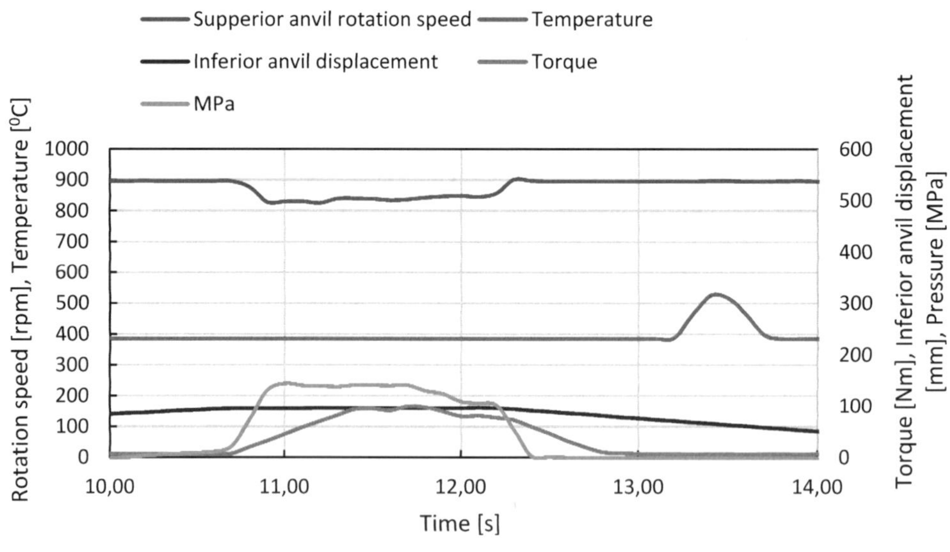

Figure 1. Schematic illustration of the HSHPT parameters of S420NL steel incorporating 1.01 degree of deformation.

IOP Conf. Series: Materials Science and Engineering **485** (2019) 012009 doi:10.1088/1757-899X/485/1/012009

3. Results and discussion

Latterly, it is of interest to enhance combination of strength and toughness of steel for advanced products [12-13], therefor the objective of this research is to improve the mechanical properties of high strength low - or ultra - low alloy (HSLA) steels through grain refinement. HSHPT can impose a shear strain far beyond the strain level attainable by conventional plastic deformation methods such as rolling or extrusion. The key parameters of severe plastic deformation are described above in this paper. Figure 1 shows the major parameters of severe plastic deformation in sample that bear 1.01 logarithmic degree. The plastic deformation lasted less than 4s. The rotational superior anvil slightly decreased during deformation. The torque attained was 100Nm. The maximum pressure required in process was only 0.17GPa. The temperature of sample reach 550^0C on the strength of friction driven by high speed rotation of superior punch.

The macrostructure analysis were done through SEM and energy dispersive X-ray spectroscopy (EDX) (Figure 2). The as-received sample has a large non-metallic compound made of MnS and Al, Ca and Si oxides formed by the dezoxidizing step (Figure 2a). In severe plastic deformed elements the nonmetallic inclusions is splitting and shrinking as can be seen from Figure 2(b and c).

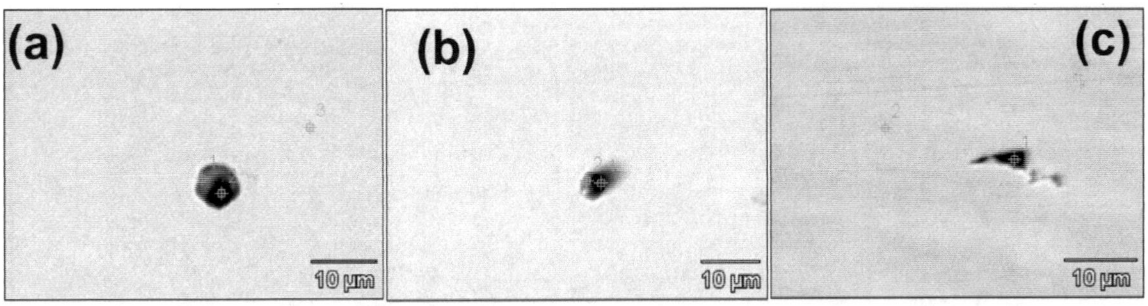

Sample a	*C-K*	*O-K*	*Mg-K*	*Al-K*	*Si-K*	*S-K*	*Ca-K*	*Ti-K*	*Mn-K*	*Fe-K*	*Ni-K*
pt1	2.58			3.33		28.39	35.30	0.08	20.05	10.26	0.01
pt2	1.31	36.77	2.71	42.59		1.01	5.65	1.04	1.03	7.84	0.06
pt3	1.12				0.40			0.07	1.73	96.59	0.10

Sample b	*C-K*	*O-K*	*Si-K*	*Mn-K*	*Fe-K*
pt1	4.05	1.08	22.93	1.13	70.81
pt2	0.98		0.34	1.92	96.76

Sample c	*C-K*	*O-K*	*Mg-K*	*Al-K*	*Si-K*	*S-K*	*Ca-K*	*Ti-K*	*Mn-K*	*Fe-K*
pt1	0.95	5.76	1.48	13.46	0.18	6.99	5.63	0.72	5.72	59.11
pt2	1.21		0.07	4.26	0.46	4.72	4.27		5.20	79.81

Figure 2. Comparison of the nonmetallic inclusions in S420NL steel: as-received (a), after HSHPT with ε=1.87 (b) and ε=2.04 (c).

The microstructure of the as-received condition has coarse ferrite colonies (bright) and few strings of pearlite phase (dark) as shown in Figure 3. The low carbon steel in rolling state consists on large volume of ferrite grains. The size of ferrite grains vary quite a bit between 4 µm and 20 µm. The volume fraction of the fine pearlite phase is observed not more than 3vol%. The strings pearlite are arranged in linear arrays, surrounded by the ferrite grains.

The alloy exhibit a process of refining the initial ferrite grains more pronounced as the application an increased degree of deformation (Figure 4a-f).

The curled and wavy morphology of the ferrite with decreased size and elongated grains confirms that the ring has deformed plastically (Figure 4a).

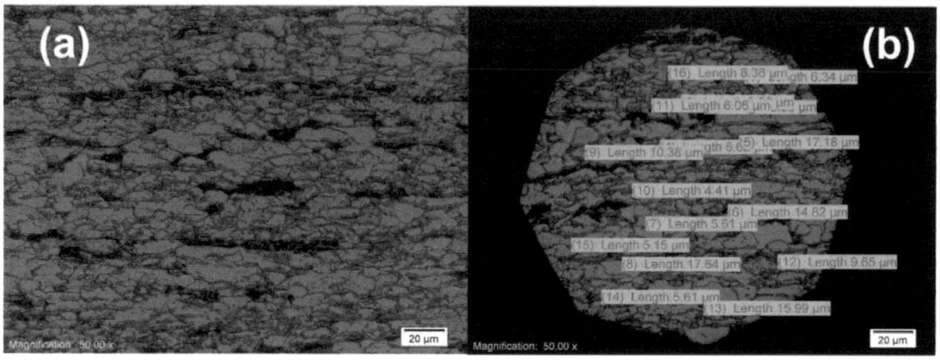

Figure 3. OM micrographs illustrate the phase volume fraction (a) and the size of the real ferrite grain (b) of the as-received samples.

At higher logarithmic degree of deformation, the ferritic structure becomes gradually aligned parallel to the shear direction.

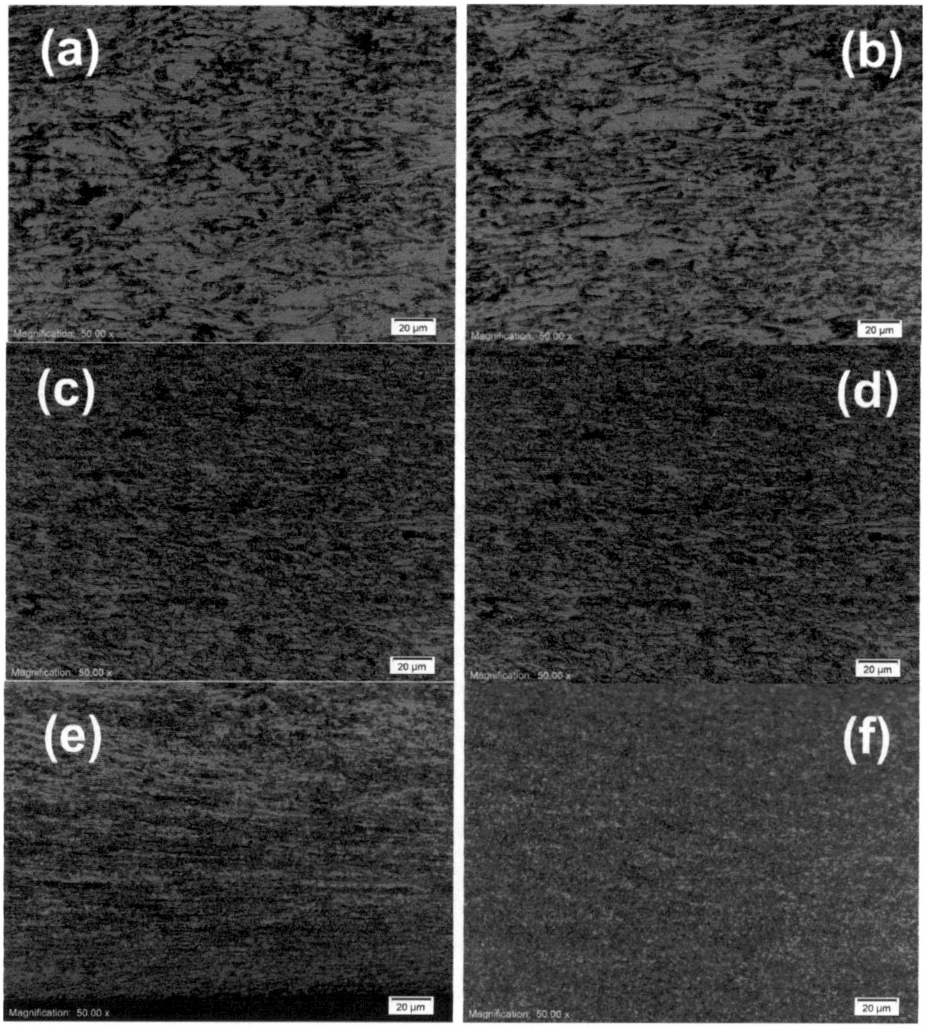

Figure 4. OM images showing the characteristic microstructures of incremental strain after HSHPT

Moreover, there is also a recognizable refinement and substructure formation within the ferrite phase (Figure 4b-d). The type, morphology, size, and density of precipitates significantly affect the overall properties of coarse grained and submicron materials [7]. Precipitates inside initial rings and tronconic elements after severe plastic deformation can be intermetallic or ceramic materials .The type of these has been described in above EDX investigation. Most of the secondary phases dispersed within and at grain boundaries of ferrite phase matrix consist on cementite, intermetallic compound. These precipitates suffers a fragmentation and distribution process along the sliding lines, as can be observed better at low deformation degrees. Besides grain refinement and fragmentation of precipitates, SPD can also induce diffusive phase transformation. This process occurs by shearing strength subsequent under high pressure concomitantly with rotating. However, the applied deformation induces strain to deform the ferrite matrix and enlarge the volume of pearlite. The increased volume fraction of fine lamellas and globular pearlite concur to the enhancing mechanical strength. The pearlite manifest higher strength because of cementite and this contributes to superior mechanical properties. The fine curved strings of pearlite are distributed and diffused in the grain boundaries of ferrite zone as shown in Figure 4e. As the shear strain increased (ε=1.3) the representative microstructural feature of the steel was uniformly distributed equiaxed submicron grains, as illustrated in Figure 4f. The bright grains under 1μm are ferrite phase surrounded by a very fine and uniformly distributed perlite.

The homogeneity of the microstructure increases with an increasing value of the plastic strain occurring during HSHPT.

The increased pearlite matrix with fine and globular morphology contributes towards improving the mechanical properties of the HSHPT'ed processed sample. In this type of processing, the sample are heated by friction inside mold. The high speed rotation of anvil (900rpm) on the initial pressed ring with 25 bars drives heating. However, a complex plastic deformation at temperature occurs. At above 500^{0}C torsion and pressing takes place simultaneously. In this moment the high pressure is applied on sample. High pressure of the order of 1GPa lead to refining the structure by concomitant shearing and sliding of crystalline grains in compact blocks.

It is well known that dislocation slip and deformation twinning are the deformation mechanisms in coarse grained metallic materials. The same two mechanisms which induced grain refinement processes take place during the SPD. However, deformation twinning has been rare in BCC materials SPD-processed and these only in case of induced by shock. The twinning process is specific to low temperature deformation. In case of HSHPT the samples are heated before the nominal pressure is applied. In the ferritic BCC phase the major deformation mode were found as dislocation activities like dislocation multiplication, accumulation, and annihilation via the formation of grain boundaries [14]. The dislocation structures are also affected by the stocking faults energy (SFE) of the materials.

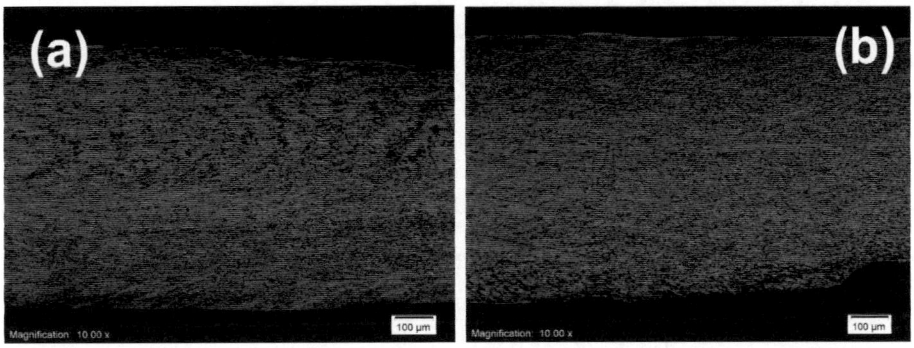

Figure 5. OM images showing the features of grain refinement and preferential slip directions on sample deformed with ε=0.29 (a) and ε=1.11 (b).

During the HSHPT more than in other SPD induced continuous grain growth process from a much smaller grain size to the final steady-state grain size which remains unchanged with further straining

[14]. The average steady-state grain size depends on material properties and more to the SPD technique and processing parameters (applied pressure, strain rate and temperature reached during deformation).

The microstructural evolution of this HSLA steel having different degree of deformation has been highlighted in Figure 6, by SEM investigation. The SEM of the sample after deformation with ε=1.87 showed the formation of annular and waved flow curves through submicron grains (Figure 6a). SEM analysis carried out on the sample incorporating 2.04 deformation degree revealed that severe plastic deformation led to the subsequent formation of a significant refinement (Figure 6b).

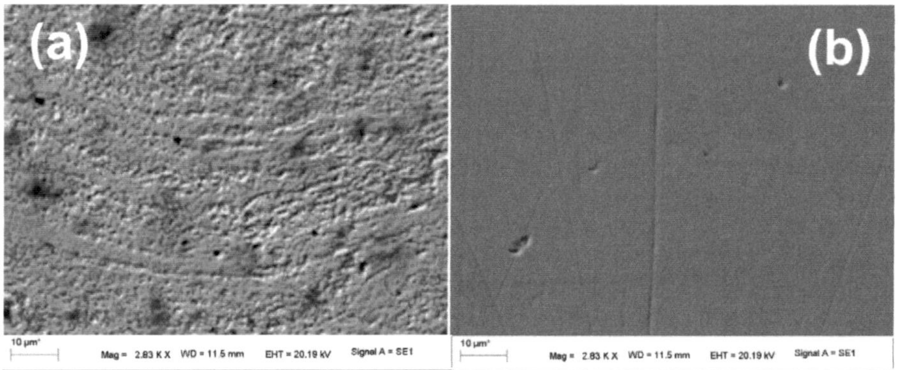

Figure 6. SEM micrographs of HSHPT'ed samples after deformation with: ε=1.87 (a) and ε=2.04 (b)

The individual grains or grains boundaries were outside the range of detection by standard SEM observation.

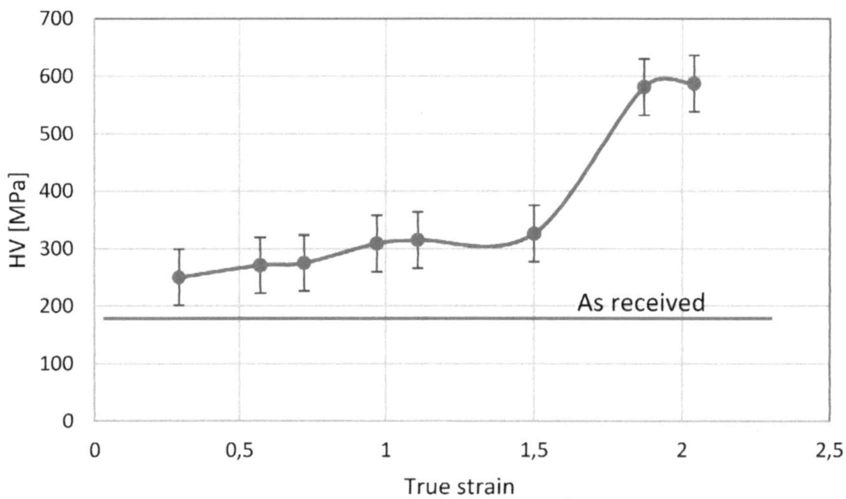

Figure 7. Vickers's microhardness evolution of HSHPT'ed samples with respect to the logarithmic degree of deformation

This method applied on bulk metallic HSLA steel manage to achieve ultrafine grained structure that possess mechanical properties that expand rolling sheet performance. The microstructural evolution, morphological changes in ferrite and pearlite colony and grain refinement of the deformed samples were correlated with the mechanical properties. In order to understand the advantage of this process, the HSHPT'ed samples were compared with the initial hot-rolled sample and the results are presented in Figure 7. The hardness of samples increased rapidly with progress of torsion deformation degree. The

IOP Conf. Series: Materials Science and Engineering **485** (2019) 012009 doi:10.1088/1757-899X/485/1/012009

Vickers hardness increases four times. The experimental tests and analysis results show that there are significant improved in the material properties of severe deformed coned rings when comparing it with the thick steel plates Standard, EN10027-1 and CR10260.

4. Conclusion

We have assessed the potential improvement of the mechanical characteristics in the S420NL low alloy (HSLA) steel by high speed high pressure torsion (HSHPT). The ultra-fine grained S420NL steel were produced by HSHPT at room temperature. Microstructure investigations show that this SPD method significantly changes the shape, distribution and grains size. The microstructure essentially refined down to the submicron level. As the shear strain increased at $\varepsilon=1.3$ the representative microstructural feature of the steel was uniformly distributed equiaxed submicron grains of ferrite phase surrounded by a very fine and uniformly distributed perlite. Beyond $\varepsilon=2$ individual grains or grains boundaries were outside the range of detection by standard SEM observation. The HSHPT-produced UFG structured S420NL steel demonstrates high hardness (587 HV). The Vickers hardness increases four times as against thick steel plates. Results obtained by this method allow further research to obtain low-weight S420NL products nanostructured by HSHPT with much improved properties.

References

[1] Gleiter H, Schimmel T and Hahn H 2014 *Nano Today* **9** pp17–68

[2] Straumal BB, Kilmametov AR., López GA, López-Ferreño I, Nó MLSan Juan J, Hahn H and Baretzky B 2017*Acta Mater.* **125** pp 274–285

[3] Edalati K.and Horita Z 2016 *Mater. Sci. Eng.* A **652** pp 325–352

[4] Kim SM, Arockiakumar R.and Park JK 2012 *Mater. Sci. Eng.* A **546** pp 53–58

[5] Toth LS and Gu C 2014 *Mater. Charact.* **92** pp 1–14

[6] Tohidi AA, Ketabchi M and Hasannia A 2013*Mater. Sci. Eng.* A **577** pp 43–47

[7] Kral P, Dvorak J, Sklenicka V, Masuda T, Horita Z, Kucharova K, Kvapilova M and Svobodova M 2018 *Mater. Sci. Eng.* A **723** pp 287–295

[8] Shi S, Zhang Z, Wang X, Zhou G., Xie G., Wang D, Chen X and Ameyama K 2017 *Mater. Sci. Eng.* A **711** pp 476–483

[9] Bagherpour E, Pardis N and Reihanian M 2018 *The International Journal of Advanced Manufacturing Technology* Springer-Verlag London Ltd., part of Springer Nature

[10] Gurău G, Gurău C, Potecaşu O, Alexandru P and Bujoreanu LG 2014 *J. Mater. Eng. Perform.* **23** pp 2396–2402

[11] Gurau G, Gurau C, Bujoreanu LG and Sampath V 2017 *IOP Conf. Ser. Mater. Sci. Eng* ICIR EUROINVENT (Iasi) vol 209

[12] Potecasu F, Marin M, Potecasu O, Marin FB and Neacsu M 2016 Proc. paper SGEM (Albena) vol 1 pp 159-166

[13] Marin M, Potecasu F, Marin FB., Potecasu O and Alexandru P 2016 Proc. paper SGEM (Albena) vol 1 pp 259-264

[14] Cao Y, Ni S, Liao X, Song M and Zhu Y 2018 **133** pp 1–59

Acknowledgments
This research was carried out within the framework of 47PCCDI/2018 project

IOP Conf. Series: Materials Science and Engineering **485** (2019) 012010 doi:10.1088/1757-899X/485/1/012010

Formation of submicrocrystalline structure in a biodegradable Mg alloy

C Gurau, G Gurau*, P Alexandru, V Basliu

Faculty of Engineering, "Dunărea de Jos" University of Galati, Domnească Street, 47, RO-800008, Galati, Romania

*gheorghe.gurau@ugal.ro

Abstract. Magnesium alloy, as biodegradable material, are attracted increasing interest in the field of bone graft substitution due to their excellent biocompatibility and good corrosion resistance. The main clinical challenges of ZK60 Mg alloy remain requirements such as: straight, fracture resistance, work hardening mechanical shock and so on, concomitant with plasticity. According to the Hall-Petch theory, an effectively enhance strength of Mg alloys may be realized by grain refinement. This paper proposes a reliable approach to fabricate Mg cast alloys down to ultrafine-grain range within just a few seconds. Ultrafine grained (UFG) Mg alloy was produced by severe plastic deformation (SPD), using high speed high pressure torsion (HSHPT) procedures. The microstructure evolution of the as-cast and as-severely deformed Mg-Zn-Zr was characterized using electron microscopy and energy dispersive spectroscopy (EDX). X-ray diffraction analyses (XRD) was performed to study the impact of severe plastic deformation on internal structure changes. Micro hardness tests indicated significant increase with the progression of true strain. Results so far have been very encouraging in developing new solution for biodegradable applications.

1. Introduction

The biomaterials has been developed to meet three different clinical requirement referring to bioactivity: virtually inert designed to no harm to tissues, surface bioactive that create tissue-bonding and temporary structures enable to biodegradation after native tissue regeneration [1-4]. Magnesium alloys are exploited as degradable biomaterials for following potential benefits features like: controllable corrosion rates, benign ability with the living body, similarity with bones concerning density and Young's modulus, high load capacity [2]. Despite their lightweight in addition to biocompability, the biomedical applications are restricted due to low ductility at room temperature. As it well known the brittleness of Mg alloys comes from their hexagonal close-packed (hcp) structure with limited slip systems. Recently, ultrahigh-pressure technique demonstrate capability to tailor microstructure for applications of Mg alloys, especially for medical Mg-based implants [5,6]. The basic requirements for using alloys in human body are high corrosion resistant properties. Severe plastic deformation is other effective approach to improve corrosion resistance and mechanical properties of alloys [5-8]. Mg alloys was severely plastic deformed using techniques, such as high pressure torsion (HPT), accumulative roll bonding (ARB) and equal channel angular extrusion (ECAE) [7]. SPDs are among the technology by which ultrafine and nanocrystalline microstructures can be produced in bulk alloys. The bimodal grain microstructure generated by the SPD processes is take into account for the improved mechanical properties of Mg alloys. Grain refinement could effective increase the combination amidst mechanical strength, low temperature superplasticity and greater corrosion resistance of bulk Mg alloys [9].

This paper focuses on a reliable approach to fabricate Mg alloy large disks with very fine grain size. In order to fabricate high mechanical properties magnesium alloys, the severe plastic deformation are adopted. Ultrafine grained (UFG) Mg alloy was produced by severe plastic deformation (SPD), using high speed high pressure torsion (HSHPT) procedures [10-13].

2. Procedure and material

Buttons of ZK60 magnesium alloy were cut from as-cast billet. In an attempt to improve mechanical resistance properties and ductility of bulk Mg alloys was applied severe plastic deformation. The biodegradable Mg alloy was processed via HSHPT. We used a variation of classic HPT technique whereby alloys are simultaneous subjected to compressive and torsion strain described in detail in [10, 11]. Specifically, our process is stir friction introduced by high speed rotation of the superior anvil concomitant with high compression and torsion. The rotational speed of the upper punch was selected at 440rpm. The deformation parameters were controlled through PLC XC 200. The degrees of deformation calculated with $\varepsilon = \ln(hi/hf)$ were: 0.57, 0.81, 1.01, 1.48 and 2.13. The achieved disks were up to 40 mm in diameter. The entire processing time lasted about 10s. To enable a better scrutiny of morphology as well as the grains refining accomplished by HSHPT process Scanning Electron Microscopy (SEM) was done via a Zeiss device. In order to evaluate elemental analysis the EDX investigations were carried out. Room temperature Vickers microhardness values were determined for alloy after HSHPT implemented with various degree of deformation.

3. Results and discussions

The effect of severe plastic deformation was investigated on ZK60 magnesium biocompatible alloy which is in possession of good biodegradability and high load capacity. The chemical composition of initial cast state alloy was evaluated by spectroscopic analysis. The results matches with the nominal component of the ZK60 alloy consisting of Mg, 5.49wt% Zn and 0.55wt% Zr. The alloy is in clinical use as degradable biomaterial and has shown good biocompatibility. The orthopedic Mg implants need a Young's modulus close to bone feature obtained by addition of Zn. The most common alloying element in Mg alloys is Zn with the benefit of enhancing their yield field. Zn and Zr alloying are appending antibacterial property, too. The studied alloy has a minor addition of calcium (0.09wt%) and aluminum (0.005wt%) chiefly add to corrosion resistance. The addition of Mn (0.025wt%) improves also corrosion behavior and suppress corrosion due to iron impurity [1]. The mechanical property and degradation properties of the Mg alloy implants should also be considered being vital in effective fracture fixations. The microstructure refinement is considered an effective way to improve corrosion properties. The corrosion rates in Mg-Zn alloys grow from amorphous conditions to larger grain sizes [9]. The degradation can be controlled by grain size achieved. In order to improve service parameter of the magnesium degradable biomaterial grain refinement was produced via HSHPT process. This severe plastic deformation technology is a top-down method, which can be used for a wide range of alloys. HSHPT is conducted on bulk samples to reduce their grain size below 1μm. The experiments were performed at RT under a pressure of 1GPa. The samples were compressed between two opposed anvils and torsion-strained by rotating the upper anvil at a rotation speed of 920rpm. The high speed of upper anvil provide heating of sample (owed by friction between sample and anvils) determinant to plastic deformation of brittle materials as magnesium alloys. The applied torque was 7Nm. Due to the lack of side constraining disks shape, up to 40 mm, was obtained within just a few seconds (5 to 9s). The thickness of disks was reduced to about 0.4mm. Overall the results presented below in Fig.1 show the representative SEM micrographs of ZK60 alloy in as-cast state and after severe plastic deformation with five increasing degree of deformation. The microstructure of as-cast Mg-Zn alloy consists of primary α-Mg matrix, $Zn_{13}Mg_{12}$ intermetallic phase placed inside the grain and bright sphere-like shaped of Zr phase (Fig.1.a). A nonuniform grain structure with an average grain size between 60-100μm is displayed in initial material.

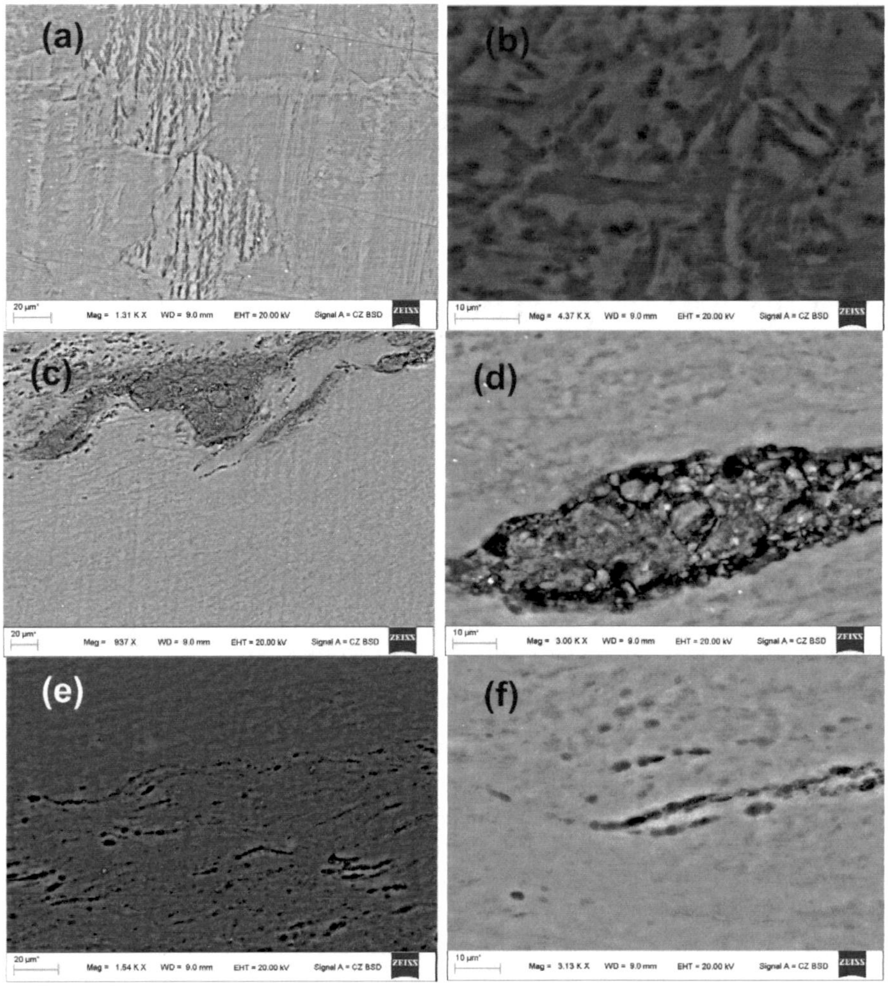

Figure 1. SEM micrographs of Mg alloy in as-cast state and after severe plastic deformation with five increasing degree of deformation.

Although, α-Mg matrix are brittle due to their hexagonal close-packed (hcp) structure with limited slip systems, submicron-grained can be successfully achieved by the SPD method selected. The HSHPT process conduct to the expected structure down to ultrafine-grain range directly from Mg cast alloys. During deformation, the imposition of an applied pressure of 1.0GPa together with the torsional straining supported by 440rpm rotation speed of superior anvil leads to extensive grain refinement.

The texture evolution and especially developed grain refinement, achieved by HSHPT on magnesium alloys follow the increasing degree of deformation. The lowest deformation degree of deformation present a crumbled structure of α-Mg matrix grains surrounded by distorted eutectic composed of α solid solution and intermetallic secondary phases $Zn_{13}Mg_{12}$ and Zn_2Zr (Fig.1b). As is apparent from Fig. 1c and d, the most important feature of the microstructures after deformation incorporating 0.81 and 1.01 logarithmic deformation degree is the presence of very fine grains arranged in parallel strings. Furthermore at these degree of deformation appears curved darker areas. These microstructures can be mainly regarded like depleted magnesium areas rich in zinc, aluminum and oxides, as can be seen in results from beneath EDX analyze (Fig.4). The most important feature brought by Fig.1e and f is certainly the clear trend of microstructural features sizes decreases. The microstructure was formed by waved flowlines which bordering equiaxed grains whose average sizes also varied with applied deformation degree. The final fine disk incorporating 2.13 true strain had significantly refined grain with an average size under 1μm (Fig.1f). HSHPT provides the capability of producing Mg alloy disks with extremely fine grain sizes which are suitable for achieving high mechanical properties, high corrosion resistant properties. The increased temperature attained by friction amid sample and the rotating speed of superior punch improve the mobility of atoms and decrease the density of imperfections. This attendant phenomenon is specific to deformation through HSHPT.

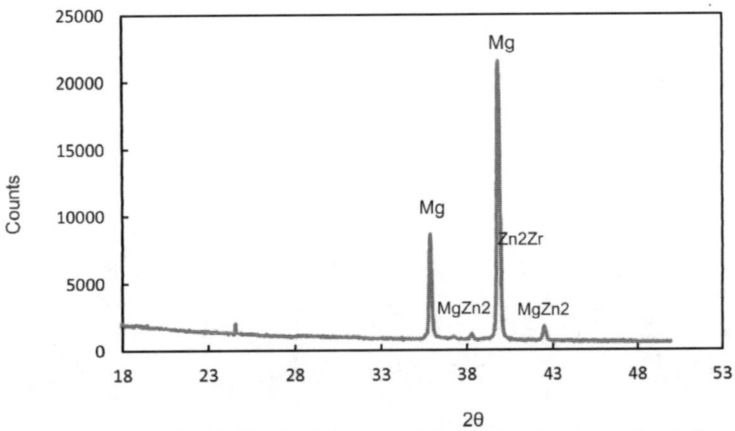

Figure 2. XRD pattern of ZK60 Mg biodegradable alloy in severely deformed state.

The XRD patterns for the Mg alloy after HSHPT severe plastic deformed with 0.81 degree of deformation is displayed in Fig.2.The alloys primarily comprised of the base material α-Mg and secondary phases such as $MgZn_2$.

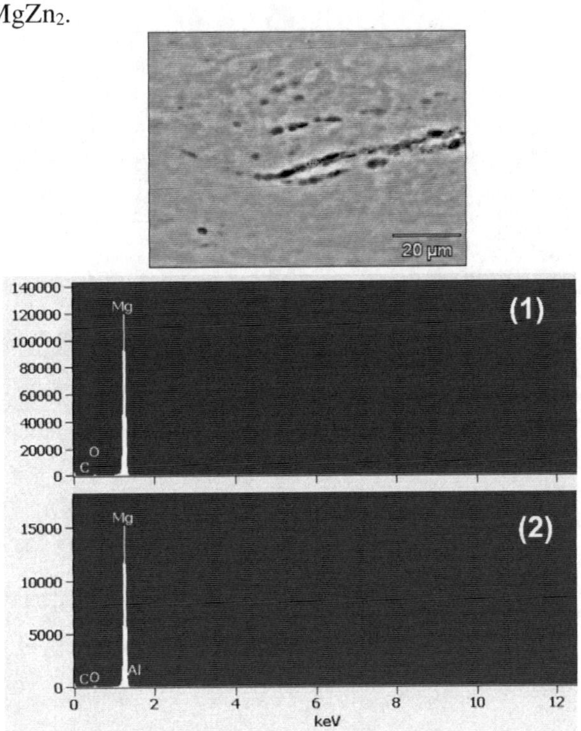

Figure 3. EDX maps of HSHPT'ed Mg alloy emphasizing curved area described by severe plastic deformation.

IOP Conf. Series: Materials Science and Engineering **485** (2019) 012010 doi:10.1088/1757-899X/485/1/012010

The corresponding EDX profile of distorted area of Mg alloy at true strain 2.13 is showing in Fig.3. The major element detected in area denoted 1 is magnesium beside traces of carbon and oxygen. The present usual formation of intermetallics phases, Mg/Zn distributing along the grain boundary or Zn/Zr precipitate, are missing. The unique condition created during HSHPT at high level of deformation leads to this unusual fact. This indicates uniform dispersion of Zn in Mg by HSHPT. Inside area exhibiting waved flow fiber (Fig.3 point 2) only an increase of cumulated aluminum oxide are noticeable.

Fig.4 shows that the element distribution of a severe deformed area Mg-Zn based alloy was determined via mapping. EDS mapping analysis revealed clearly distinguishable two different areas inside the alloy microstructure. The white section mainly consisted of Mg and black section is rich in Zn. The EDX results also emphasis the presence of some very small and uniform distributed Zr-rich precipitates in the microstructure of the alloy.

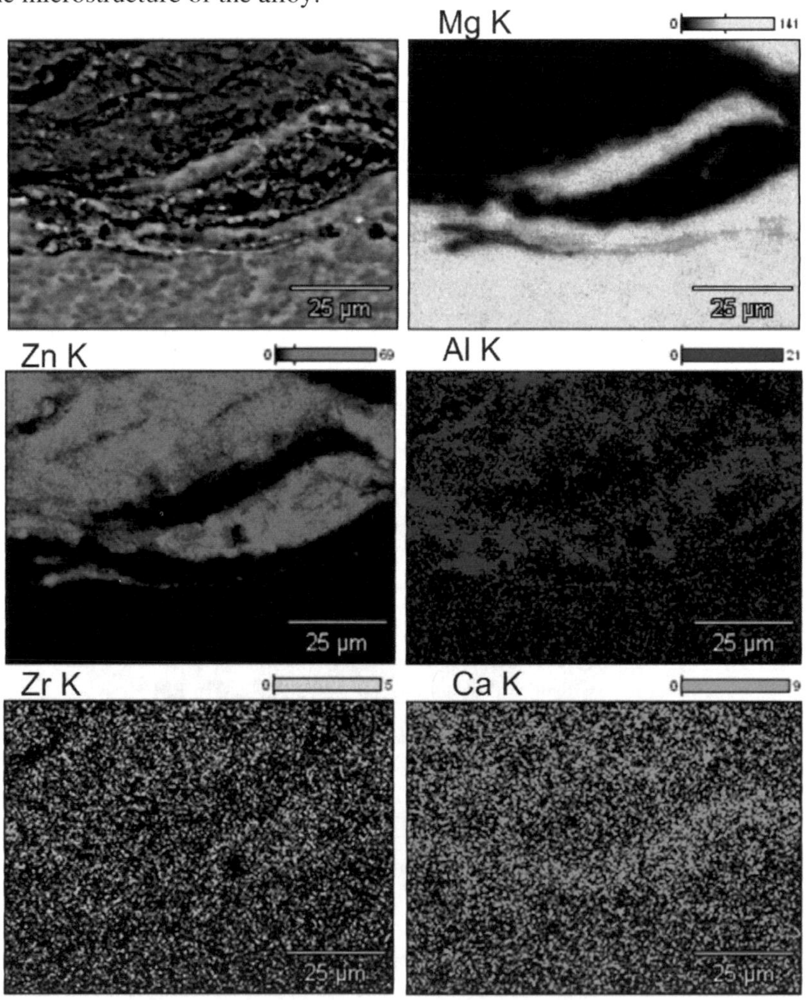

Figure 4. Scanned area and variation of chemical element distribution after HSHPT of ZK60 alloy

Fig.5 presents 3D confocal images of Mg alloy after severe plastic deformation highlighting drop and rise in the distribution of Mg concentrations on curved mixed areas from Fig. 5. The severe plastic deformation by HSHPT can cause the rotation and movement of large blocks of material due to the high-speed rotation of the upper punch.

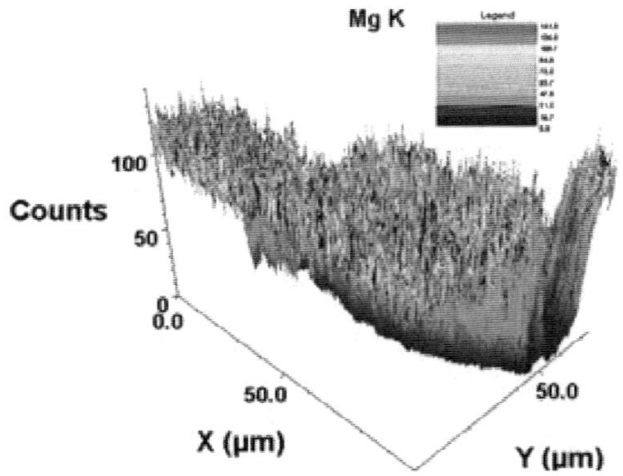

Figure 5. EDX 3D confocal images of Mg alloy after severe plastic deformation

The average grain diameter in a polycrystalline alloy influence the mechanical properties. The fine-grained alloy is harder and stronger than one that is coarse grained on the strength of a greater grain boundary area that obstruct dislocation motion. The increment in micro hardness was achieved along with grain size decreased with an increase in true strain applied. The Vickers micro hardness of the Mg-Zn-Zr alloy as a function of degree of deformation is presented in Figure 7. The hardness values for the SPD Mg alloy incorporating 0.57 degree of deformation was 337HV, which increased with increasing true strain and reached to 493HV at 2.13 degree of deformation. The hardness of the initial Mg alloy is 79HV which is significantly lower than that for HSHPT'ed Mg alloy. The high hardness of the severe deformed alloys was attributed to grain refinement < 1μm. The overall results showed that microhardness evolution was correlated to grain size according to Hall-Petch relation [14-18].

4. Conclusions

The primary findings of the current study can be summarized as follows:

1. Buttons of ZK60 alloy in cast state were successfully severe plastic deformed for the first time by HSHPT.
2. The structure of biodegradable magnesium alloy after SPD has been explored with SEM-EDX and XRD devices.
3. Formation of submicrocrystalline structure in a biodegradable Mg alloy that takes place in severe deformed was highlighted in investigation.
4. Results so far have been very encouraging in developing new solution for biodegradable applications useful in the bone reconstruction field.
5. This SPD process require straightforward fabrication strategies for large-scale production being desirable to develop large rotation shape products.

Acknowledgments

This research was carried out within the framework of 47PCCDI/2018 project

References

[1] Chen Q and Thouas GA 2015 *Mater. Sci. Eng. R Reports* **87**, pp 1-57

[2] Feng Y, Zhu S, Wang L, Chang L, Hou Y and Guan S 2018 *Fabrication and Characterization of Biodegradable Mg-Zn-Y-Nd-Ag Alloy: Microstructure, Mechanical Properties, Corrosion Behavior and Antibacterial Activities*

IOP Conf. Series: Materials Science and Engineering **485** (2019) 012010 doi:10.1088/1757-899X/485/1/012010

[3] Wang M, He BB, and Huang MX 2018 *Strong and Ductile Mg Alloys Developed by Dislocation Engineering*
[4] Wang X, Xu S, Zhou S, Xu W, Leary M, Choong P, Qian M, Brandt M, and Xie YM, 2016 *Biomaterials* **83**, 127-141
[5] Ge B, Fu H, Deng K, Zhang Q, and Peng Q, 2018 *Bioact. Mater.* **3**, 250-254
[6] Feng J, Li H, Deng K, Fernandez C, Zhang Q, and Peng Q, 2018 *Unique Corrosion Resistance of Ultrahigh Pressure Mg-25Al binary alloys*
[7] Alizadeh R, Mahmudi R, Ngan AHW, Huang Y, and Langdon TG 2016 *Mater. Sci. Eng. A* **651**, pp 786-794
[8] Oliveira JP, Duarte JF,. Inácio P, Schell N, Miranda RM, and Santos TG 2017 *Mater. Des.* **113**, pp 311-318
[9] Chen J, Tan L, Yu X., Etim IP, Ibrahim M and Yang, K 2018 J. Mech. Behav. *Biomed. Mater.* **87**, pp 68
[10] Gurău G, Gurău C, Potecaşu O, Alexandru P, and L. G. Bujoreanu, 2014 *J. Mater. Eng. Perform.* **23**, pp 2396–2402
[11] Gurau G, Gurau C, Bujoreanu LG, and Sampath V, 2017 *IOP Conf. Ser. Mater. Sci. Eng.* ICIR EUROINVENT (Iasi) vol 209
[12] Gurau C, Gurau G, and Sampath V, 2018 *I.J.Eng.Mater.Sci.* **25** pp 5-10
[13] Bagherpour E, N. Pardis, and M. Reihanian, 2018 *An Overview on Severe Plastic Deformation : Research Status , Techniques Classification , Microstructure Evolution , and Applications* The International Journal of Advanced Manufacturing Technology Springer-Verlag London Ltd., part of Springer Nature
[14] Edalati K and Horita Z 2015 *Mater. Sci. Eng. A* **652** pp 325-352
[15] Panigrahi A, *et al.* 2016 *Biomed. Mater.* **62**, pp 93
[16] Kang JY, Kim JG, S. K. Kim, K.-G. Chin, S. Lee, and H. S. Kim 2016 *Scr. Mater.* **123**, p 122
[17] Zhang W, Wen YH, Li N and Huang SK 2007 *Mater. Sci. Eng. A* **19** pp 454-455
[18] Valiev RZ, Enikeev NA, Murashkin MY, Kazykhanov VU, and Sauvage X, 2010 *Scr. Mater.* **63**, pp 949-952

IOP Conf. Series: Materials Science and Engineering **485** (2019) 012011 doi:10.1088/1757-899X/485/1/012011

Quantitative image analysis in some iron powder metallurgy materials

M Marin and F B Marin
Faculty of Engineering "Dunărea de Jos" University of Galati, Domnească Street, 47, RO-800008, Galati, Romania

mihaela.marin@ugal.ro

Abstract. The goal of the present paper is to analyze the porosity in some powder metallurgy (P/M) alloys by two different methods, using the conventional method and by image processing analysis method. Quantitative image processing is a new and useful tool which is able to determine the pore size, pore size distribution and porosity of parts obtained by P/M route. One important and significant disadvantage of P/M processing is the presence of porosity due to their activity as crack initiators and due to their presence as the stress distribution in compacts is nonhomogeneous in the cross-section. Two atomized iron powders obtained by powder metallurgy with particles of different sizes (< 45, 45-63, 63-100, 100-150, >150 μm) are the based materials studied in this paper. The analyzed powders were subjected to uniaxial pressing with 500 MPa as the applied pressure. The compacts disc dimensions obtained are $\phi\, 8 \times 6$ mm in dimensions. The compacts were sintered in a laboratory furnace at a temperature of 1.120 °C. The sintering time was 60 and 90 minutes. The sintered specimens obtained were analyzed regarding their density, microstructure and porosity. Porosity of iron-based P/M materials was measured using the conventional and image processing technique. A correlation between the experimental and software data analysis could be established.

1. Introduction

The industry with the most powder metallurgy's applications is represented by the automotive and aerospace industry, where 90% of the total world powder production is represented by iron and alloy powder [1, 2]. In the P/M conventional technology, the products are made by the process of blending, compacting and sintering ferrous powder elemental or alloyed. The limitation of this technology is represented by the small voids present in the parts after sintering, known as pores or porosity. Porosity plays a major influence on the mechanical properties of the sintered parts obtained by P/M due to their actions as stress concentrators and as a consequence, low values for strength and toughness [3-9]. Due to the progress in the field of the materials science concerning the application of image analyse [10-16] it has become an important method used for the investigation of the porosity and particle size in sintered powder metallurgy materials. There are several image processing software available to analyse images, such as: ImageJ, MATLAB, Avizo, PerGeos, Image Pro, OMERO, ENVI, ICY and others. Image J [17] is one of free and useful image analysis software to analyze porosity in an image by applying filters and adjustments using Threshold binary function, the pore areas are recognized, so the software generates a separate image only with porosity and can calculate it as percentage. The goal of this paper is to study the porosity in some sintered powder metallurgy materials by applying an image processing software tool and to compare the obtained results with the values obtained by conventional method.

2. Experimental details

The specimens used in this study are represented by two comercial powders produced from atomized iron powders and iron-based pre-alloyed powder with Cu, Ni and Mo as raw material. In table 1 is presented the chemical composition of the powder samples.

IOP Conf. Series: Materials Science and Engineering **485** (2019) 012011 doi:10.1088/1757-899X/485/1/012011

Table 1. Chemical composition of analyzed powders.

Powder type	Cu	Mo	Ni	C
P_1	0.10	0.03	0.05	<0.01
P_2	1.50	0.50	4.00	<0.01

The particle size distribution of analyzed powders was ranged in 45 and 150 μm. In table 2 are presented the particle size distribution of analyzed powders. The metal powders were homogenized with a lubricant, respectively zinc stearate -1% and then compacted into cylindrical shape with the dimensions of 8 mm diameter x 6 mm height at a pressure of 500 MPa.

Table 2. Particle size distribution of analysed powders.

Powder type	Particle size distribution, [%]				
	<45 μm	45-63 μm	63-100 μm	100-150 μm	>150 μm
P_1	25	21	29	19	5
P_2	21.3	20.2	32.6	23	3.1

For both powders, zinc stearate was used as die lubricant during pressing. In P/M process lubricants are indispensable due to theirs action in reducing friction between powder particles and the die. Also, the lubricants are added because can improve significantly the powder's compressibility. Afterward, the green compacts were subjected to sintering at temperature of 1120 °C for 60 and 90 minutes in a laboratory furnace. The microstructural characterization and porosity behaviour were performed on unetched specimens using an optical microscope OLYMPUS BX51M. Regarding the determination of porosity characteristics, 100x magnification for samples was used. Following sintering, the sintered density of the samples was measured using conventional method and by applying an image processing software tool.

3. Results and discussion

Concerning the density of the samples, green and sintered densities were calculated by using geometrical method.

Density of a green part (*ρg*) obtained by P/M route is calculated using the follow relation:

$$\rho g = (mg / vg), \tag{1}$$

where mg = mass of the sintered part, vg = volume of the sintered part.

Density of a sintered part obtained by P/M route (*ρs*) is calculated using the follow relation:

$$\rho s = (ms / vs), \tag{2}$$

where ms = mass of the sintered part., vs = volume of the sintered part.

The total porosity (*Pt*) of the sitnered compact obtained by P/M route, in volume percent, is calculated using the follow equation:

$$Pt = 100\ (1 - \rho s/\rho t)\ [\%], \tag{3}$$

where ρs, ρg and ρt are the sintered density, green density and theoretical density.

The green and sintered densities of the analyzed specimens calculated using the geometrical method are represented in table 3.

IOP Conf. Series: Materials Science and Engineering **485** (2019) 012011 doi:10.1088/1757-899X/485/1/012011

Table 3. Green and sintered densities of the analyzed specimens.

Powder type	Green density, (g/cm^3), [ρg]	Sintered density, (g/cm^3), [ρs]	
		Sintered at 1120° C and 60 minutes	Sintered at 1120 ° C and 90 minutes
P_1	6.78	6.83	6.84
P_2	6.81	6.85	6.89

Concerning the microstructure characterization of analyzed samples, the images acquisition was carried out on unetched samples using a digital camera coupled to an optical microscope (Olymps BX51M), at 100x magnification. Examination of figures 1-4 shows the optical micrographs of the sintered specimen P_1 and P_2 with different sintering time, 60 and respectively, 90 minutes and the images obtained after rendering using Image J.

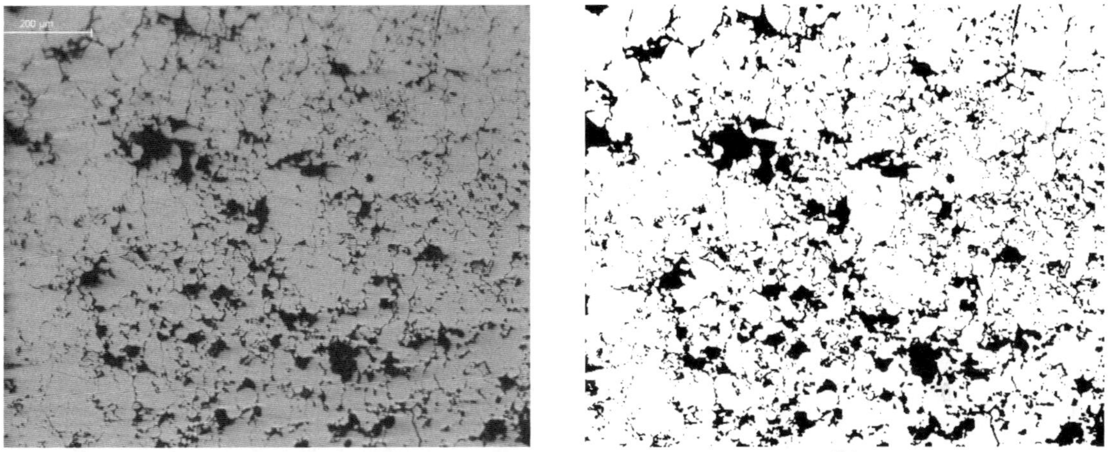

(a) (b)

Figure 1. Microstructure of sample P_1 at 1120° C for 60 minutes: (a) initial image and (b) after software rendering.

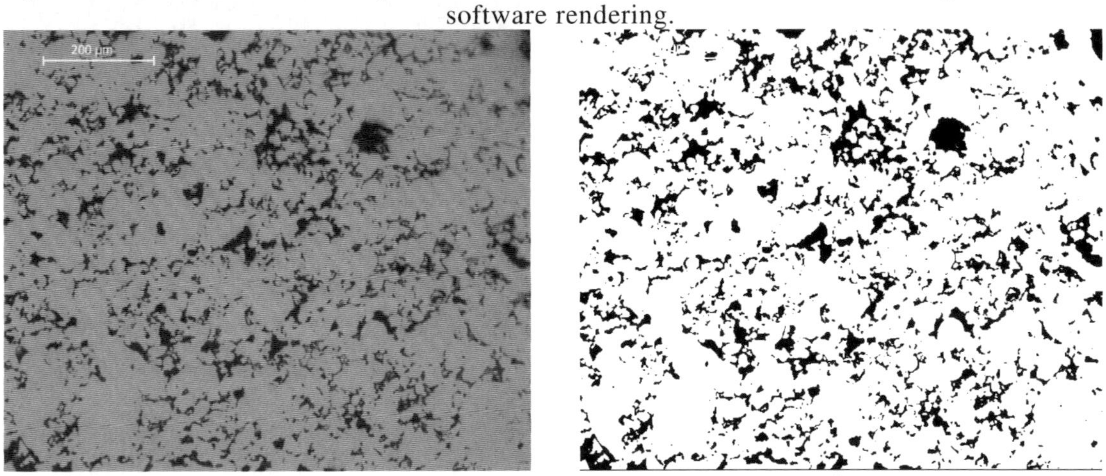

(a) (b)

Figure 2. Microstructure of sample P_2 at 1120° C for 60 minutes: (a) initial image and (b) after software rendering.

IOP Conf. Series: Materials Science and Engineering **485** (2019) 012011 doi:10.1088/1757-899X/485/1/012011

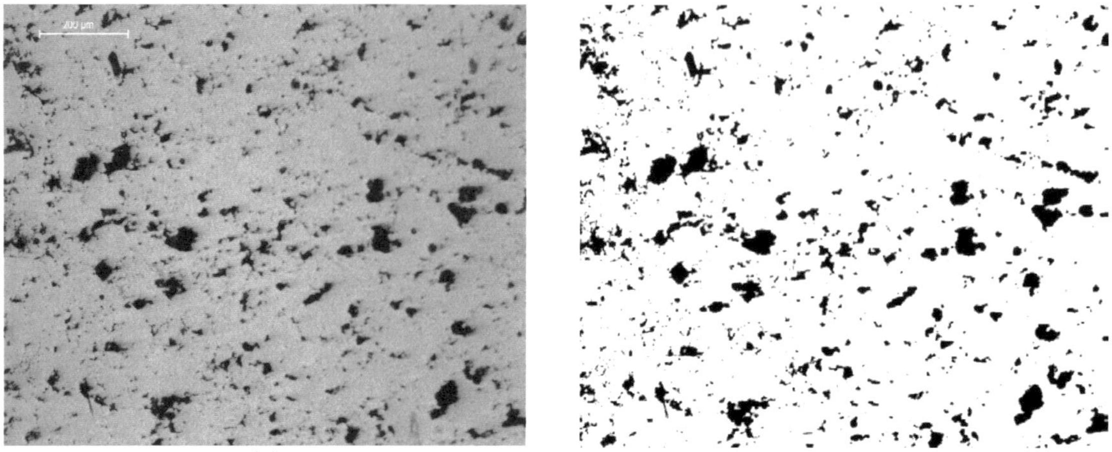

(a) (b)

Figure 3. Microstructure of sample P_1 at 1120° C for 90 minutes: (a) initial image and (b) after software rendering.

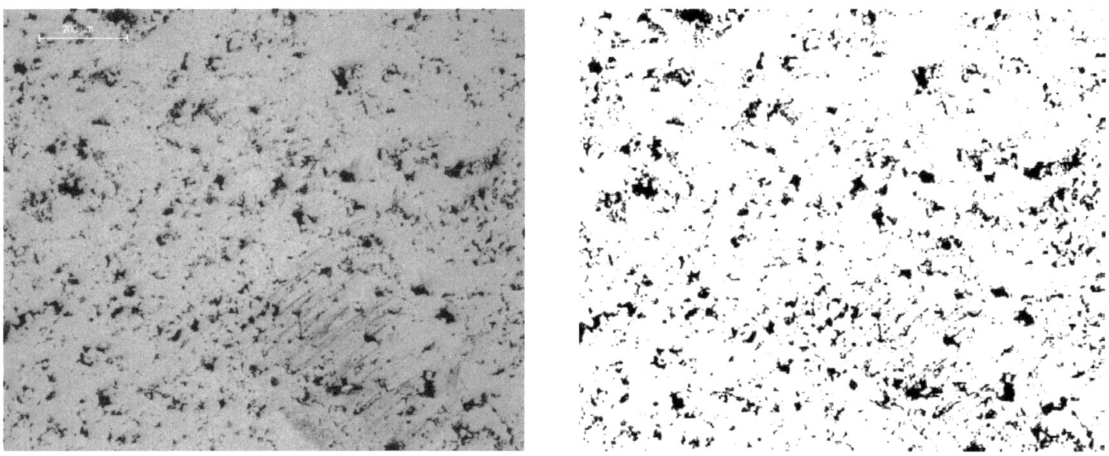

(a) (b)

Figure 4. Microstructure of sample P_2 at 1120° C for 90 minutes: (a) initial image and (b) after software rendering.

By using the quantitative image analysis method, a comparative study of porosity was performed. Also, using the Image J software it was possible to generate a 3D image of the surface for the analysed sintered samples and is presented in figures 5 and 6.

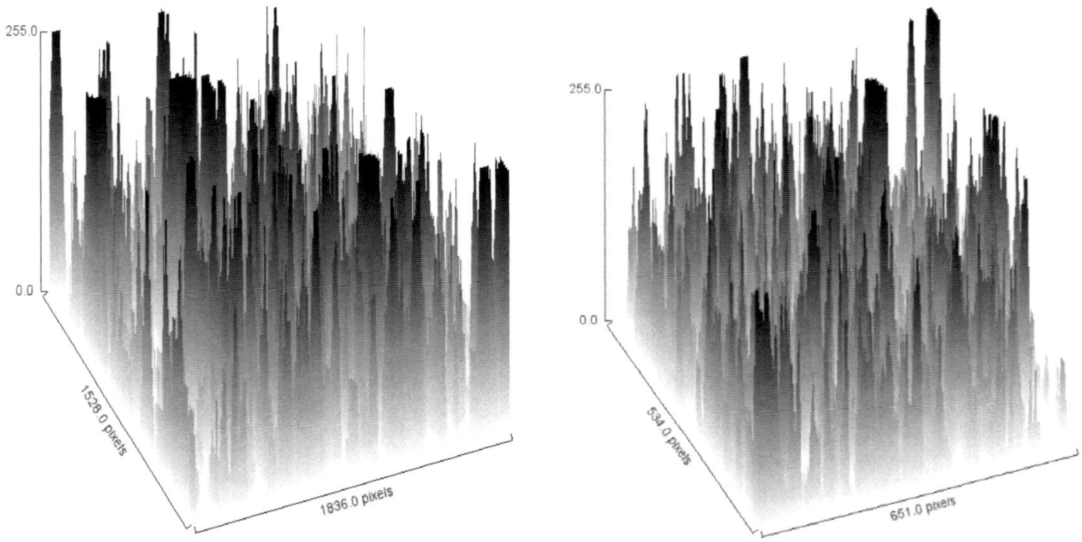

(a) **(b)**

Figure 5. 3D image of the surface for the samples sintered at 1120° C for 60 minutes, obtained using Image J software: (a) P_1 and (b) P_2.

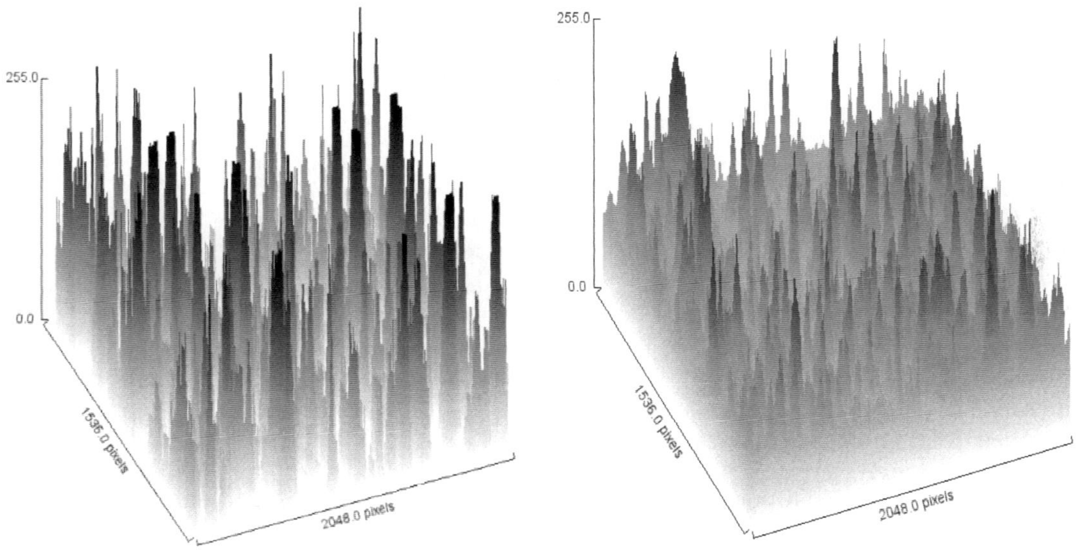

(a) (b)

Figure 6. 3D image of the surface for the samples sintered at 1120° C for 90 minutes, obtained using Image J software: (a) P_1 and (b) P_2.

Concerning the porosity measurements, in table 4 are presented the values obtained using conventional method by density technique and by image analysis technique using an image processing software, Image J.

The results showed that the longer sintering time gives a higher sintered density values and smaller porosity content. This can be attributed to the diffusion enhancement, which leads to reduce pore size and close porosity. By increasing the sintering time, the porosity decreases [18-20]. Also, a low values in porosity in sintered state for sample P_2 is observed due to the alloying elements, Cu, Ni and Mo [21-26].

Table 4. Percentage porosity of analyzed alloys obtained from conventional method and from image analysis using Image J.

Powder type	Porosity from convential method, (%)		Porosity from image analysis, (%)	
	Sintered at 1120° C for 60 minutes	Sintered at 1120° C for 90 minutes	Sintered at 1120°C for 60 minutes	Sintered at 1120° C for 90 minutes
P_1	13.11	13.00	13.83	13.40
P_2	12.90	12.40	13.97	13.94

During sintering process, Cu forms a liquid phase, acting as a semi-permeable film around the Fe particles, which improves the bond between the powder particles, while Ni and Mo remain in a solid state, so a partial diffusion of Ni in Fe is obtained.

4. Conclusions

The porosity measurements of analyzed samples using image analysis software - Image J was correlated to the density of the sintered samples obtained by geometrical method. A correlation between higher density and a decreasing in porosity was established. The sample P_2 had a lower porosity and a higher density. The measurements of porosity resulted from density technique are ranging from 13.11 % to 13.00% for sample P_1 and from 12.90% to 12.40% for sample P_2. The porosity measurements resulted by using the Image J software are ranging from 13.83% to 13.40% for sample P_1 and from 13.97% to 13.94% for sample P_2. A correlation between the experimental and software data analysis was established.

References

[1] Marini D, Cunningham D and Corney J 2018 *Proc IMechE Part B: J Engineering Manufacture* **232** pp 650–69
[2] Akhtar S, Saad M, Misbah M and Sati M 2018 *Materials Today: Proc.* vol 5 (Amsterdam: Elsevier), issue 9 part 3 pp 18649-55
[3] Marin M, Potecaşu F, Marin F B and Potecaşu O 2017 *Conf. Proc. of 17th Int. Multidisciplinary Scientific GeoConf. SGEM* vol 17 (Sofia: STEF92 Technology) pp 125-132
[4] Mohan P, Elshalakany A, Osman T, Amigo V and Mohamed A 2017 *Journal of Alloys and Compounds* **729** pp1215-25
[5] Srijampan W et al. 2015 *Materials & Design* **88** pp 693-701
[6] Gurau C, Gurau G, Sampath V and Silva R C 2016 *Conf. Proc. of 16th Int. Multidisciplinary Scientific Geoconf. SGEM* vol 1 (Sofia: STEF92 Technology) pp 175-182
[7] Sulowski M 2010 *Powder Metallurgy* **53** 2 pp 125-140
[8] Chawla N and Deng X 2005 *Materials Science and Engineering A* **390** pp 180-88
[9] Mann R et al. 2017 Wear **386–387** pp 1-16
[10] Almanstotter J 2016 *Metal Powder Report* **71** 5 pp 333-38
[11] Gacsi Z 2003 *Materials Science Forum* **414-415** pp 213-220
[12] Marin M, Potecaşu F, Marin F B, Potecaşu O and Alexandru P 2016 *Conf. Proc. of 16th Int. Multidisciplinary Scientific Geoconf. SGEM* vol 1 (Sofia: STEF92 Technology) pp 259-264
[13] Kabir M R and Richter M 2017 Materials **10** pp 1-15
[14] Gurau C, Gurau G, Sampath V 2018 *Indian Journal of Engineering and Materials Sciences* **25** 1 pp 5-10
[15] Vasilescu E, Florea B, Semenescu A, Chivu O and Babis C 2016 *Rev Chim* **67** 7 p 1329
[16] Ciripova L, Hryha E, Dudrova E, Vyrostkova A 2012 *Materials & Design* **35** pp 619-25
[17] Image J software- https://imagej.nih.gov/ij/

[18] Yan Y, Nash G and Nash P 2013 *International Journal of Fatigue* **55** pp 81-91
[19] Marin M, Drugescu E, Drugescu, Elena, Potecasu O, Potecasu F, Cordeiro R S 2010 *Metalurgia International* **15** pp 95-100
[20] Martin F, Garcia C and Blanco Y 2011 *Materials Science and Engineering: A* **528** 29–30 pp 8500-11
[21] Angela W D, Tellez L, Alcala J, Martinez E and Cedeno V 2014 *Materials and Design* **58** pp 12–18
[22] Radchenko P, Radchenko O and Panichkina V 2004 *Powder Metallurgy and Metal Ceramics* **43** pp 132-138
[23] Gherghescu I A, Ioniţă D, Ciucă S and Dumitrescu R E 2018 *Rev Chim* **69** 1 pp 86 - 90
[24] Marin M, Potecasu F, Potecasu O, Istrate G 2015 *Conf. Proc. of 15th Int. Multidisciplinary Scientific Geoconf. SGEM* vol 1 (Sofia: STEF92 Technology) pp 181-186
[25] M Brânzei, Gherghescu I A and Ciucă S 2016 *Rev Chim* **67** 6 pp 1068-1072
[26] Narayanasamy R, Anandakrishnan V and Pandey K S 2009 *Materials Science and Engineering: A* **517** 1–2 pp 30-36

IOP Conf. Series: Materials Science and Engineering **485** (2019) 012012 doi:10.1088/1757-899X/485/1/012012

Bearing failure prediction using audio signal analysis based on SVM algorithms

F B Marin, C Solomon , M Marin

Faculty of Engineering, "Dunărea de Jos" University of Galati, Domnească Street, 47, RO-800008, Galati, Romania

E-mail: mihaela.marin@ugal.ro

Abstract. Bearings are machine elements used in a wide variety of applications including transportation. The accurate prediction of a bearing failure is important to sensitive applications to secure its safety during the service life. Bearing failure prediction is useful both in bearing testing phase as well as in case of lifetime use. Real-time Audio signal analysis and advanced algorithms are able to identify the incipient failure, caused by defects, fatigue, overload or poor maintenance. Audio signal analysis and processing remains a domain where technique and algorithm needs to be developed. In this paper is presented a proof-of-concept technique and equipment developed to predict failure of bearings in case of testing phase. For this study, acoustic emission signals were measured and analyzed during life testing of bearing while other sound source are also recorded. Correlation between the acoustic emission patterns were identified in order to identify noise signal and identify the signal associated with bearing degradation. The developed solution to isolate other sounds signals means that the technique could be used while lifetime of the bearings. The results of this study provide evidence that accurate estimation of the failure of various bearings is possible by processing the vibration signal acquired from a single point, even in case of multiple sound sources are present and introduce noise in signal processing. The SVM classifier provides at least 92% mean accuracy. The influence of model on prediction accuracy has also been discussed in the work.

1. Introduction

Prognostics and health management (PHM) domain of study [1] is important and has proven the techniques highly effective in improving reliability. PHM refers to the techniques concerning condition monitoring, fault diagnostics and fault prognostics. Remaining useful life (RUL) prediction aims at assessing the degradation of bearings and to predict failure. Stimulating low maintenance costs, increasing plant availability, increasing productivity and safety have led to a growing interest in monitoring machine conditions.

Bearings are a common component of machines in industrial applications. This is the reason they have received tremendous importance from researchers in the field of PHM [2]. Different techniques are used to detect and diagnose bearing defects[3,4]. These could be classified as vibration and acoustic measurements, wear analysis and temperature measurement. Vibration monitoring and analysis in rotary machines is the most prevalent technique used to monitor, detect and analyze the state of the structure in real time or at predetermined time intervals due to rapid data collection and interpretation [5].

Vibration signatures of the machine inform the operator to make decision for maintenance such as replacing the bearings [5]. Early detection of bearing defects is therefore essential to prevent damage to the alternate parts of a machine. The deformations of the bearing can be sorted as localized and distributed. Localized defects include bursts, cracks and pits caused by fatigue on tread surfaces. These defects could be caused by manufacturing errors and operating environments. Thus, monitoring the condition of bearings has been considered as a fundamental and basic part of any modern production facility [6]. Appropriate monitoring predicts the probability of a malfunction before it really happens. Several vibration monitoring methods and signal processing techniques have been explored to monitor conditions [7]. Bearing testing machine are important in confirming numerical simulation of bearings, that the bearings RUL prediction is correct.

The SVM [8] is an artificial intelligence algorithm able to identify and predict RUL based on trained data. In this paper, the SVM is used to predict the bearings degradation process. Different strategies have been applied to measure the vibrations and acoustic responses of faulty bearings; sound and sound pressure techniques, shock pulse method, time and frequency vibration measurements and acoustic emission technique [9,10]. Several statistical indicators that can be used as part of the real environment are the peak, the mean square root, standard deviation, peak factor, rebound factor, asymmetry, impulse factor, flattening, and shape factor [11]. It has been found that different statistical measures may indicate defective bearings. In order to test bearing materials, lubricants and bearing design the bearing manufacturer are using machine fatigue testing of bearing components [12]. The investigator compares theoretical life, measured in hours, resulted from design of bearing to the results outputted by testing machines.

2. Experimental details

In order to determine the occurrence of bearing surface defects, a real-time real time acoustic measurement system has been developed. The statistical parameters that are extracted from the acoustic signatures obtained from the bearing via the microphones are largely in the temporal domain. Correct analysis of signal time series can produce data measurements. Time domain statistics are used as irregular and trending parameters in an attempt to detect the presence of incipient bearing damage. In this paper, statistical and frequency analyzes of the acoustic signals for the detection of the defects of the bearing were performed. The bearing condition monitoring was implemented by analyzing the data of the captured acoustic signal. The functional information flow of the acoustic measurement system is shown in figure 1.

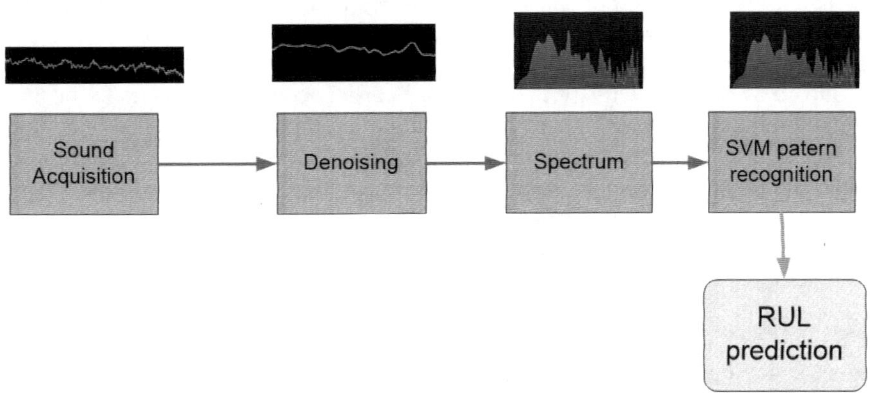

Figure 1. Architecture for bearing fault detection and prediction.

A test device has been built to record the acoustic signals of good and defective bearings in the working environment. We have used the MA-910 bearings test machine, as shown in figure 2, a laptop, a microphone and 60 bearings. CAW33 type 22205 bearings were used for experimental work. Acoustic signals of good and defective bearings were recorded at 3700 rpm.

For this study, acoustic emission signals were measured and analyzed during life testing of bearing while other sound source are also recorded. Correlation between the acoustic emission patterns were identified in order to identify noise signal and identify the signal associated with bearing degradation.

The architecture describing flow of information is depicted in figure 1 with several steps: *i*) sound acquisition have been performed with a microphone on the testing machine; *ii*) denoising phase is very important in testing facilities as in the same room several testing machines might work in the same time; iii) spectrum analysis is the phase where spectrum is extracted from data and is described by 5 descriptors that we indicate bellow; *iv*) SVM pattern recognition of the sound pattern is able to predict remaining life time based on the 5 descriptors defining shape of spectra.

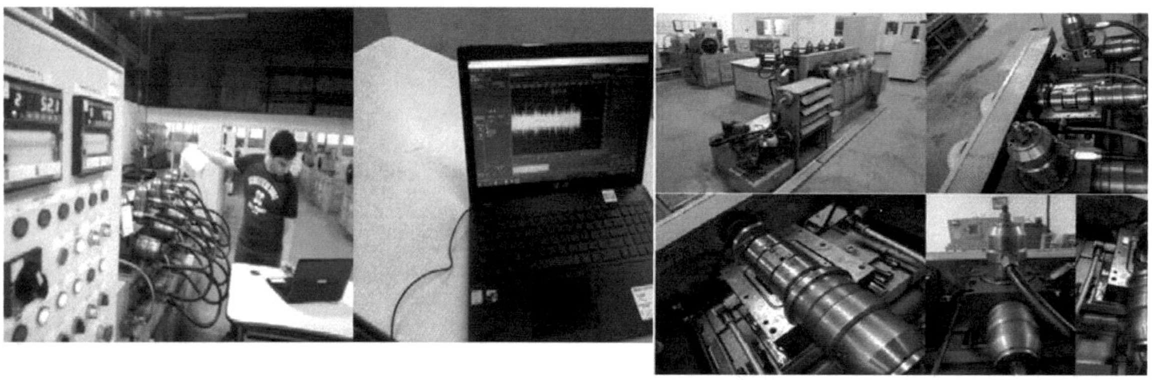

Figure 2. MA-910 Bearing Testing Machine.

A test device has been built to record the acoustic signals of good and defective bearings in the working environment. In order to securely mount the microphone a custom foam mount was fabricated. This 3D printed component ensures the microphone is unobstructed and allows for minimizing vibrations from testing machine. The cavity of the mount ensures no Helmholtz resonances can build up. A microphone was used to capture the acoustic signals from the bearings. It requires a maximum 3.6V supply and draws only 120μA and it is quoted as having a sensitivity of -38dB. Bearing acoustic signals under different conditions were recorded using MS Windows 7.0 Sound application. Furthermore, the audio converter is used to convert .wma files into .wav format to import them into FLSTUDIO using the wav-read function. The audio converter was also used to convert the sampling rate to 22.05 kHz. The statistical and frequency analysis techniques were then applied using the FLSTUDIO functions and tools.

Frequency spectra obtained for different bearings that have been categorized in frequency ranges based on power spectra. It has been found that the acoustic signals obtained from defective bearings have the highest power. The signals obtained from a good bearing had the lowest power.

The spectra for each case have exported as points coordinates defining the curves of spectra. As we notice in the figure 3 and figure 4, the comparing spectra shape for good condition bearing versus spectra shape for defective bearing, we noticed a distinct pattern. In figure 6 we can easily notice how the spectra shape for defective bearing assembles to each one, as well as spectra shape defining good condition bearings looks similarities.

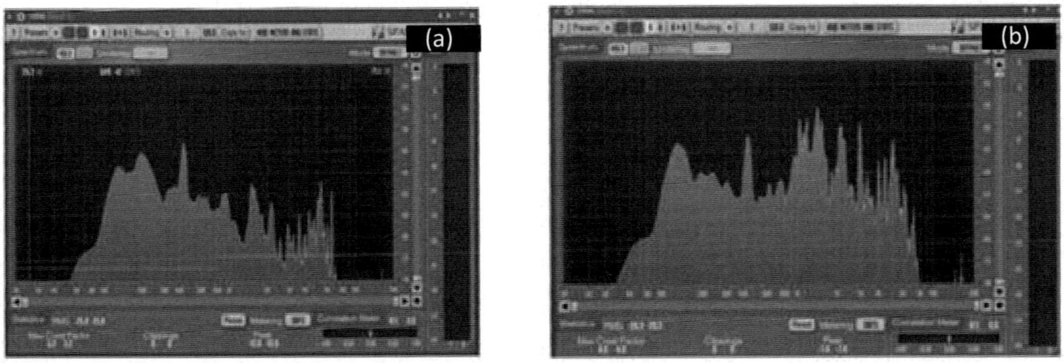

Figure 3. Spectra shape: (a) for good condition bearing and (b) for defective bearing.

To determine the sound frequency content of the good and defective bearings, the spectral power density of the captured sound signals was analyzed using the FLSTUDIO software.

The Fast Fourier Transform (FFT) method was used for spectral analysis for which a window function was selected to remove the captured signal heads to eliminate discontinuities.Since the acoustic signals are not deterministic in nature, the Welch method is able to compute the spectral power density of the acoustic signals. The spectral power density graphs of the acoustic signals of good and defective bearings obtained during experimentation are shown in figure 4.

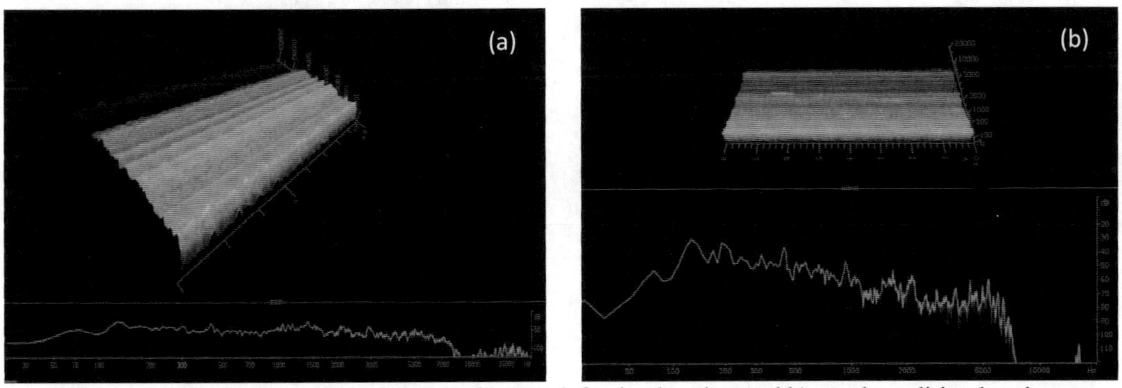

Figure 4. 3D representation of spectra for: (a) defective bearing and b) good condition bearing.

The acoustic bearings were recorded at a rotational speed of 3700 rpm. The experiment was performed several times, each of the acoustic signals for good condition and defective bearings were captured and used to calculate the parameters of the spectra. Acoustic signals of good and defective bearings are shown in figure 5.

IOP Conf. Series: Materials Science and Engineering **485** (2019) 012012 doi:10.1088/1757-899X/485/1/012012

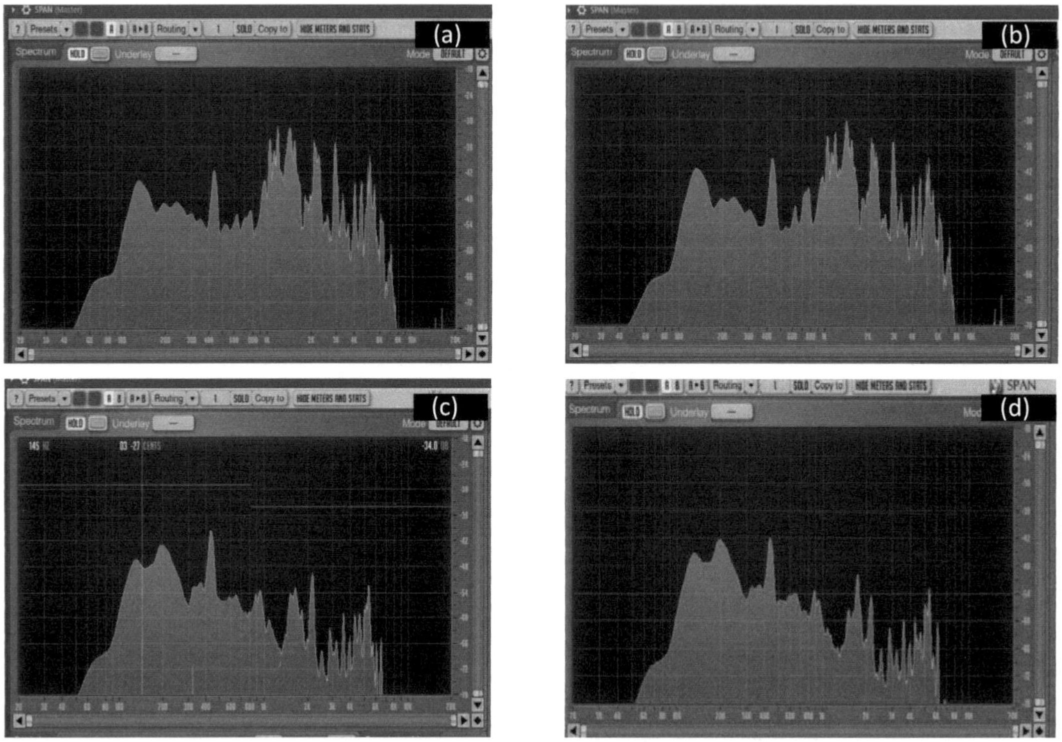

Figure 5. Acoustic signals of different bearing conditions at different speeds: (a) and (b) for good condition bearing spectra shape, (c) and (d) for defective bearing spectra.

SVM regression model was implemented using Rapidminer 5.0 software. The parameter evaluation technique is used to select five features that are used as predictors in multi-class support vector machine (SVM) classification. The same as we, humans describe shapes, we identifiy five parameters as inputs for SVM. SVM algorithm is concerning drawing a hyperplane between "good solutions" and "bad solutions", as depicted in figure 6.

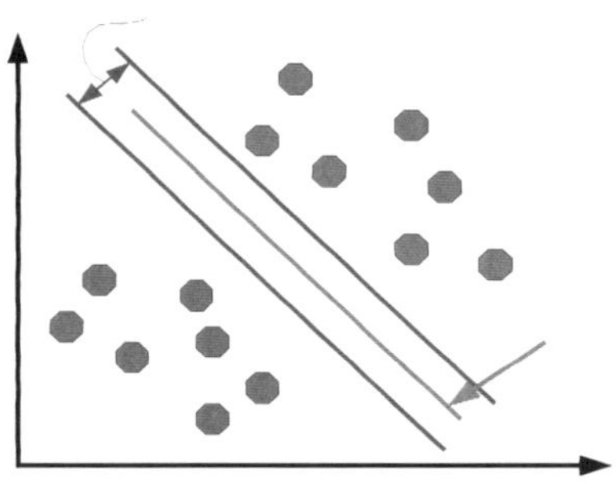

Figure 6. SVM Concept.

IOP Conf. Series: Materials Science and Engineering **485** (2019) 012012 doi:10.1088/1757-899X/485/1/012012

As described in figure 7, several parameters have been defined using spectra of the sound registered. P1 defines the angle of the first peek, P2 defines the length between first two peeks, P3 defines is the angle of the second peek, P4 the distance between second and third peek and P5 is the surface of the third peek.
By using this similitude that defines difference between sound patterns in the two cases, the SVM is trained for 50 known data sets of acoustic data.

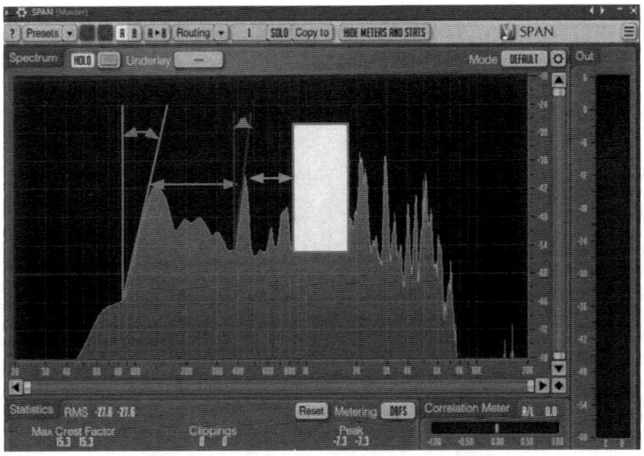

Figure 7. Parameters of the SVM algorithm defined by the spectra shape

3. Results

The diagnosis was performed with the help of Support Vector Machine using Rapidminer 5.0 software where parameters are those five defined above. Then the SVM model is trained using training data of 50 test and 20 test samples are used for testing the technique. The developed solution to isolate other sounds signals means that the technique could be used while lifetime of the bearings as well on bearing testing machines. In figure 9 is depicted the results using the technique described in this paper predicted versus real values. The results of this study provide evidence that accurate estimation of the failure of various bearings is possible by processing the vibration signal acquired from a single point, even in case of multiple sound sources are present and introduce noise in signal processing. The SVM classifier provides at least 89% mean accuracy, as shown in figure 8.

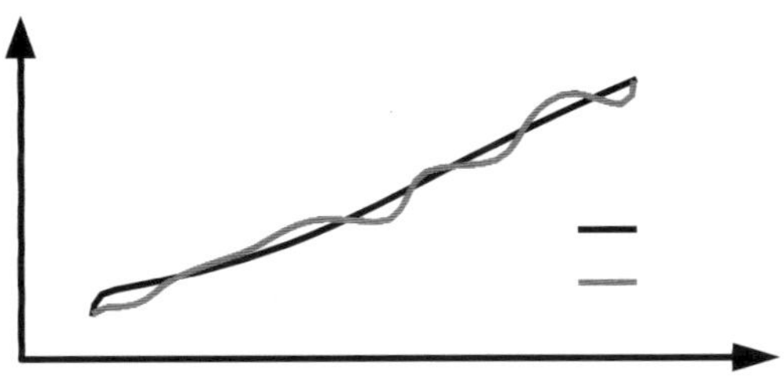

Figure 8. Actual vs predicted data for RUL

IOP Conf. Series: Materials Science and Engineering **485** (2019) 012012 doi:10.1088/1757-899X/485/1/012012

4. Conclusions

Bearings are vital components in rotary machines that have received increased attention in the field of condition monitoring. This paper aimed at monitoring the condition of bearings on the basis of acoustic analysis based on SVM and predicting remaining useful life of bearings. Spectral shapes have clearly shown that bearings with different physical conditions emit acoustic signals with different spectral power densities. SVM algorithm using five defined parameters are able to describe spectra pattern and thus enabling to predict remaining useful life on bearing testing machine.

References

[1] Rodrigues L 2018 *IEEE/ASME Trans. Mechatron* **23** 1 pp 141–150

[2] Shao H, Jiang H, Zhang X and Niu M 2015 *Meas. Sci. Technol* **10** pp 57-63

[3] Tandon N ans Choudhury A 1999 *Tribology International* **32** (8) pp 469–480

[4] Houser D and DrosjackM 1973 Vibration Signal Analysis Techniques Technical Report U.S. Army Air Mobility Research and Development Laboratory

[5] Soualhi A, Razik G, Clerc G and Doan D 2014 *IEEE Transactions on Industrial Electronics* **61** pp 74

[6] Vachtsevanos G, Lewis F, Roemer M, Hess A and Wu B 2006 Intelligent fault diagnosis and prognosis for engineering systems Wiley

[7]Williams T, Ribadeneria X, Billington S, Kurfess T 2001 MSSP **15**(5) pp 979–93

[8]Baillie D and Mathew J A 2009 *Mechanical Systems and Signal Processing* **10** 1 pp 1-17

[9] Cortes C and Vapni N 1995 *Machine Learning* **20** (3) pp 273–97

[10]Gryllias K and Antoniadis 2012 *Eng. Appl. Artif. Intell* **25(**2) pp 326–344

[11] Debnath R, Takahide1N and Takahashi H 2004 *Pattern Analysis & Applications* 7 pp 164-77

[12] Sadeghi F, Jalalahmadi B, Slack T, Raje N and Arakere N 2009 *ASME Trans* **131** 4 pp 110-16

IOP Conf. Series: Materials Science and Engineering **485** (2019) 012013 doi:10.1088/1757-899X/485/1/012013

Photovoltaic panels cost optimization with flexible polymer semiconductor cells using fuzzy logic

M Miron[1], D I Morariu[1], I Cucos[2], C I Alecu[3], L Judele[2], D Lepadatu[4]

[1]Phd student at Technical University "Gh. Asachi" Iaşi, Faculty of Civil Engineering and Building Services, Romania
[2]Senior Lecturer at Technical University "Gh. Asachi" Iaşi, Faculty of Civil Engineering and Building Services, Romania
[3]Senior scientific researcher at Institute of Economic and Social Research Gh. Zane - Iasi, Romanian Academy Branch
[4]Associate Professor at Technical University "Gh. Asachi" Iaşi, Faculty of Civil Engineering and Building Services, Romania

daniel.lepadatu@gmail.com

Abstract. This paper presents the research undertaken by the authors to determine the parameters of electric power produced by a photovoltaic panel using flexible semiconductor polymer cells, aiming a comparative study between this panel and other types of conventional photovoltaic panels, with monocrystalline, polycrystalline and amorphous silicon, based on technical and economic indicators, established under uncertainty conditions. We will also present the variation of these dimensions according to the incidence angle of solar radiation. Thus, researches will show that for its value between 36-38 degrees, you get very good conversion yields. The economic study presented also reveals that under uncertainty there is a correspondence between the panel type and its physical and economic characteristics.

1. Introduction

The energy economy and its production of renewable resources is a major concern of researchers lately [1-8] and the use of solar energy as a polluting and virtually inexhaustible energy by capturing it with different methods is a the current topic that responds specifically to the development requirements of the CE mentioned in Directive 2010/31 and transposed in the concept of E4 house [9-11]. In this paper we accomplished a comparative study on the electricity production of four types of photovoltaic panels, for 1 KW installed power, as follows: monocrystalline silicium-PVGIS panels; polycrystalline silicium-PVGIS simulation panels; amorphous silicium-PVGIS simulation panels; semiconductor polymer-PVGIS simulation panels and semiconductor polymer panels-experimental research. PVGIS (Figure 1) is an application for geographic assessment of solar energy resources and simulation of photovoltaic system performance. It provides an estimation of daily and monthly average solar irradiation, as well as simulations of the daily average electricity production, monthly and annual average of photovoltaic systems according to their type, geographic position, inclination angle and orientation. For using the program the geographic coordinates of the position of the simulated panel will be introduced and also the functional characteristics of the simulated photovoltaic panel (Figure 1). Also, you can also choose the file type where to save the calculated data.

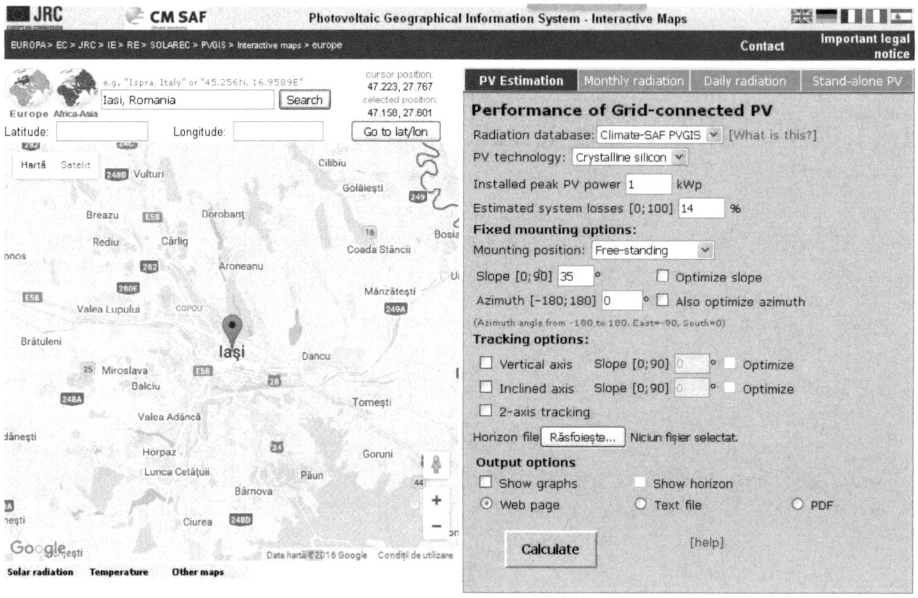

Figure 1. Photovoltaic Geographical Information System (PVGIS) - application

2. Materials and methods

2.1. *Flexible photovoltaic cells with semiconductor polymers*

Significant advances have been made in the field of photovoltaic energy over the last few years by using cells based on organic semiconductor materials. Conventional solar cells are made of inorganic materials such as silicium. Although the conversion efficiency of these cells is relatively high, they require expensive materials and energy-intensive manufacturing technologies. The main reasons for the increased interest in organic semiconductor materials are:

- ease of manufacturing and low cost of processing
- flexibility due to their ability to mold on any type of surface
- ability to be produced from recyclable materials and using renewable energy sources.
- keeps the solar energy conversion capacity regardless of the amount of solar radiation transmitted.

Conjugated polymers are particularly attractive in comparison with silicium because they are powerful light dampers visible and can be deposited in flexible substrates on large surfaces using wet processing technologies such as printing or layer coating.

In recent years, several types of organic photovoltaic cells have emerged, including: solar polymer cells, polymer-fullerene solar cells, small molecule cells and hybrid solar cells.

3. Results and discussions

This paper presents the research undertaken by the authors to determine the parameters of electric power produced by a photovoltaic panel using flexible semiconductor polymer cells, aiming a comparative study between this panel and other types of conventional photovoltaic panels, with monocrystalline,

polycrystalline and amorphous silicon, based on technical and economic indicators, established under uncertainty conditions.
The application (PVGIS) used performs simulation and provides tabular results in the specified file format. Average daily electricity production Ed, average monthly production Em, average daily sum of total solar irradiation per square meter of the studied panels (Figures 2-5).

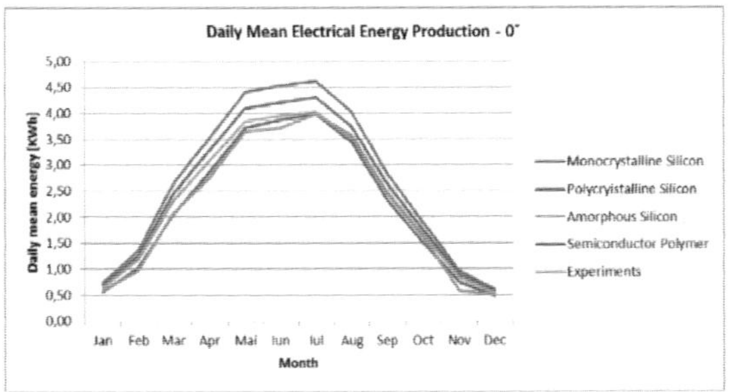

Figure 2. Daily mean energy production variation – angle of 0^0

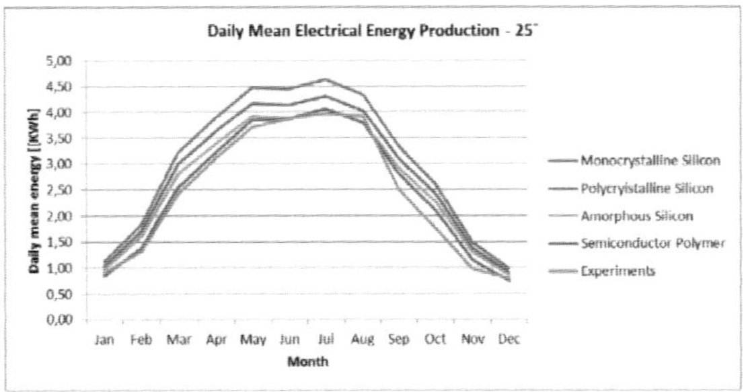

Figure 3. Daily mean energy production variation – angle of 25^0

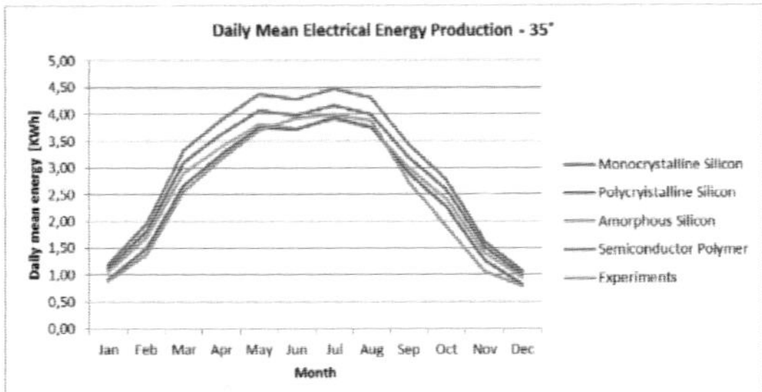

Figure 4. Daily mean energy production variation – angle of 35^0

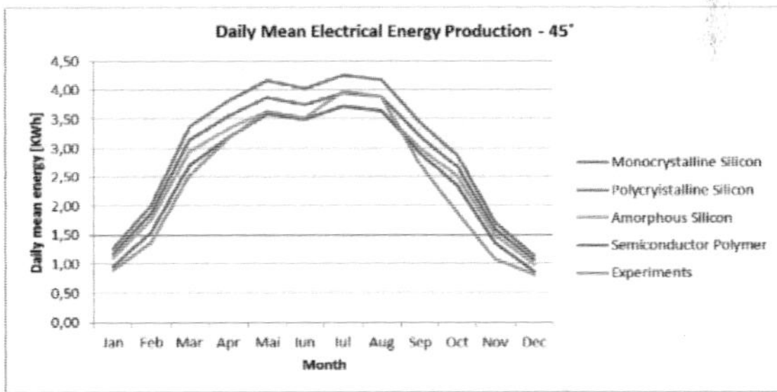

Figure 5. Daily mean energy production variation – angle of 45^0

The analysis of the results presented above confirms that there is a dependence between the solar radiation absorption capacity and the orientation angle of this panel being optimal for a value in the range 36-38 depending on the nature of the photo absorbent material used. The experimental data were obtained on a panel made up of organic semiconductor photovoltaic cells [14].

3.1. *Economic comparisons between the four types of photovoltaic panels under uncertainty conditions*

An economic study was also carried out using an expert system for uncertainty decision based on fuzzy logic [12-14] methods for the four types of photovoltaic panels (figures 6 - 9) using a technical and economical specific indicators. The four indicators considered were the following:

- Specific price in two
- Specific power and

In order to make management decisions under uncertainty, we can use the following techniques that use the decision matrix {Rij}:

- Decision technique Max-Min or pessimistic method:

$$V_{optim} = Max_i \; min_j\{R_{ij}\} \tag{1}$$

IOP Conf. Series: Materials Science and Engineering **485** (2019) 012013 doi:10.1088/1757-899X/485/1/012013

- Decision technique Max-Max or optimistic method:

$$V_{optim} = Max_i \, Max_j\{R_{ij}\} \quad (2)$$

- The optimality technique or Hurwicz method:

$$H_i = \alpha \times A_i + (1 - \alpha) \times \alpha_i \quad (3)$$

where:

α – optimist coefficient ($0 < \alpha < 1$);

A_i – the most favorable element of the line i;

a_i – the worst element of the line i;

In this case the optimal alternative is:

$$V_{optim} = Max_i \, \{H_i\} \quad (4)$$

- Proportionality technique or Bayes-Laplace method:

$$V_{optim} = Max_i \frac{1}{n} \sum_{j=1}^{n} \{R_{ij}\} \quad (5)$$

- The technique of minimizing regrets or Savage method:

The regrets matrix is determined:

$$\{r_{ij}\} = Max_i \, \{R_{ij} - Max_i \, R_{ij}\} \quad (6)$$

$$V_{optim} = min_i \, Max_j\{r_{ij}\} \quad (7)$$

The specific price and annual output per unit of power are relevant when a certain installed power is required and the specific square matrix price and specific power per square meter are relevant for estimating a certain amount of available panels.

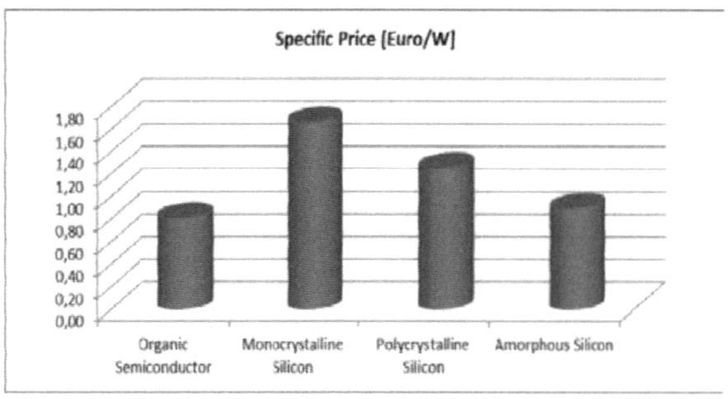

Figure 6. Specific prices in euro per watt for the four types of photovoltaic panels

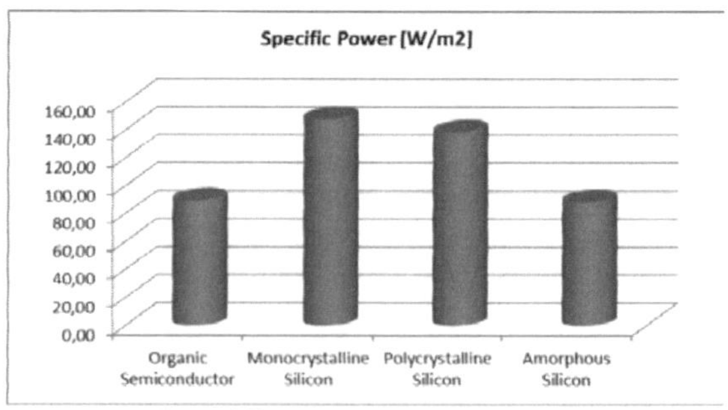

Figure 7. Specific power per square meter for the four types of photovoltaic panels

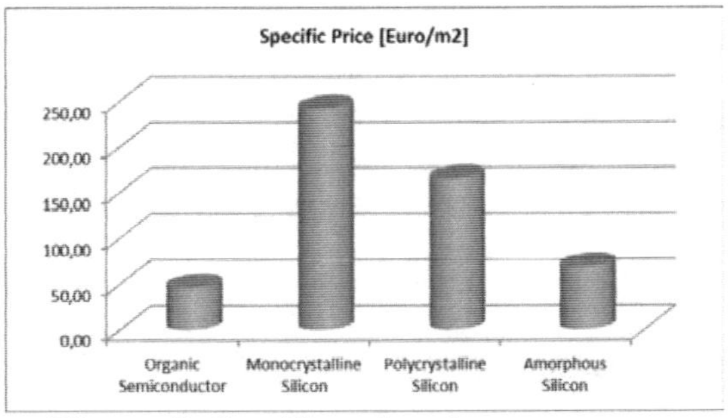

Figure 8. Specific prices in euro per square meter for the four types of photovoltaic panels

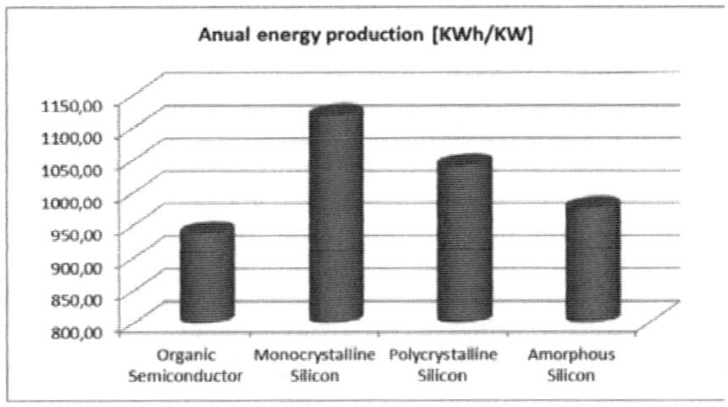

Figure 9. Annual energy production installed for the four types of photovoltaic panels

We also present a comparison between simulation results and experiments resulting in a good fit (figures 10, 11).

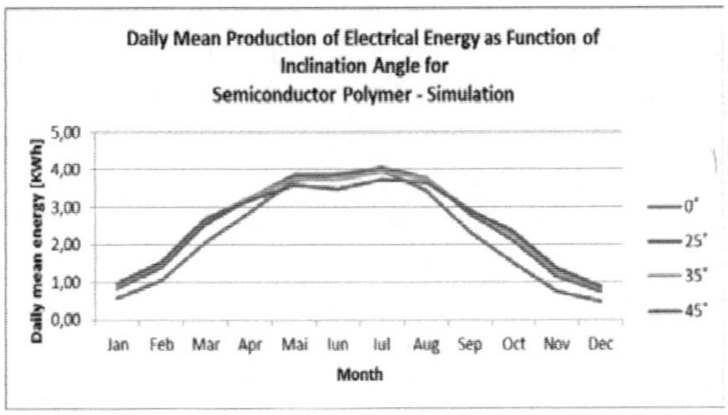

Figure 10. Variation of average daily electricity production depending on the angle of inclination resulting from simulations

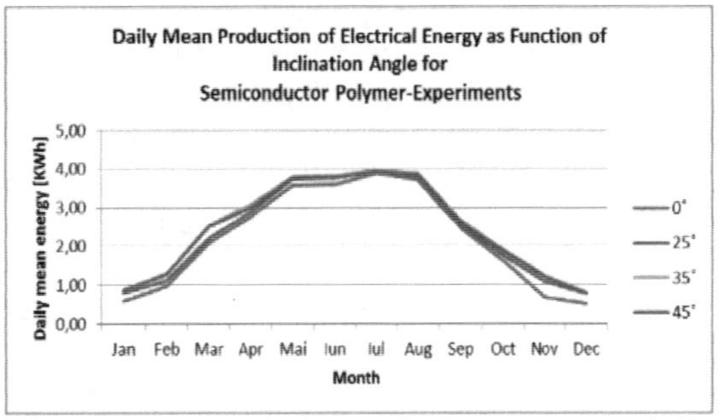

Figure 11. Variation of the average daily electricity production depending on the angle of inclination resulting from the experiments

The economic study shows that monocrystalline and polycrystalline silicium panels, although expensive the energy production and the specific power are high. Compared to the organic semiconductor panels and amorphous silicium that can reach up to 25% of the above-mentioned value and their power output exceeds 50%. It is found that the optimal option in most cases is that of semiconductor polymer panels.

4. Conclusions

The paper presented confirms the attractiveness of the development of technologies based on semiconductor polymer in photovoltaic energy production and has the following advantages:

- low production costs due to manufacturing technology at low temperatures
- ease of application to support surfaces.

The economic studies presented reveal the market value of the new semiconductor polymer panels and the correspondence between their type and the energy performance that converge to a clear conclusion that they provide a viable alternative to quality.

References

[1] Yilmaz F 2018 Thermodynamic performance evaluation of a novel solar energy based multigeneration system *Applied Thermal Engineering* **143** pp 429-437
[2] Bofu C Craciun I Giurma-Handley C R Antonescu I Telisca M Boariu C Alecu I 2014 Correlation

between green energy cadastre and environmental monitoring *Journal of Environmental Protection and Ecology* **15** (4) pp 1751–1758

[3] Cucos I 2014 Research on the impact of technical, economic and environmental impact of energy management systems for integrated waste *The Yearbook of the Gh. Zane Institute of Economic Research*, Romanian Academy Publishing House **23** (1) pp 63-72

[4] Gelsor N Gelsor N Wangmo T Yi-Chun C Frette O Jakob J Stamnes Hamre B 2018 Solar energy on the Tibetan Plateau: Atmospheric influences, *Solar Energy* **173** pp 984-992

[5] Giurma I Craciun I Giurma-Handley C.R Boariu C Cercel P Antonescu I 2014 Considerations concerning geothermal energy storage using a heat accumulator, *Journal of Environmental Protection and Ecology - Environmental management* **15** (4) 1759–1766

[6] Cucos I 2015 Systems management methods of waste biomass energy administration in mountain areas, *Journal of Montanology*, **III** (3) pp 147-156

[7] Mohamed Nfaoui, Khalil El-Hami, 2018 Extracting the maximum energy from solar panels, *Energy Reports* **4** pp 536-545

[8] Cucos I 2014, Research on the impact of technical, economic and environmental impact of energy management systems for integrated waste, *The Yearbook of the Gh.Zane Institute of Economic Research* **23** (1) pp 63-72

[9] Vilas G Pol 2016 Editorial overview: Energy and Environmental Engineering: Emergent electrical energy efficacy (E4) *Current Opinion in Chemical Engineering* **13** pp 6-8

[10] Ferrante A 2016 Chapter 2 - *Nearly Zero Energy Urban Settings (ZEUS)*, Editor(s): Annarita Ferrante, Towards Nearly Zero Energy, Butterworth-Heinemann pp 65-168

[11] Cucos I Antonescu I Miron M 2016 Recherches sur la gestion de l'energie des batiments civiles et industrielles, *Acte du II ème Seminaire doctoral international de la francophonie* 02-04 november 2016 Faculty of Urbanism and Architecture Chisinau, Republic of Moldova Ed. PRIMEXcom Chisinau pp 50-67

[12] Antonescu I 2015 Premises for Creating the Database and Knowledge Base of an Expert System for Dynamic Management of Renewable Energy Resources under Uncertainty Conditions *The Yearbook Of The Gh.Zane Institute Of Economic Research* **24** (1) pp 87-94

[13] Antonescu I 2015 Architecture of an Expert System for Dynamic Management of Renewable Energy Resources under Uncertainty Conditions *Buletin of Politechnic Institute of Iasi Electronic Section* Tom LXI (LXV) **2** pp 9-17

[14] Antonescu I Miron M Cucos I 2016 Considérations concernant l'utilisation de panneaux photovoltaïques avec cellules flexibles au polymères semi-conductrices, *Acte du II ème Seminaire doctoral international de la francophonie* 02-04 november Facultaty of Urbanism and Architecture Chisinau, Republic of Moldova Ed PRIMEXcom Chisinau pp 36-41.

Acknowledgements

This paper was realized with the support of project EFECON – ECO-INNOVATIVE PRODUCTS AND TECHNOLOGIES FOR ENERGY EFFICIENCY IN CONSTRUCTION, POC/71/1/4 - Knowledge Transfer Partnership, Cod MySMIS: 105524, ID: P_40_295, Project co-financed by the European Regional Development

IOP Conf. Series: Materials Science and Engineering **485** (2019) 012014 doi:10.1088/1757-899X/485/1/012014

Advanced tools for optimization and efficiently monitoring behaviour in service stage of buildings using renewable energy

D I Morariu[1], I Serbanoiu[2], M Miron[1], D Lepadatu[3]

[1] Phd Student at Technical University "Gh. Asachi" Iasi, Faculty of Civil Engineering and Building Services, , Romania
[2] Professor at Technical University "Gh. Asachi" Iaşi, Faculty of Civil Engineering and Building Services, Romania
[3]Associate Professor at Technical University "Gh. Asachi" Iaşi, Faculty of Civil Engineering and Building Services, Romania

daniel.lepadatu@gmail.com

Abstract. In the topographical surveying works devices have got among the required accessories also tools called tripod. They have the role of correctly installing and settling the equipment above the station point in order to perform the measurements. Various types of tripods are known: the classic tripod (made from aluminum or wood), the adjustable laser tripod, the elevator tripod, the tripod for supporting and adjusting the picket. The aim of this scientific paper is to present my invention that is concerned on a new type of tripod device with photovoltaic cells called mobile solar tripod which has the purpose of providing the energy needed for the equipment used and thus extending the duration of survey measurements by using renewable energies. The tripod being an almost universal accessory along with other topographic instruments, it can be used for some topo-geodetic activities such as monitoring behavior in service stage of buildings, marking the points on the ground, determining ground elevations and other works that require a long time for the objects studied. This present work will describe the advantages and disadvantages of using the existing tripods compared to the new proposed device.

1. Introduction

The evolution of terrestrial measurements and topography is directly correlated with the development and continuous improvement of topographic instruments. The progress of the equipment area for determining geometric elements, means of calculation and reporting data influences the upgrade of methods and procedures in topographic measurements. This automatically involves making more complex jobs and improving their quality at a faster pace, with cost savings. Monitoring behaviour in service stage of buildings is accomplished with different methods and techniques. The progression of technology is characterized by the ways used for efficiently monitoring the behaviour of buildings in time, used to optimize the costs and duration of their execution. The objective of this scientific paper is to present a modern device that can be universally used for several types of topographical equipment that assures their own autonomy by accumulating and then converting the solar energy into electrical energy called mobile solar tripod. Also, we will present the advantages and disadvantages of existing tripods compared to the device proposed by the authors.

2. Material and method

Together with making of first maps and topographic plans, the requirement to measure the size and shape of land appeared. This led to the materialization of the first tools for drawing graphical representations. Topographic devices have progressed in terms of construction and technologies due to the achievements of electronics [1]. Surveying instruments have among the essential accessories, also the use of tripod. This tool has the role of supporting many types of devices such as: total stations, level, laser scanner, static Global Positioning System (GPS) and others [2]. Such instruments have been used since ancient times, but the components were separate and their assembly was needed. The tripod is made from three legs provided with three spikes for fixing in the ground, three clamps to be able to slide, made of metal (Figure 1b) or wood (Figure 1a). At the top of the instrument, it is found the tripod table on which it is placed the survey device fixed with the screw pump. It can be placed above the point station in a convenient place and can be adjusted to a height that is favourable for the operator [3]. The role of this device is to support and maintain the stability of the instruments against horizontal movements and forces. At first they were made from wood then made from aluminium greatly which reduced their weight. Tripods can sustain the equipment (Figure 2) and its accessories [3].

a) Wood b) Aluminium

Figure 1. Types of classic tripods

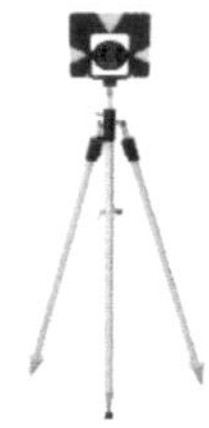

Reflector Holder

Figure 2. Tripod for accessories

Furthermore, the evolution is found also in the change of design, material and use of modern tripods. Contemporary tripods are manufactured for supporting the survey equipment, but also for topographic accessories. In the next paragraphs, we will present some of the latest tripods which are used in the survey work. The tripod for supporting and adjusting the staff according to the patent created by Voicu F. and Domilescu V.M in 1989 consists of a tripod case with screws on each leg which modifies vertically and allows adjustments of rod at different heights with the help of a bar and a staff support. Keeping in position the device is done with by the means of fixing screws and a bridle clamp [4]. Another tool is the elevator tripod. This invention is made from a body to which a crank mechanism and a wheel are attached, used to raise the equipment to the desired level by the user. The device is generally applied in civil engineering to elevate the level to the operator`s height in order to save time for setting and adjustments [4]. The tripod for supporting the surveying peg regardless of the kind of land without the help of an operator. The patent developed by Vargas R. in 1983 is a two-legged topographic tripod and a central pole. A telescopic tube is found inside the central pole and is extended to a suitable height with the help of compression clamps at the upper end. The goal of the telescopic tube is to increase the height of the tripod [5]. Starting from the classic tripod that performs only support, settling on the station point and stability over the entire length of measurements, reaching at many types of instruments that have one or more uses (tripod for staff/peg support, tripod with elevator). On the other hand, there are types of tripods used in different areas besides topography. We can enumerate the tripod (Figure 4) used in meteorology, which is a part from the weather station that consists from lighting rod, solar radiation sensor, wind speed direction sensor, RT/cellular antenna, solar panel, relative humidity sensor with radiation shield, enclosure houses datalogger, power supply and modem, raingage and grounding rod [6]. Furthermore, we can add the solar powered led tripod beacon (Figure 5) used in road traffic for allowing easier visibility on the streets or highways, made from a tripod stand and reflector with flashing lights on the left and right side [7]. Also, in the construction area tripods are used for levels, mostly for measuring interiors and tripods (Figure 5) are found in photography for supporting cameras when taking pictures.

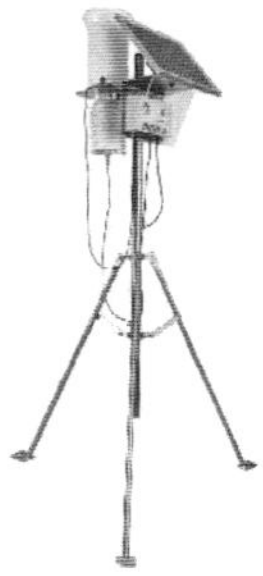

Figure 3. Weather station

Figure 4. Solar powered led tripod beacon

Figure 5. Tripod-photography

This is due to the evolution of technologies that radically changed the range of land measurements, facilitating the way in which works are done and achieving technically and economically feasible results.

3. Results and discussions

The modern system built for contemporary topographic instruments, presented in this scientific paper can be used in supervising the activity of buildings. In fact, through this modern tripod it is created the autonomy needed for the measurement process efficiently utilizing the renewable solar energy. Thus, we can use electronic devices for a long time without being conditioned by traditional batteries that are limited to a certain number of hours while operating. The proposed invention called Mobile Solar Tripod (Figure 6a) keeps the classic principle of a building tripod: three spikes that have the function of fixing the three legs in the ground and three clamps which allow the feet to be adjusted at different heights convenient to the operator. The innovation of this device consists in fixing photovoltaic panels on each leg of the tripod (Figure 6c) to provide continuous energy to the instrument that performs the topographic measurement activity. The layout of photovoltaic cells on each side of the tripod is necessary due to the movement of the Sun, thus offering the possibility of catching the light source from any position. Panels transfer the energy through a cable connected to the solar energy conversion system in electricity. The converter is placed in the tripod table (Figure 6b), located at the top of the legs. The solar energy conversion device is connected to a battery that operates as an external accumulator. The transmission of the electrical energy to the measuring instrument's battery is made by a cable whose length allows the device to rotate in the set of directions.

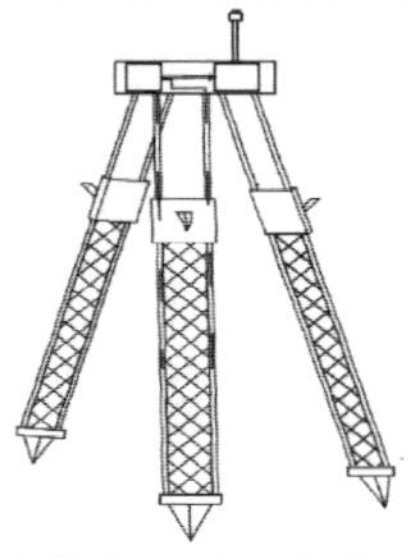

a) Solar tripod view

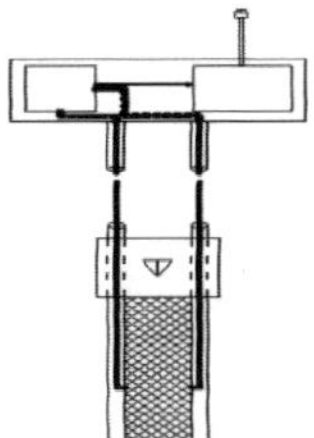

b) Connection view panel-battery

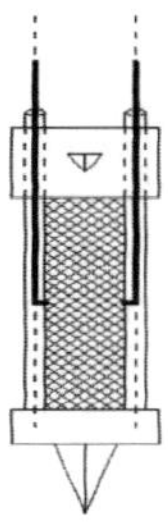

c) Tripod leg view

Figure 6. Mobile solar tripod

Photovoltaic systems represent the future of technology [8-10], especially in areas where access to an electrical power is impossible. Its components include: solar cells, accumulator batteries, load regulators, static converters and other components useful for the achievement of a solar energy system (protections against electrical discharges, circuit breakers and fuses). The photoelectric cell is assimilated to a photosensitive diode, its process is relying on the properties of semiconductor materials.

The operating principle is called the photoelectric effect. There are also other researchers who have studied the renewable energy conversion phenomenon of a various forms [11-14]. For the production of solar cells, silicon is often used as a semiconductor material. There are several types of photovoltaic cells: monocrystalline cells, polycrystalline cells, amorphous cells, Cadmium telluride photovoltaics cells (CdTe), Copper Indium Selenium solar cells (CIS), Copper indium gallium selenide solar cells (CIGS). Monocrystalline cells are the first to have been made on the basis of a silicon block crystallized in a single crystal. Their capacity is between 12-16%. They have two disadvantages: the high price and long payback time of the supplied energy [14-18]. Polycrystalline cells are built on the basis of a silicon block crystallized in several crystals having different orientations. Their yield is 11-13%, but they imply lower production costs than monocrystalline ones [14-18]. On the other hand, amorphous cells have lower production, 5-10% than crystalline and costs are low. They consist of a glass support or synthetic material on which a thin layer of silicon is deposited. Their advantage is good behavior in diffuse or fluorescent light, having good performance during high temperatures (Renewable energy course). Cadmium telluride photovoltaics (CdTe), Copper Indium Selenium (CIS), Copper indium gallium selenide (CIGS) technologies are under industrialization, but their development may be stopped due to the high levels of cadmium in their composition being highly toxic. The cell generates a voltage of 0.5, regardless of its type (monocrystalline or polycrystalline). Depending on their size, they produce a current of up to 8 amperes (A). The battery of the total station has 6 V, so for a good charging performance it is necessary:

$$x = 6\ V \div 0.5\ V = 10 \tag{3.1}$$

where x represents the number of photovoltaic cells that are useful for recharging the battery in the device. The most advisable alternative from the point of view of productivity and the cost of manufacturing the device could be polycrystalline cells. Photovoltaic cells have a length of life for 25 years, with very good resistance to temperatures of - 40 ° Celsius [14-18].The glass of the top layer is self-cleaning during rain and also periodic dust removal is recommended. It is advisable to position the feet from the ground at an angle between 30 ° and 45 ° to better capture the sun's energy.

Table 1.Advantages and disadvantages of tripods

Types of tripods	Advantages	Disadvantages
Classic tripod	- performs the centering and supporting during the measurement;	- it does not contain other features that facilitate the performance of terrestrial measurements;
Tripod for supporting and adjusting the staff	- allows only fixing and centering the staff at different heights for reading the topometric point;	- it cannot support the equipment during the measurements;
Elevator tripod	- performs only centering, riding and adjusting the machine at the operator's height;	-is only useful for construction works (yard) or for building interiors;
Tripod for supporting the surveying peg	- performs only the support of the peg in vertical position, regardless the land`s composition; - there is no need for the operator to hold the peg each point;	- the device only performs the tasks of supporting the peg in the upright position; - it should be placed every time on the point after the end of the sighting;
Mobile solar tripod	- performs centering, setting and supporting the device during measurements; - it can be used continuously without interrupting the measurement cycle for changing battery; - the layout of the panels on each leg of the device is useful for retaining solar energy from any angle;	- the high cost of the instrument due to the photovoltaic cell construction; - the replacement of accumulator batteries or electronic equipment (battery inverter, solar charger) may occur periodically.

IOP Conf. Series: Materials Science and Engineering **485** (2019) 012014 doi:10.1088/1757-899X/485/1/012014

The proposed device by the authors, called „Mobile Solar Tripod" keeps the classic construction of a simple tripod, but instead it uses renewable energy to charge the battery of the topographic instruments during measurements. The advantages of using the mobile solar tripod are: centering, setting and assuring stability during topographical determinations, its use provides a continuous rhythm of the measurement cycle without the need for changing or charging the battery to complete the operations and the arrangement of panels on each leg of the equipment provides the ability to collect solar energy regardless of the position of the Sun (table 1). The price for building this device can be between 2000 – 3000 euros depending on the type of the solar panels used, generator and accumulator. A solution for reducing costs would be adding only one solar panel outside the tripod like in the weather station case, but it depends on the autonomy the tools used during topographic measurements need to fully charge. Although, in this scientific paper we presented a novelty in the topographic area by making this solar tripod, it can be improved by attaching extra solar panels on the tripod`s table or by enclosing a extra device like an external battery if more power is demanded. We can add that especially for more complex operations like monitoring buildings deformations in service stage, the necessity of the continuous process of monitoring instruments and consequently the need for a continuous supply of the battery accumulators can occur and with this modern device this inconvenience completely disappears. The only disadvantages could be the value of the instrument`s construction due to photovoltaic panels and materials used, but also the situation where accumulators, battery inverters or solar chargers can be damaged and they need to be changed.

4. Conclusions

Topography has a well-defined utility and role in land surveys. This field is divided into two extended categories: general topography (elevations in the plan and field profiles) and engineering topography, which includes the monitoring of buildings behavior over time. Both use contemporary tools and methods to achieve goals. Currently, in many fields the tripod is used: meteorology, civil engineering, photography, road traffic for supporting different equipment in vertical position. On the other hand, the decisive elements that have transformed the concept of survey measurements consist of modern digital technologies and tools. In addition to digital equipment, the mobile solar tripod is an optimal solution for long-term topography operations, such as monitoring of building`s behavior in time, providing them with the necessary energy autonomy. It ensures the efficiently monitoring behavior in service stage of buildings using renewable energy with photovoltaic panels. The novelty of this instrument is that it provides continuous power to the device's accumulator during the process of survey measurements without the need to change the battery or interrupt topographic operations. Also, the device is universal, applying to several types of topo-geodetic equipment: total station, static Global Positioning System (GPS), laser scanner, digital level. Our future perspective is to build the device, showing that the instrument works, can provide energy through solar cells and continuous use of measuring tools during the monitoring of buildings in service stage. Also, in order to put in practice the solar tripod, we need an expert in electricity and solar panels to guide us during the fabrication of the patent. In conclusion, we can state that the proposal made is indispensable by the autonomy provided to the building monitoring devices in the service stage because they can ensure an efficient functioning of these instruments which will allow them to obtain data with high accuracy and relatively low cost. Thus, by introducing the use of these innovative tools from technical point of view we obtain an optimization of time, precision and especially the rational use of renewable energy is achieved.

References

[1] Morariu D.I Lepădatu D 2017 Topo-geodetic modern methods and techniques for building monitoring process *International Agriculture and Food Engineering Symposium – Section Water and Soil* **60** (2) Iasi Romania pp 139-144
[2] Boş N C Iacobescu O Boş N 2015 *Digital Topography* C H Beck Publishing

[3] Brisan M.C 2006 Topography Matrix Rom Publishing Technical University of Civil Engineering Bucharest
[4] Voicu F Domilescu V.M 1989 105853 Patent, RO (Romania) available on-line at: http://www.osim.ro.
[5] Vargas R 1983 4366940 Patent US (United States) available on-line at: http://www.espacenet.com
[6] https://www.researchgate.net
[7] https://www.uniformswarehouse.com
[8] Antonescu I Miron M Cucoş I 2016 Considérations concernant l'utilisation de panneaux photovoltaïques avec cellules flexibles aux polymères semi-conductrices, *IIème Seminaire Doctoral International Francophone* Chisinau Editura PRIMEXcom, Republica Moldova Chisinau pp 36-41
[9] Cucoş I Antonescu I Miron M 2016 Recherches sur la gestion de l'energie des batiments civiles et industrielles, *Iième Seminaire Doctoral International Francophone* EdituraPRIMEXcom, Chisinau Republica Moldova pp 50-67
[10] Lepădatu D Judele L Săndulache G Roşu A.R Diac M 2014 Evolution of distance measurement techniques by optical or optoelectronics way in applied topography, *Proceedings of Symposium on Agriculture and Food Engineering, Scientific Works* Agronomy Series **57**(2) 107-112
[11] Bofu C Craciun I Giurma-Handley C R Antonescu I Telisca M Boariu C Alecu I 2014 Correlation between green energy cadastre and environmental monitoring *Journal of Environmental Protection and Ecology 15* Environmental management **4**
[12] Giurma I Craciun I Giurma-Handley C R Boariu C, Cercel P, Antonescu I 2014 Considerations concerning geothermal energy storage using a heat accumulator, *Journal of Environmental Protection and Ecology* **4** Environmental management 1311-5065
[13] Antonescu I 2015 Premises for Creating the Database and Knowledge Base of an Expert System for Dynamic Management of Renewable Energy Resources under Uncertainty Conditions *The Yearbook of The Gh.Zane* Institute Of Economic Research **24** (1)
[14] Cazacu N 2015 Renewable energy course University of Dunărea de Jos Galaţi Romania
[15] Cucos I Munteanu C March-April 2015 The experimental researches of advanced systems for managing thermosensitive heat treatments *Journal of Optoelectronics and Advanced Materials* **17** (3-4) 409-420
[16] Cucoş I 2014 Research on the impact of technical, economic and environmental impact of energy management systems for integrated waste *The Yearbook of the Gh.Zane Institute of Economic Research* **23** (1) 1453-4061 Romanian Academy Publishing House pp 63-72
[17] Antonescu I 2015 Architecture of an Expert System for Dynamic Management of Renewable Energy Resources under Uncertainty Conditions *Bulletin of Polytechnic Institute of Iaşi* **2** pp 9-17
[18] Miron M Antonescu I Cucoş I 2016 Contributions à l'étude de la production d'électricité d'un panneau de cellules photovoltaïques avec des polymères semi-conducteurs *Iième Seminaire Doctoral International Francophone* Editura PRIMEXcom Chisinau Republica Moldova pp 15-20

Acknowledgments

1. *The patent developed by Phd. student Morariu Diana Ioana, presented in this scientific paper has been accepted by the State Office for Inventions and Trademarks (OSIM), Romania under the patent application number A/00465 in 2017.*
2. *This paper was realized with the support of project EFECON – ECO-INNOVATIVE PRODUCTS AND TECHNOLOGIES FOR ENERGY EFFICIENCY IN CONSTRUCTION, POC/71/1/4 - Knowledge Transfer Partnership, Cod MySMIS: 105524, ID: P_40_295, Project co-financed by the European Regional Development Fund.*

IOP Conf. Series: Materials Science and Engineering **485** (2019) 012015 doi:10.1088/1757-899X/485/1/012015

Tensile characteristics for a class of polymeric blends (PP + PA6)

A E Musteaţă[1], M Botan[2], G Pelin[2], L Deleanu[1], D Constantinescu[3]

[1]Department of Mechanical Engineering, Faculty of Engineering „Dunarea de Jos" University, Galati
[2]National Institute for Aerospace Research „Elie Carafoli" INCAS, Bucharest, Romania
[3]ICEFS COM Savinesti, Romania

lorena.deleanu@ugal.ro

Abstract. This paper presents the influence of component concentration of a class of polymeric blends with different concentration of PP and PA6 and a constant concentration of additives ($CaCO_3$, LDPE and an adhesive) Young's modulus does not have a clear tendency of depending on PA6 concentration and testing speeds, but performs in a band between 1500...1900 MPa, except for the value for PA6 at the lowest testing speed (v=10 mm/min), that is 1454.8 MPa. Under large testing speeds, there were cracks or voids towards the loading direction, forming large parallel 'empty channels'. Investigations are done on possible causes of these longitudinal voids in order to clear if they occur during molding or/and during testing and to eliminate them. Except for values at low testing speed, the average values for tensile stress at break perform in a band of less than 8 MPa. The elongation at break is less than 10% for all materials, except for PA6 at the lowest testing speed, when the highest value for this characteristic was obtained. The lowest values were obtained for material with 20% PA6. For blends with 60% and 80% PA6, the difference between values, at the same testing speed, is the smallest. Energy at break has a slightly increase with the concentration of PA6. From the mechanical point of view, among the formulated blends, material D (with 80% PA6) is the most promising as has the highest stress and energy at break (11.43 J at v=10 mm/min and 8.7 J at v=1000 mm/min), but values are still less than those for PA6.

1. About polymeric blends

Polymers have complex mechanical behavior and their blends could exhibit even more unexpected properties. This is the reason of testing and comparing data and having analyses that help engineers to avoid failure and to select the appropriate material for a particular application.

Some relevant works [1-5] underlined the close relationship between morphology of polymeric blend and its properties. Polymer blends could tailor the mechanical and thermal properties of polymeric materials, because the resulting mixtures keep some advantages of expensive polymers and shadow the non-desired features of low-cost one [6, 7].

Immiscible polymers blends exhibit sea-island morphologies with large droplet sizes and a lack of adhesion between components, as a result of the high interfacial tension between the two immiscible phases. Thus, these blends generally exhibit poor mechanical properties, since mechanical performance is a function of blend morphology and interfacial interactions. To overcome these issues, several compatibilization strategies that include adding block copolymers, compatibilizing agents and nanoparticles have been developed in the last few decades [8].

Polypropylene (PP) and polyamide (PA6) are polymers still in the focus of research due to their properties. This introduction presents several studies that underline the interest in the blend containing one or both of these polymers. Polyamide/polypropylene (PA/PP) blends or composites are interesting because both components are relatively cheap (in 2018, PP was 1.50...3 $/kg and PA6 was 1...5 $/kg), with advantageous properties, and are processable by melt blending. The compatibilisation of binary polymer compounds may be made by adding graft copolymer, segments of which have physical or chemical affinity with two immiscible homopolymers.

The way to improve interface interaction of immiscible polymer blend is adding lowmolecular weight compatibilizers with "soft" nature, unfortunately with negative effects on the final mechanical strength. Shi [9] proposed an interfacial interlocking design strategy for immiscible polypropylene (PP)/polyamide (PA) blend.

Ruehle et al. [10] provided data on the physicochemical properties of a promising biorenewable blend system which exhibits the novel feature of thermodynamic miscibility Biorenewable polyamide-11 (PA11) is blended with partially biorenewable polyamide-6,10 (PA610) to produce thermoplastics of different carbon content. The two polyamides are fully miscible in the melt; the blends exhibit crystallization induced phase separation. Mechanical properties of the blends show intermediate values compared to the homopolymers; significant improvements in the properties of PA11 with the incorporation of 25wt% PA610 are pointed out.

Important dilatation is produced by plastic deformation under tension of neat PP and PP/PA6/POE blends, for which the POE to PA6 concentration ratio equals 1/2. The detailed mechanisms of this volume change are investigated by Bai [11]. At low alloy content, dilatation results from decohesion of PA6 particles from PP matrix. As the amount of PA6 and POE increases, voids are nucleated preferentially in the thicker POE interphase making a shell around the PA6 particles, and secondarily in isolated POE particles.

Recycled polyamide 6 (PA6) from post-industrial waste fibers (PIW) and carpet waste (PCW) indicate the presence of polypropylene (PP) in PCW. X-ray fluorescence and X-ray diffraction show that inorganic contaminants of PCW and PIW are calcium carbonate (CaCO3) and titanium dioxide (TiO2), respectively. Due to higher inorganic filler content, PCW exhibits higher melt viscosity and higher storage modulus than that of PIW. PIW has 20% higher tensile strength than that of PCW [12].

Yousfi [13] reported that for talc filled PP systems, the incorporation of both natural and synthetic talc improves their thermal stability. For talc-filled PP systems, the addition of synthetic talc slightly improves the elastic modulus, but induced an improvement in thermal stability of polypropylene matrix, the mechanical and thermal-related properties being positively affected by this surface treatment. If the highest mechanical reinforcement is obtained in the presence of natural talc, the best ductility is reached in the presence of synthetic talc. For PA6/talc systems, the mechanical properties varied depending on the surface free energy of talc. The strongest improvement in strength properties was for synthetic talc-filled PA6 nanocomposites. Its highly polar character and its ability to create strong acid–base interactions with the functional groups of polyamide 6 are the main reasons to explain the strong interfacial interactions generated between the synthetic talc and the polar matrix.

Wang et al. [14] studied the morphology of polypropylene (PP)/polyamide 66 (PA66) blend, with organoclay and maleic anhydride grafted polypropylene (MAPP). Without MAPP, a transition from typical sea-island morphology to coarse co-continuous morphology is observed when organoclay add relative to PA66 phase reaches 10 wt%. With 10 wt% of organoclay, the coarse co-continuous morphology is transformed to fine fibrillar and then to rod-like morphology with increasing MAPP contents. For PP/PA66 blend mixed at an asymmetric composition (70/30w/w), coarse co-continuous morphology can be formed by intermediate feeding PP into premelted PA66/10 wt% organoclay compound, instead of typical sea-island morphology for the case without organoclay. With organoclay content of 10 wt%, the coarse co-continuous morphology can be transformed to fine fibrillar and then to fine rod-like morphology by adding increasing amounts of MAPP.

Palacios et al. [15] investigated silica nanoparticles (NS) and polypropylene grafted with maleic anhydride (PPgMA) to stabilize morphology and improve properties of an immiscible PP/Nylon 6 blends. Blends of 80% PP and 20% PA6 with a fixed amount of PPgMA and different NS content were prepared by melt mixing. Two PPgMA compatibilizing agents with different grafting level and two modified NS, one hydrophilic and other hydrophobic, were employed. Despite the PA6 drop reduced size, the blends exhibited unsatisfactory mechanical performance. In the end, a PP/PA6 blend stabilized with 1.6% hydrophobic NS and compatibilized with high grafting level PPgMA exhibited a fine-tuned morphology with enhanced ductility and barrier properties. NS nanoparticles act as outstanding stabilizers but do not promote interfacial adhesion between the phases.

The nanocomposites PA/PP/PP-g-MA (polypropylene grafted with maleic anhydride)/graphene oxide were characterized by Alexandrescu et al. [16]. The blends with initial proportion PA6/PP of 270/30 were produced by injection, for bearing seals, contact plates and components for railways, with shock resistance higher than 5-8 kJ/m², wear resistance, resistance to temperatures of -40...240°C, with environmental temperatures from -40 to 60°C, in rain, snow or sunshine. The value of tensile strength of nanocomposites increases compared to the value of polyamide and PA/PP-g-MA/PP blend. Graphene oxide concentrations higher than 1% lead to decrease the shock resistance (Izod method). Percentages in the range 0.1-1% lead to maximum values of physical mechanical parameters.

Properties of PA + PP blends have also been reported by [17-19].

Immiscible blends of 80 wt% polypropylene (PP) with 20 wt% polyamide (PA) or polycarbonate (PC) were prepared by melt mixing with or without the addition of 5% nanosilica. The nanosilica produced a strong reduction of the disperse phase droplet size, because of its preferential placement at the interface. PP developed spherulites, whose growth was unaffected by blending, while its overall isothermal crystallization kinetics was strongly influenced by nucleation effects caused by blending. Addition of nanosilica resulted in an enhancement of the strain at break of PP+PC blends whereas it was observed to weaken PP+PA blends [20].

The morphology of immiscible 80/20 polypropylene/polyamide 6 blends that contain different types of nanosilica and a compatibilizer agent was correlated with their linear and non-linear rheological behavior. Polypropylene grafted with maleic anhydride (PPgMA) was used as compatibilizer agent. Two types of modified silica nanoparticles were added in blends, one hydrophilic (NSE) and the other hydrophobic (NSH). The size of PA droplets was reduced 12 times when the compatibilizer agent was added; and 25 times when hydrophobic nanosilica (NSH) was additionally included in the formulation [21].

2. Materials

All samples were prepared by melt extrusion in molding machine at ICEFS Savinesti Romania. Details of the molding technology is presented in another work [6], [7]. The component concentration of tested materials are given in Table 1.

Table 1. Symbols of formulated materials

	PA6	PP	LDPE	$CaCO_3$	POLYBOND adhesive
	Composition [wt%]				
PA6	100	-	-	-	-
A	20	65	5	7	3
B	40	45	5	7	3
C	60	25	5	7	3
D	80	5	5	7	3
PP	-	100	-	-	-

IOP Conf. Series: Materials Science and Engineering **485** (2019) 012015 doi:10.1088/1757-899X/485/1/012015

3. Tensile properties of the formulated blends

Bone samples made of neat polymers (PP and PA6) and blends with different concentrations of these polymers were tested in traction (see Figure 3 for samples tested at speed of 1000 mm/min). Figures 1 and 2 give the aspect of stress–strain curves for each tested polymeric material, for two different testing speeds.

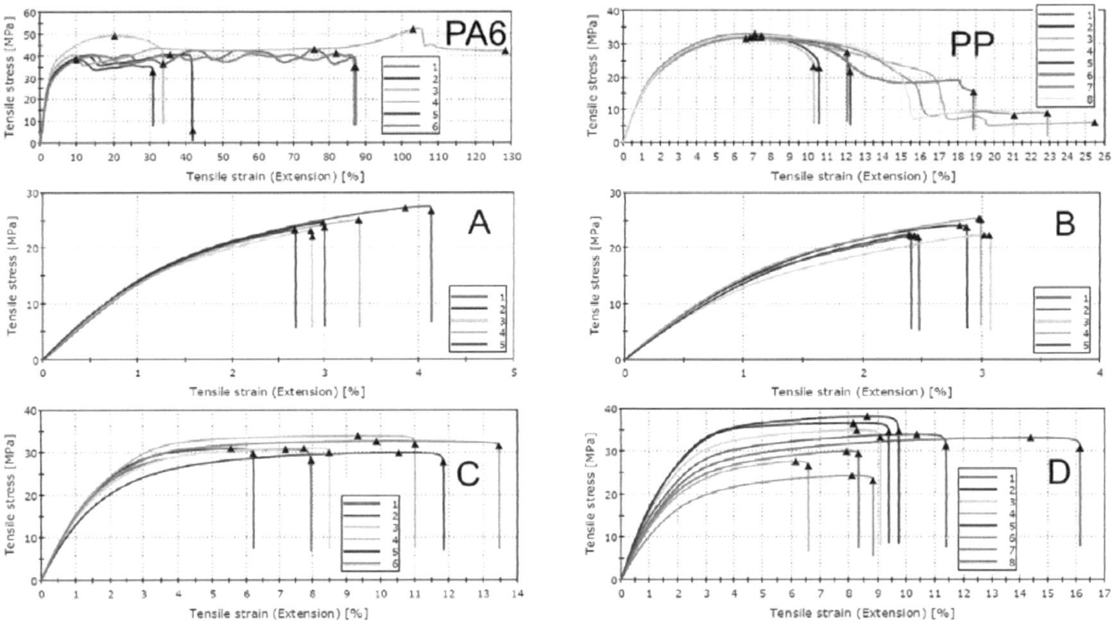

Figure 1. Strain - stress curves for all materials, tested at speed v=10 mm/min

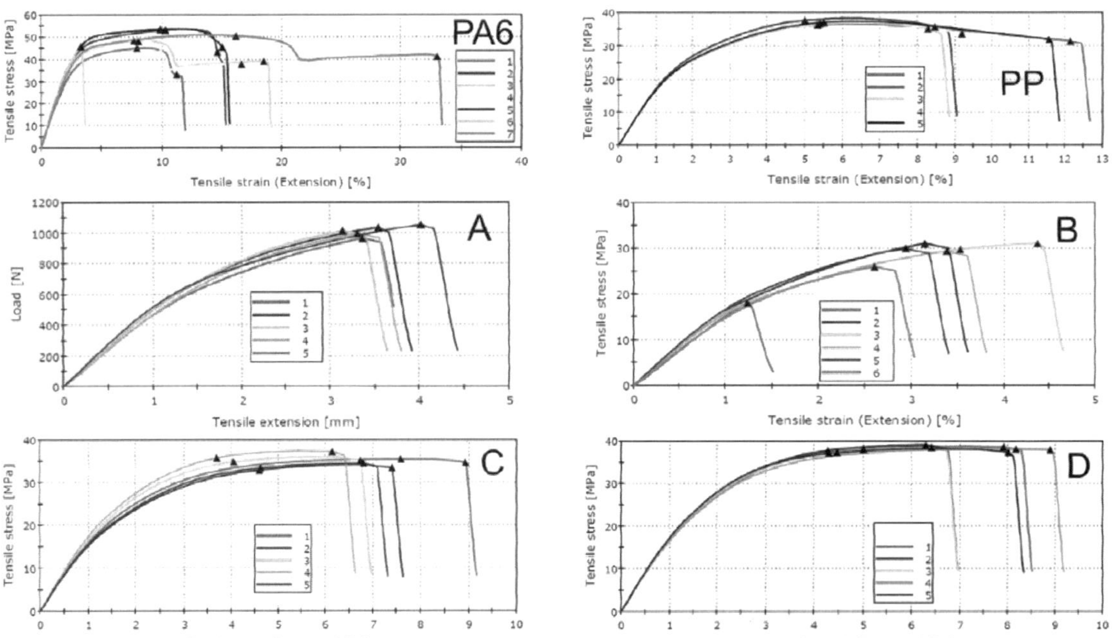

Figure 2. Strain - stress curves for all materials, tested at speed v=500 mm/min

IOP Conf. Series: Materials Science and Engineering **485** (2019) 012015 doi:10.1088/1757-899X/485/1/012015

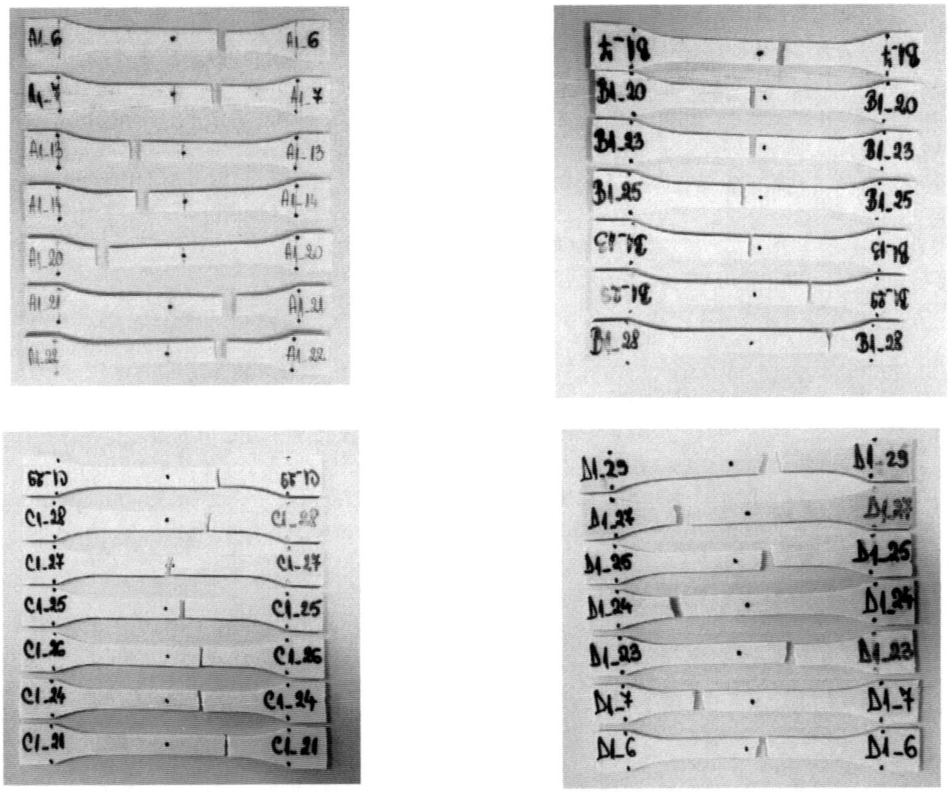

Figure 3. Samples broken in tensile tests, at testing speed of 1000 mm/min (The capital letter designates the material as in Table 1.)

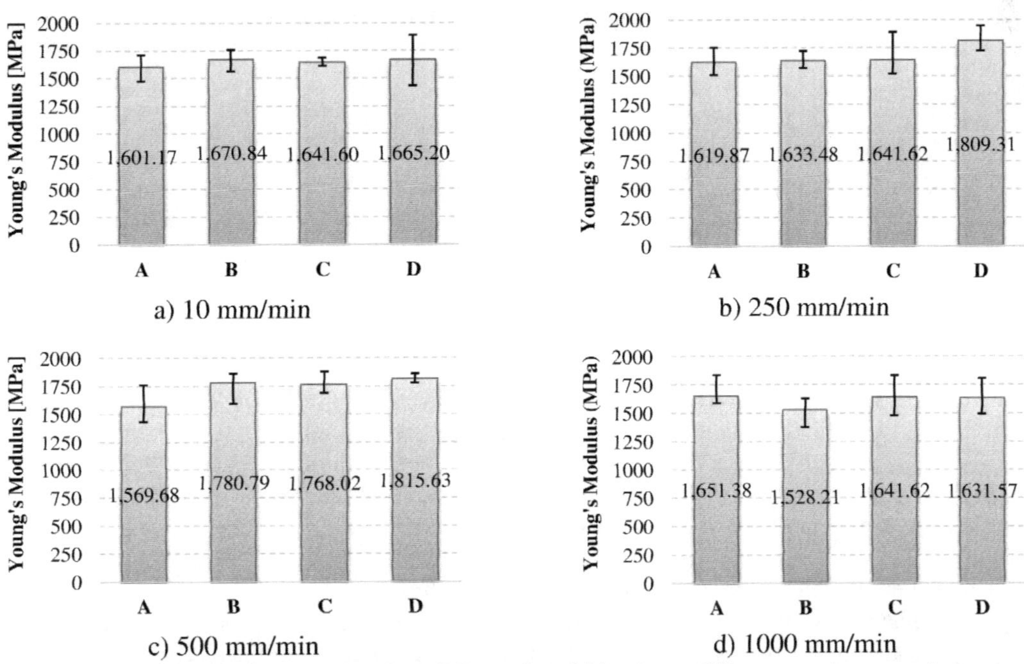

Figure 4. Young modulus for all formulated blends at different testing speeds

Taking into account the stress–strain curves, both neat polymers have exhibit a dual behavior at the lowest testing speed (v=10 mm/min). For PA6, some samples were broken at 30...40% strain but others were broken after 87...130%. PP has several samples broken at 10...15%, others at higher values (till 17...25%) with an intermediate plateau of stress between 10...20 MPa. At low concentration of PA6 (20...40%wt, materials A and B), the curves are very close and the strain at break has values in a narrow range (1.5% for 20% PA6 and less than 1% for 40% PA6). For higher contents of PA6 (materials C and D), the curves are similar, but spread on a larger space. At the testing speed of v=500 mm/min (Figure 2), PP have the curves almost overlapping and PA6 exhibit a distribution more distinct for each sample, with plateaux of stress at approximately 40 MPa.

The tensile properties of polymeric blends, obtained from averaging at least 5 valid tests, are summarized in Figures 4 to 7.

Young's modulus does not spectacularly depends on blend concentration in PP and PA6, but, unexpectedly, this characteristic is less sensitive for v=10 mm/min and v =1000 mm/min. For the intermediate test speeds (v=250 mm/min and v=500 mm/min) there is a slight tendency of increasing it with the PA6 content. Thus, the concentration of components has no evident influence on this mechanical characteristic. For the same testing speed, the greatest difference in this parameter was obtained when comparing material A to material D at v=500 mm/min (15.67%).

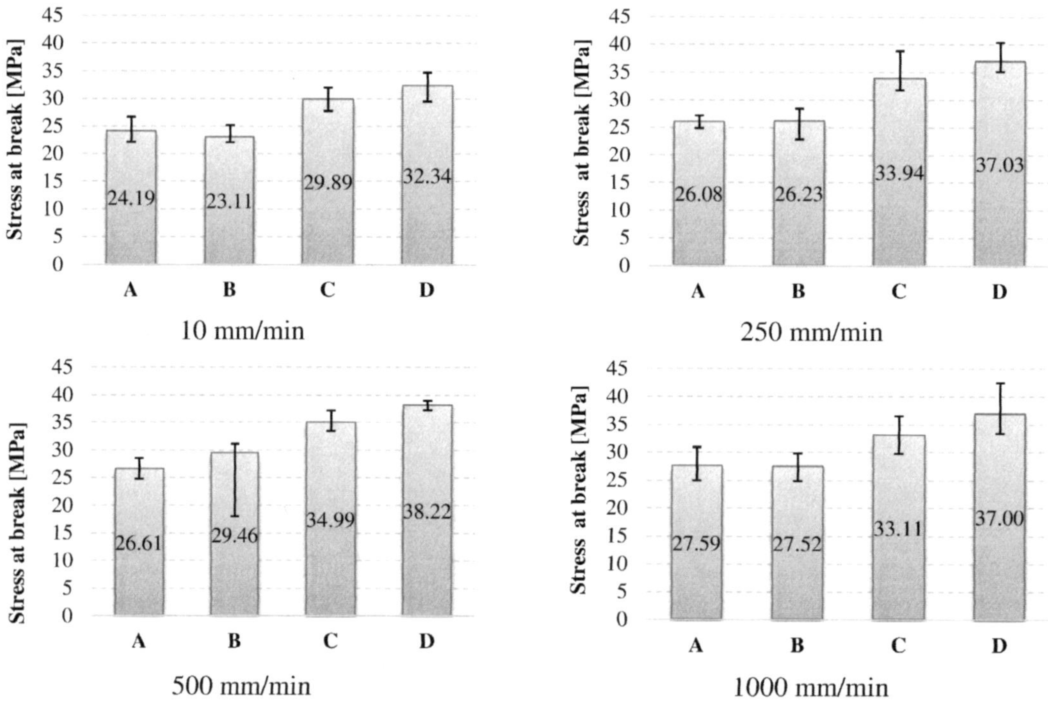

Figure 5. Stress at break for all formulated blends at different testing speed

Stress at break (Figure 5) grouped the results. Blends A and B (with low PP content) have lower values and blends with high content of PA6 (C and D) are higher. Comparing A to D, this characteristic is higher for the latest with 26% at v=10 mm/min, 41.9% at v=250 mm/min and v=500 mm/min, 34.1% for the highest testing speed (v=1000 mm/min). But the differences between C and D for this characteristic are smaller. D has the same value (37...38 MPa) for all testing speeds, except for the smallest one.

Bai et al [11] explained the behavior of their blends of PP/PA6/POE by: the increasing contribution of PA6 that intrinsically deforms with less cavitation than PP, the post-cavitation rubber-like stretching of POE particles and the early formation of a percolating network of shear bands from the diffuse array of voids formed after the yield point. These mechanisms explained the gradual increase of the resistance to impact of the PP/PA6/POE as their alloying content is increased.

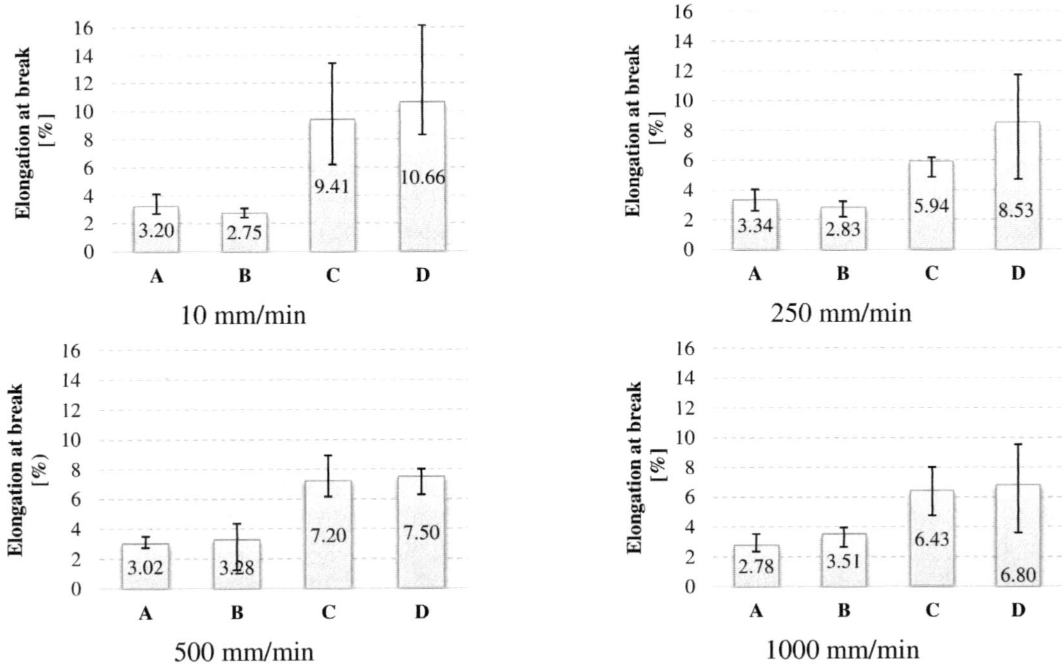

Figure 6. Elongation at break for all formulated blends at different deformation rate

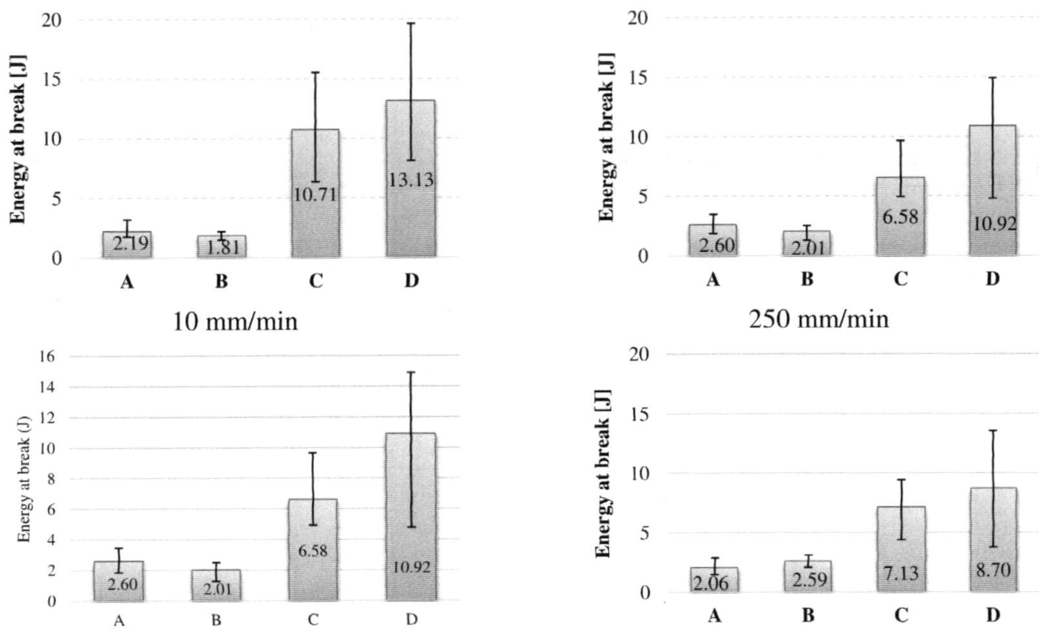

Figure 7. Energy at break for all formulated blends at different testing speeds

Elongation at break also grouped the blends in two classes (Figure 6):

- materials A and B have low values (2.75%...3.51%), slightly decreasing with the testing speeds,
- materials C and D with higher values.

IOP Conf. Series: Materials Science and Engineering **485** (2019) 012015 doi:10.1088/1757-899X/485/1/012015

Energy at break (Figure 7) also makes distinctive the same groups of materials: A and B with lower values (2.75 J... 3.51J) and the group of materials C and D with higher values and a more pronounced dependence on testing speed.

For PA6, at testing speeds of 250 mm/min to 1000 mm/min, the Young's modulus is insensitive to deformation rate (Figure 8). Young's modulus does not have a clear tendency of depending on PA6 concentration and testing speeds, but performs in a band between 1500...1900 MPa, except for the value obtained for PA6 at the lowest testing speed (v=10 mm/min), that is 1454.8 MPa. For PP and PA6 there were obtained the lowest values at testing speed v= 10 mm/min. For the blends, the dependence of Young's modulus on the testing speed and PP/PA6 concentrations is weak.

Under large testing speeds, the cracks are turned towards the loading direction, forming large parallel 'empty channels'. Investigations are needed on possible causes of these longitudinal voids in order to clear if they occur during molding or/and during testing and to eliminate them.

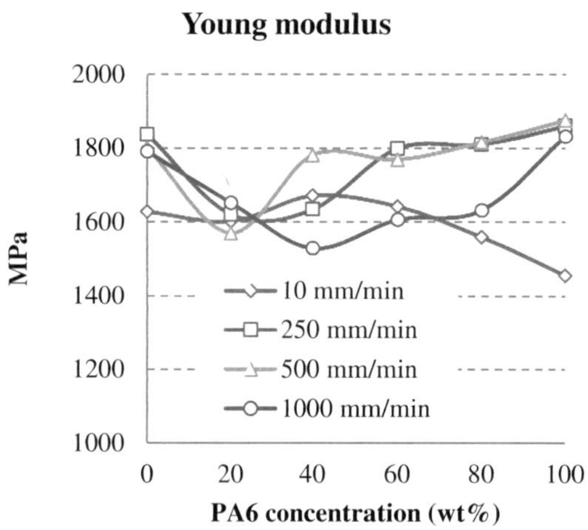

Figure 8. The Young's modulus as function of PA6 concentration and testing speed

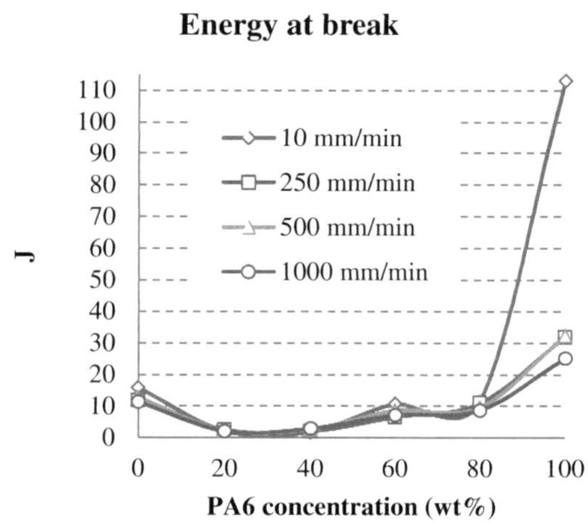

Figure 9. The energy at break as function of PA6 concentration and testing speed

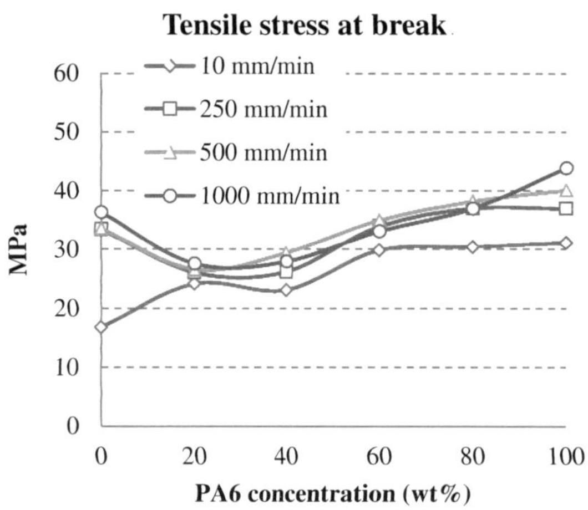

Figure 10. The tensile at break as function of PA6 concentration and testing speed

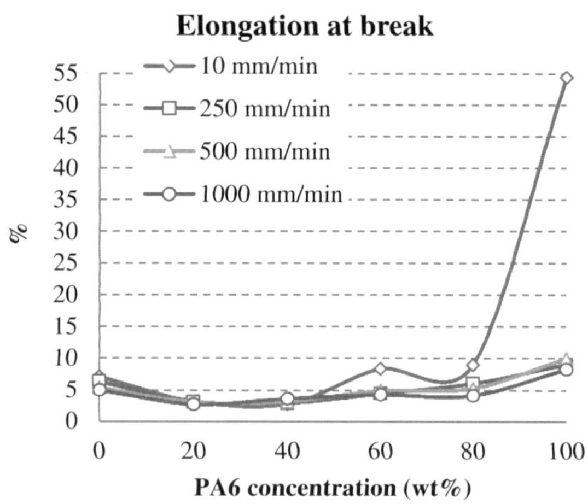

Figure 11. The elongation at break as function of PA6 concentration and testing speed

Energy at break is high only for PA6 at the lowest testing speed (v=10 mm/min). For PP, this characteristic has only a variation of 28.2% having as reference the value for v=10 mm/min. Each of designed materials has the values for the energy at break very concentrated, in spite of very different testing speeds. Analyzing Figure 9, one may notice a slightly increase of this characteristic with the concentration of PA6. The value of energy at beak for material D (80% PA6), which has the highest values of this characteristic among the formulated blends, is three or four times smaller that those obtained for PA6.

Except for values at the lowest testing speed (v=10 mm/min), the average values for tensile stress at break perform in a band of less than 8 MPa, these being hard to be ranked based on testing speed (Figure 10).

The elongation at break is less than 4% for materials with 20% and 40% PA6, between 6 and 10% for the materials with 60% and 80% PA6. The lowest values were obtained for A (with 20% PA6). For C and D, the difference between values, at the same testing speed, is the smallest.

4. Conclusion

From the mechanical point of view, among the formulated blends, material D (with 80% PA6) is the most promising as has the highest stress and energy at break (11.43 J at v=10 mm/min and 8.7 J at v=1000 mm/min), but values are still less than PA6.

Acknowledgments
The authors address thanks to Doctorale School for Mechanical and Industrial Engineering, "Dunarea de Jos" University of Galati and INCAS Bucharest.

References

[1] Utracki L A, 2002 Polymer Blends Handbook (Kluwer Academic Publishers, Dordrecht)
[2] Utracki L A 1995 History of commercial polymer alloys and blends (From a.perspective of the.patent literature) Polym Eng Sci **35** 1–17
[3] Sharma K R 2012 Polymer Thermodynamics. Blends, Copolymers and Reversible Polymerization (CRC Press Taylor & Francis Group)
[4] Robeson L M 2007 *Polymer Blends: A Comprehensive Review* (Hanser. Gardner, Munich)
[5] Cîrciumaru A 2013 *Caracterizarea şi testarea materialelor polimerice* (Europlus)
[6] Botan M, Georgescu C, Dobrea D, Deleanu L 2012 Mechanical properties for blends of PBT and PTFE, *7th Intern Symposium KOD 2012, Machine and Industrial Design in Mechanical Engineering*, Balatonfured, Hungary 531–536
[7] Botan M, Georgescu C, Pirvu C, Deleanu L 2014 Influence of aramid fibers on mechanical properties of two polymeric blends, *22nd International conference on materials and technology* Portoroz Slovenia, 53
[8] Sui G, Jing M, Zhao J, Wang K, Zhang Q, Fu Q, 2018 A comparison study of high shear force and compatibilizer on the phase morphologies and properties of polypropylene/polylactide (PP/PLA) blends Polymer 154 119-127
[9] Shi S, Huang P, Nie M, Wang Q 2017 Polypropylene/polyamide blend featuring mechanical interlocking via controlled interfacial diffusion and recrystallization Polymer, 132, 23-30
[10] Ruehle D. A, Perbix C, Castañeda M, Dorgan J R, Mittal V, Halley P, Martin D 2013 Blends of biorenewable polyamide-11 and polyamide-6,10 *Polymer* **54** 6961-6970
[11] Bai S, G'Sell C, Hiverc J-M, Mathieu C 2005 Polypropylene/polyamide 6/polyethylene-octene elastomer blends. Part 3. Mechanisms of volume dilatation during plastic deformation under uniaxial tension *Polymer* **46** 6437–6446
[12] Ghasemi H, Mirzadeh A, Bates P J, Kamal M R 2015 Characterization of recycled polyamide 6: Effect of polypropylene and inorganic contaminants on mechanical properties *Polymer Testing* **42** 69-78

IOP Conf. Series: Materials Science and Engineering **485** (2019) 012015 doi:10.1088/1757-899X/485/1/012015

[13] Yousfi M, Livi S, Dumas A, Le Roux C, Crépin-Leblond J, Greenhill-Hooper M, Duchet-Rumeau J 2013 Use of new synthetic talc as reinforcing nanofillers for polypropylene and polyamide 6 systems: Thermal and mechanical properties *Journal of Colloid and Interface Science* **403** 29–42

[14] Wang L, Shi C, Guo X-Z, Yu J, 2014 Tailoring the morphology of polypropylene/polyamide 66 blend at an asymmetric composition by using organoclay and maleic anhydride grafted polypropylene *Journal of Industrial and Engineering Chemistry* **20** 259–267

[15] Palacios J K, Sangroniz A, Eguiazabal J I, Etxeberria A, Müller A J, 2016 Tailoring the properties of PP/PA6 nanostructured blends by the addition of nanosilica and compatibilizer agents *European Polymer Journal* **85** 532–552

[16] Alexandrescu L, Sönmeza M, Georgescu M, Niţuică A, Ficai R, Trusca R, Gurău D,Tudoroi L, 2017 Polyamide/Polypropylene/graphene oxide nanocomposites with functional compatobilizers. Morpho-structural and physico-mechanical characterization *Procedia Structural Integrity* **5** 675-682

[17] Suter J L , Coveney P V, Greenwell H C, Thyveetil M-A 2007 Large-scale molecular dynamics study of montmorillonite clay: Emergence of undulatory fluctuations and determination of material properties *J. Phys. Chem. C*, **111**, 8248–8259

[18] Zhang B, Wong J S-P, Yam R C-M, Li R K-Y 2010 Enhanced wear performance of nylon 6/organoclay nanocomposite by blending with a thermotropic liquid crystalline polymer *Polymer Engineering and Science*, **50** 900–910

[19] Chow W S, Mohd Ishak Z A, Karger-Kocsis J et al 2003 Compatibilizing effect of maleated polypropylene on the mechanical properties and morphology of injection molded polyamide 6/ polypropylene/organoclay nanocomposites *Polymer* **44** 7427–7440

[20] Laoutid F, Estrada E, Michell R M, Bonnaud L, Müller A J, Dubois P 2013 The influence of nanosilica on the nucleation, crystallization and tensile properties of PPePC and PPePA blends Polymer 54 3982-3993

[21] Sangroniz L, Palacios J K, Fernández M, Eguiazabal J I, Santamaria A, Müller A J, 2016 Linear and non-linear rheological behavior of polypropylene/polyamide blends modified with a compatibilizer agent and nanosilica and its relationship with the morphology, *European Polymer Journal* **83** 10–21

IOP Conf. Series: Materials Science and Engineering **485** (2019) 012016 doi:10.1088/1757-899X/485/1/012016

Mathematical modelling and optimization of the thermomechanical treatment process applied to AlZn4.5Mg1Cu alloy

M I Neacsu[1*], G Coman[1], A M Adam[2]

[1] Faculty of Engineering, "Dunarea de Jos" University of Galati, Street No 47 Domneasca cod 800008 Galati, Romania
[2] Faculty of Medicine and Pharmacy, "Dunarea de Jos" University of Galati, Street No 47 Domneasca 800008 Galati, Romania

*uscaeni@yahoo.com

Abstract: The paper presents the results of the experimental researches as a result of applying the thermo-mechanical treatment for the AlZn4.5Mg1Cu alloy used in the aeronautical industry, in order to optimize the processing parameters. The research is based on a laborious experimental thermomechanical processing program applied to AlZn4.5Mg1Cu aluminium alloy to obtain certain imposed values of the mechanical properties with the lowest possible expense. The objective function in the optimization of the investigated thermomechanical treatment regime is the energy consumption "Q = f (t, τ, ε)" taking into account some restrictions regarding the values of the investigated mechanical properties. The best value for the objective function is obtained by determining values for the independent variables of the thermo mechanical treatment process, for which the conditions for obtaining values imposed on variables dependent on minimum energy consumption Q are met.

1. Introduction

The development of the aeronautical industry is very much related to the development of materials science and, in particular, of performance materials with special physical-mechanical properties. It is well known that the field of aircraft industry is a cutting edge field and therefore requires high performance materials. The AlZn4.5Mg1Cu alloy is part of the Al-Zn-Mg-Cu special alloy class. These alloys have the great advantage of having low density and high mechanical characteristics. This is why they are of particular interest to the aeronautical industry and the machinery industry [1-6].

Generally, these alloys have a deformability and a low mechanical strength which greatly changes with the application of thermal treatments [7, 8]. As a result of the aging process, natural or artificial, precipitations are formed in the structure of these alloys which lead to the hardening of the alloys. The mechanical properties of resistance become higher as the precipitations formed are finer, more dispersed and more numerous in the mass of the basic solution (solid solution) [1, 2, 4]. The complex interactions between base matrix dislocations and precipitated particles in the alloy structure as a result of aging lead to increased mechanical characteristics. Thermal processing takes place in solid state transformations which, if permitted in the alloy balance diagram, are an essential condition for carrying out a heat treatment, by quenching in solution and artificial or natural aging on an aluminium alloy [9, 10].

The increased resistance of Al-Zn-Mg-Cu alloys is directly proportional to the Zn or Zn + Mg content increase, thus generating Zn and Mg-rich fine metastable precipitated areas, which are the so-called GPs [1-4, 6]. The stages of the structural hardening mechanism following the application of thermal aging treatments are: formation of the supersaturated solid solution, by application of solution hardening, formation of Guinier - Preston zones during aging processes; first the precipitates form a metastable zone (η') , then as the temperature gradient is higher, these areas become stable, e.g. formation of precipitates $MgZn_2$ [6, 10-13].

2. Experimental details

The chemical composition of the alloy from which samples for experimental research were made is shown in Table 1.

Table 1. Chemical composition of the alloy under investigation [14]

Alloy \ Elements	Zn (%)	Mg (%)	Cu (%)	Si (%)	Fe (%)	Pb (%)	Mn (%)	Al (%)
AlZn5,7MgCu	4.5	1.4	0.2	0.35	0.4	0.35	0.5	rest

In order to use these materials in the aeronautical industry, especially in the machinery industry, after the thermomechanical processing, this alloy must acquire the mechanical properties prescribed in EN 485-2-2013 [14], which are shown in Table 2.

Table 2. Alloy properties [14]

Alloy \ Mechanical property	Rm (MPa)	$Rp_{0,2}$ (MPa)	A_5 (%)	HB
AlZn4,5Mg1Cu	350	250	10	104

This paper investigated the influence of thermomechanical treatment on the mechanical properties of the AlZn4,5Mg1Cu alloy.

Figure 1 presents schematically the thermomechanical processing to which the studied samples were subjected.

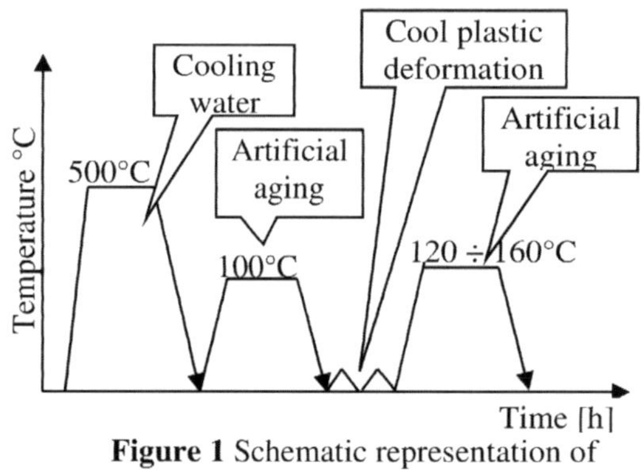

Figure 1 Schematic representation of thermomechanical processing

Thermomechanical processing consisted of the following:

- quenching in solution at 500 ° C for 2 hours;
- preliminary artificial aging at 100 ° C for 1 hour;
- cold plastic deformation with three degrees of deformation $\varepsilon_1 = 10\%$, $\varepsilon_2 = 20\%$ and $\varepsilon_3 = 30\%$ in order to ensure the dimensions set and to study the effect it has on the mechanical properties.
- final artificial aging at the following temperatures: $T_1 = 120$°C, $T_2 = 140$ ° C, $T_3 = 160$ °C with a holding time: $\tau_1 = 8$ hours, $\tau_2 = 12$ hours, $\tau_3 = 16$ hours, for each temperature.

Cold lamination has been done to increase the hardness of the material, as well as to create favourable conditions for the occurrence of precipitations around dislocations in the subsequent final artificial aging.

At the end of thermomechanical processing, the sample sizes are: length L = 200 mm, width l = 60 mm, thickness h = 5 mm.

3. Experimental results and discussion

After performing the thermomechanical treatment, the samples were subjected to mechanical tests where the values of the properties Rm, $Rp_{0,2}$, A5, HB which are presented in Tables 3, 4, 5 are set. The values in the tables are the average of 5 measurements.

Table 3 Values of the mechanical properties after thermomechanical treatment with a degree of plastic deformation ε = 10 /%

Artificial aging time (h)	The temperature of artificial aging °C											
	120				140				160			
	Rm (MPa)	Rp_{02} (MPa)	HB	A_5 (%)	Rm (MPa)	Rp_{02} (MPa)	HB	A_5 (%)	Rm (MPa)	Rp_{02} (MPa)	HB	A_5 (%)
8	302	216	103	11,8	294	218	98	12	264	205	95	12,4
12	324	245	113	11,5	312	229	112	10,5	280	225	98	11,3
16	361	271	118	11	341	267	117	10	294	241	103	10,5

Table 4 Values of the mechanical properties after thermomechanical treatment with a degree of plastic deformation ε = 20 %

Artificial aging time (h)	The temperature of artificial aging °C											
	120				140				160			
	Rm, (MPa)	Rp_{02}, (MPa)	HB	A_5 , (%)	Rm, (MPa)	Rp_{02}, (MPa)	HB	A_5, [%]	Rm, (MPa)	Rp_{02}, (MPa)	HB	A_5, (%)
8	366	329	124	9,6	333	291	120	10	326	274	109	10,2
12	381	336	137	9	362	298	133	9,7	356	285	125	10
16	392	353	149	8,6	377	316	143	9.2	365	303	132	9,5

Table 5 Values of the mechanical properties after thermomechanical treatment with a degree of plastic deformation ε = 30 %

Artificial aging time (h)	The temperature of artificial aging °C											
	120				140				160			
	Rm, (MPa)	Rp_{02}, (MPa)	HB	A_5 , (%)	Rm, (MPa)	Rp_{02}, (MPa)	HB	A_5, (%)	Rm, (MPa)	Rp_{02}, (MPa)	HB	A_5, (%)
8	385	330	143	7.8	354	282	131	8	329	261	122	8,3
12	398	350	154	7.5	383	325	148	7,4	362	282	131	8,4
16	423	365	160	7.0	415	348	152	7,1	384	315	135	8,7

The determination of the mechanical properties values revealed that values of the mechanical properties studied which correspond to the requirements of the specific norms in force [14] were not obtained for all combinations of the considered thermomechanical processing parameters (time, temperature, degree of plastic deformation).

The values of the studied mechanical properties are based on the values of the three parameters of the thermo-mechanical treatment regime and for this reason the mechanical properties were considered as functions of three variables: Rm = Rm (T, τ, ε), Rp0,2 = Rp0, 2 (T, τ, ε), A5 = A5 (T, τ, ε), HB = HB (T, τ, ε).

By using an interp3 specific function from the MATLAB program package, it was obtained the interpolation of the functions of three variables Rm = Rm (T, τ, ε), Rp0, 2 = Rp0, 2 (T, τ, ε), A5 = A5 (T, τ, ε), HB = HB (T, τ, ε).

Interpolation was performed in turn for all four mechanical characteristics studied according to the three parameters: T - the final artificial aging temperature, τ - the final artificial aging time and ε - the degree of cold plastic deformation preceding the final artificial aging [15].

Aluminium alloys are "sensitive" to small variations in heat treatment temperatures and for this reason we set a five to five degree variation for the final artificial aging temperature. Thus, nine interpolated values for the treatment temperature were obtained in the range of: 120 ° C and 160 ° C.

The artificial aging time from 8 hours to 16 hours was discretized with an hour's step, resulting in 9 interpolation values.

For the degree of plastic deformation in the range of 10% - 30%, a number of 21 incremental values having a 1% step were interpolated.

Interpolation resulted in a volume of 1701 interpolated values for each property.

The optimization of the thermo-mechanical process meant in these conditions, in the first stage, finding out of these 1701 data corresponding to each property, only those data that simultaneously fulfil the conditions imposed by EN_485-2-2013, and in the second stage the calculation of the energy consumption required for thermomechanical processing for all the situations identified in the first stage and choosing the variant for which Q energy is minimal [15, 16].

Calculation of energy consumption in the form of heat (thermal energy) was made by calculating the total Q_{total} energy consumed by the heat treatment furnace and the rolling mill used for plastic deformation. These calculations were made using the relationship below:

$Q_{total} = Q_{total\ furnace} + Q_{lam}$, [17], (1)

where: - $Q_{total\ furnace}$ - the amount of heat required to reach and maintain the treatment temperature throughout the heat treatment;

- Q_{lam} - the amount of energy consumed for sample rolling.

$Q_{total\ furnace} = Q_A + Q_B$, [17], (2)

Q_A - the amount of heat (energy) consumed during the furnace heating period;

Q_B - the amount of heat (energy) consumed during the maintenance period at the heat treatment temperature.

$Q_{lam} = U \bullet t_{lam} \bullet I$ kWh [17], (3)

Q_{lam} - energy consumed for cold plastic deformation;

U - the power supply voltage of the rolling mill motor;

U = 380 V;

tlam - the rolling time of a sample, hours;

I - the intensity of the electric current used for rolling, A.

The program developed in MATLAB with the help of the interp3 function and the mathematical equations of thermal balance (1), (2), (3) shows that for any value imposed from the total of 1701to any of the four properties, a number of variants possible was obtained for which the optimum in terms of total energy consumption can be determined.

With the help of the program developed in MATLAB, a graphical interface was created, a graphical interface allowing visualization (simulation) of those possible situations and choosing a large number of values for any of the four properties.

Figure 2 shows an example of such a simulation.

IOP Conf. Series: Materials Science and Engineering **485** (2019) 012016 doi:10.1088/1757-899X/485/1/012016

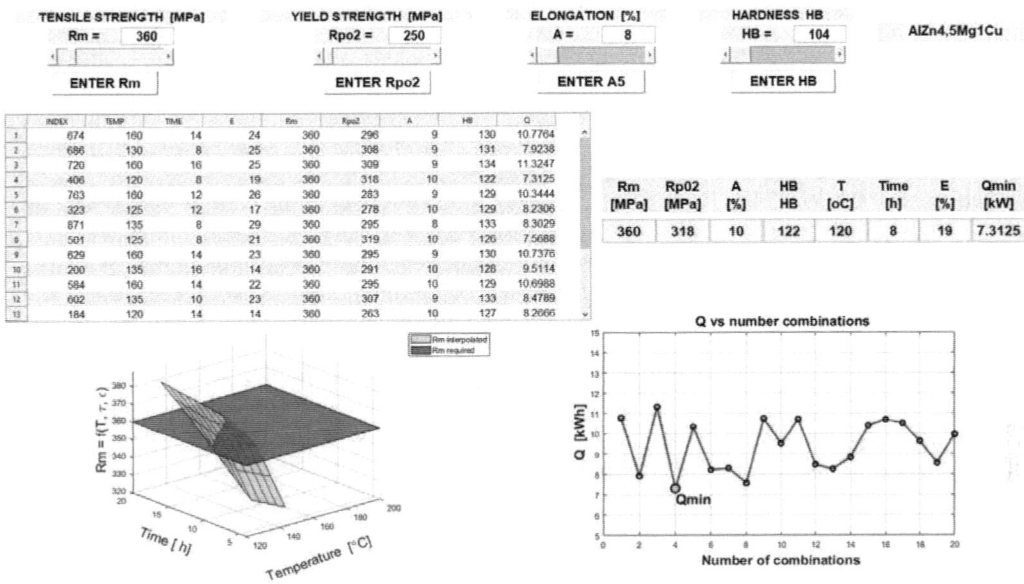

Figure 2 Graphical interface for simulating possible situations from the point of view of the imposed mechanical characteristics and energy calculation Q

4. Summary and conclusions

The conclusions drawn from the research described in this paper were as follows:

- the existence of a dependence between the values of the mechanical strength properties and the degree of cold plastic deformation before the final artificial aging;

- the resistance properties values are inversely proportional to the thermal treatment temperature and directly proportional to the values of the artificial aging time;

- the mechanical strength of the alloy processed records the highest values for the artificial aging time of 16 hours and the temperature of 120ºC and the lowest resistance values are recorded for the 8 hour aging and the temperature of 160ºC;

- With increasing the final artificial aging time and lowering the aging temperature as well as increasing the degree of plastic deformation, the alloy's elongation value is decreasing;

- the MATLAB graphical interface allows simulation and identification based on mathematical computation of Qtotal's minimum value, which is the optimum for thermomechanical processing;

- with this graphical interface, the values of the thermomechanical processing parameters can be highlighted in tabular form for those situations in which a certain value of one or more of the studied properties is to be obtained.

References

[1] Z.W. Du, Z.M. Sun , B.L. Shao , T.T. Zhou b, C.Q. Chen, Quantitative evaluation of precipitates in an Al–Zn–Mg–Cu alloy after isothermal aging, Materials Characterization 56 (2006) 121–128;

IOP Conf. Series: Materials Science and Engineering **485** (2019) 012016 doi:10.1088/1757-899X/485/1/012016

[2] J.C. Werenskiold, A. Deschamps, Characterization and modeling of precipitation kinetics in an Al–Zn–Mg alloy, Materials Science and Engineering: A, Structural Materials: Properties, Microstructure and Processing, 2000; 293: 267–74;

[3] A. Deschamps, G. Fribourg, Y. Brechet, J.L. Chemin, C.R. Hutchinson, In situ evaluation of dynamic precipitation during plastic straining of an Al–Zn–Mg–Cu alloy, Acta Materialia Volume 60, Issue 5, (March 2012), 1905–1916;

[4] C.R. Hutchinson, F. de Geuser, Y. Chen, A. Deschamps, Quantitative measurements of dynamic precipitation during fatigue of an Al–Zn–Mg–(Cu) alloy using small-angle X-ray scattering, *Acta Materialia*, Volume 74, (2014), Pages 96–109;

[5] M.Rosso, I. Peter, C.Castella, et al., Optimization of composition for self-hardening AlZn10Si8Mg alloys, International Conference on Extrusion and Benchmark (ICEB), *Materials today-proceedings,* Volume: 2 Issue: 10 Pages: 4949-4956 Part: A, 2015;

[6] Kh. A. Ragab, M. Dellah, M.Bournane, et al., Influence of thermal treatment media on characteristics of Al-Zn-Mg alloys, International journal of cast metals research, Volume: 28 Issue: 5 Pages: 257-262, 2015;

[7] B. M. Afshari, S. S. Mirjavadi, Y.A. Dolatabad, et al., Effects of pre-deformation on microstructure and tensile properties of Al-Zn-Mg-Cu alloy produced by modified strain induced melt activation, *Transactions of nonferrous metals society of china* Volume: 26 Issue: 9 Pages: 2283-2295 Published: SEP 2016;

[8] J. A. Oesterreicher, G.Kirov, S. A Gerstl, et al., Stabilization of 7xxx aluminium alloys, Journal of alloys and compounds, Volume: 740, Pages: 167-173, 2018;

[9] D.K. Pratiwi, N.P.E. Utami, Effect of ageing time 200 degrees C on microstructure behaviour of AlZn-Cu-Mg cast alloys, Sriwijaya international conference on engineering, science and technology (SICEST 2016), 2016;

[10] W.Yang, L. Liu, J. Zhang, et al, Insight into the partial solutionisation of a high pressure die-cast Al-Mg-Zn-Si alloy for mechanical property enhancement , Materials science and engineering a-structural materials properties microstructure and processing, Volume: 682, Pages: 85-89, Published: 2017;

[11] Kh. A. Ragab, M. Dellah, M.Bournane, et al., Influence of thermal treatment media on characteristics of Al-Zn-Mg alloys, *International journal of cast metals research*, Volume: 28, Issue: 5, Pages: 257-262, 2015;

[12] J.Zhang, L.Yang, R. Zuo, Effects of solution and aging treatments on the microstructures and mechanical properties of 7055 al alloys, Rare metal materials and engineering, Volume: 44 Issue: 4 Pages: 956-960, 2015;

[13] J. S. Robinson, D. A. Tanner, S.Van Petegem, et al., Influence of quenching and aging on residual stress in Al-Zn-Mg-Cu alloy 7449, *Materials science and technology*, Volume: 28 Issue: 4 Pages: 420-430, 2012;

[14] xxx EN 485-2-2013;

[15] U. A Curle, L. A. Cornish, G. Govender, Predicting yield strengths of Al-Zn-Mg-Cu-(Zr) aluminium alloys based on alloy composition or hardness, *Materials & design* Volume: 99 Pages: 211-218, 2016;

[16] K. Ural, A study of optimization of heat-treatment conditions in retrogression and reageing treatment of 7075-t6 aluminum-alloy, *Journal of materials science letters*, Volume: 13 Issue: 5. Pages: 383-385, 1994;

[17] B. Popa, E. Man, M. Popa – Termotehnică, agregate şi instalaţii termice, Editura tehnică, Bucureşti, 1979;

IOP Conf. Series: Materials Science and Engineering **485** (2019) 012017 doi:10.1088/1757-899X/485/1/012017

Thermodynamic properties of BCC interstitial alloy FeH under pressure

T H Nguyen [1], Q H Nguyen[2], G Coman[3], V Dinh Quang[2] and C Tran Dinh[2]

[1]University of Transport and Communications, 3 Cau Giay, Dong Da, Hanoi, Vietnam
[2]Hanoi National University of Education, 136 Xuan Thuy, Cau Giay, Hanoi, Vietnam
[3]Dunarea de Jos University of Galati, 47 Domneasca, Romania

nthoaly@utc.edu.vn

Abstract. The analytic expressions of the mean nearest neighbor distance, the free energy and the thermodynamic quantities for interstitial alloy AB with body-centered cubic (BCC) structure under pressure are derived by the statistical moment method. The theoretical results are applied to interstitial alloy FeH. Our calculated results for Fe as main metal are in good agreement with experiments.

1. Introduction

Thermodynamic and elastic properties of metals and interstitial alloys are very interesting to many theoretical and experimental researchers [1,2]. There are some recent investigations on the density [3], the compressibility [4] and the sound velocity [5] of solid FeH at high pressures.

In this paper, we built a thermodynamic model for interstitial alloy AB with bcc structure, under pressure, by using the statistical moment method (SMM) [6-8] and applied these results to the interstitial alloy FeH.

2. Method

In the model of interstitial alloy AB with the body-centered cubic (BCC) structure, the atoms A with large size stay in the corners and center of the unit cell, the interstitial atoms B with smaller size stay in the center of the faces of unit cell. In [6-8], we derived the analytic expressions of the nearest neighbor distance r_1, the cohesive energy u_0 and the alloy parameters $k, \gamma_1, \gamma_2, \gamma$ for atoms B, A, A_1 (the atom A in the center) and A_2 (the atom A in the corners). The equation of state for interstitial alloy AB with BCC structure at temperature T and pressure P is written in the form

$$Pv = -r_1\left(\frac{1}{6}\frac{\partial u_0}{\partial r_1} + \theta x \operatorname{cth} x \frac{1}{2k}\frac{\partial k}{\partial r_1}\right). \quad (1)$$

At 0K, this equation has the form

IOP Conf. Series: Materials Science and Engineering **485** (2019) 012017 doi:10.1088/1757-899X/485/1/012017

$$Pv = -r_1\left(\frac{1}{6}\frac{\partial u_0}{\partial r_1} + \frac{\hbar\omega_0}{4k}\frac{\partial k}{\partial r_1}\right). \tag{2}$$

If knowing the form of potential interaction φ_{i0}, eq. (2) permits us to determine the nearest neighbor distance $r_{1X}(P,0)$ $(X = B, A, A_1, A_2)$ at 0K and pressure *P*. After knowing $r_{1X}(P,0)$, we can determine the alloy parameters $k_X(P,0), \gamma_{1X}(P,0), \gamma_{2X}(P,0), \gamma_X(P,0)$ at 0K and pressure *P*. After that, we can calculate the displacements $y_{0X}(P,T)$ as in [6-8]. From that, we derive the nearest neighbor distance $r_{1X}(P,T)$ at temperature *T* and pressure *P*

$$r_{1B}(P,T) = r_{1B}(P,0) + y_{0A_1}(P,T), r_{1A}(P,T) = r_{1A}(P,0) + y_{0A}(P,T),$$

$$r_{1A_1}(P,T) \approx r_{1B}(P,T), r_{1A_2}(P,T) = r_{1A_2}(P,0) + y_{0B}(P,T). \tag{3}$$

Then, we calculate the mean nearest neighbor distance in interstitial alloy AB by the expressions as follows

$$\overline{r_{1A}(P,T)} = \overline{r_{1A}(P,0)} + \overline{y_0(P,T)},$$

$$\overline{r_{1A}(P,0)} = (1-c_B)r_{1A}(P,0) + c_B r'_{1A}(P,0), r'_{1A}(P,0) = \sqrt{3}r_{1B}(P,0),$$

$$\overline{y_0(P,T)} = (1-7c_B)y_{0A}(P,T) + c_B y_{0B}(P,T) + 2c_B y_{0A_1}(P,T) + 4c_B y_{0A_2}(P,T), \tag{4}$$

where $\overline{r_{1A}(P,T)} \equiv a_{AB}(P,T)$ is the mean nearest neighbor distance between atoms A in interstitial alloy AB at pressure *P* and temperature *T*; $\overline{r_{1A}(P,0)} \equiv a_{0AB}(P,0)$ is the mean nearest neighbor distance between atoms A in interstitial alloy AB at pressure *P* and 0K; $r_{1A}(P,0)$ is the nearest neighbor distance between atoms A in clean metal A at pressure *P* and 0K; $r'_{1A}(P,0)$ is the nearest neighbor distance between atoms A in the zone containing the interstitial atom B at pressure *P* and 0K and c_B is the concentration of interstitial atoms B.

The free energy of alloy AB with BCC structure and the condition $c_B << c_A$ has the form

$$\psi_{AB} = (1-7c_B)\psi_A + c_B\psi_B + 2c_B\psi_{A_1} + 4c_B\psi_{A_2} - TS_c,$$

$$\psi_X \approx \frac{N}{2}u_{0X} + \psi_{0X} + 3N\left\{\frac{\theta^2}{k_X^2}\left[\gamma_{2X}X_X^2 - \frac{2\gamma_{1X}}{3}\left(1+\frac{X_X}{2}\right)\right] + \right.$$

$$\left. + \frac{2\theta^3}{k_X^4}\left[\frac{4}{3}\gamma_{2X}^2 X_X\left(1+\frac{X_X}{2}\right) - 2\left(\gamma_{1X}^2 + 2\gamma_{1X}\gamma_{2X}\right)\left(1+\frac{X_X}{2}\right)(1+X_X)\right]\right\},$$

$$\psi_{0X} = 3N\theta\left[x_X + \ln(1-e^{-2x_X})\right], X_X \equiv x_X \coth x_X, \tag{5}$$

where ψ_X is the free energy of atom *X*, ψ_{AB} is the free energy of interstitial alloy AB, S_c is the configuration entropy of interstitial alloy AB.

The isothermal compressibility of interstitial alloy AB has the form

$$\chi_{TAB} = \frac{3\left(\frac{a_{AB}}{a_{0AB}}\right)^3}{2P + \frac{\sqrt{3}}{4a_{AB}}\frac{1}{N}\left(\frac{\partial^2 \psi_{AB}}{\partial a_{AB}^2}\right)_T}. \quad (6)$$

The thermal expansion coefficient of interstitial alloy AB has the form

$$\alpha_{TAB} = -\frac{k_B \chi_{TAB}}{3}\left(\frac{a_{0AB}}{a_{AB}}\right)^2 \frac{a_{AB}}{3V_{AB}} \frac{\partial^2 \psi_{AB}}{\partial \theta \partial a_{AB}}. \quad (7)$$

The heat capacity at constant volume of interstitial alloy AB is determined by

$$C_{VAB} = (1 - 7c_B)C_{VA} + c_B C_{VB} + 2c_B C_{VA_1} + 4c_B C_{VA_2},$$

$$C_{VX} = 3Nk_B \left\{ Y_X^2 + \frac{2\theta}{k_X^2}\left[\left(2\gamma_2 + \frac{\gamma_{1X}}{3}\right) X_X Y_X^2 + \frac{2\gamma_{1X}}{3}\left(1 + Y_X^2\right) - \gamma_{2X}\left(Y_X^4 + 2X_X^2 Y_X^2\right)\right]\right\}, Y_X \equiv \frac{x_X}{\sinh x_X}. \quad (8)$$

The heat capacity at constant pressure of interstitial alloy AB is determined by

$$C_{PAB} = C_{VAB} + \frac{9TV_{AB}\alpha_{TAB}^2}{\chi_{TAB}}. \quad (9)$$

3. Numerical results

For the interaction between two Fe atoms, we use the n - m potential as follows

$$\varphi(r) = \frac{D}{n-m}\left[m\left(\frac{r_0}{r}\right)^n - n\left(\frac{r_0}{r}\right)^m\right], \quad (10)$$

where the potential parameters are shown in Table 1.

Table 1. The m - n potential parameters [9].

Interaction	m	n	$D(\text{eV})$	$r_0(\mathring{A})$
Fe – Fe	7.0	11.5	0.4	2.4775

For the interaction between the Fe atom and the H atom, we use the Morse potential as follows

$$\varphi(r) = D\left[e^{-2\alpha(r-r_0)} - 2e^{-\alpha(r-r_0)}\right], \quad (11)$$

where the potential parameters are shown in Table 2.

Table 2. The Morse potential parameters [10].

Interaction	$D(\text{eV})$	$\alpha(\mathring{A})^{-1}$	$r_0(\mathring{A})$
Fe – H	0.32	1.34	1.73

Our numerical results for FeH alloy are described in Figure 1 to Figure 12 .When the concentration $c_H \to 0$, we obtain the thermodynamic quantities of Fe. Our calculated results in Figure 3, Figure 4, Figure 8 and Figure 12 are in rather good agreement with experiments and other calculations.

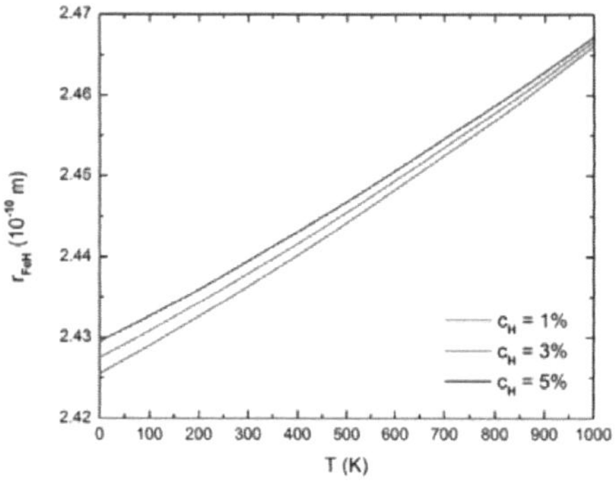

Figure 1. $r_{FeH}(T)$ of alloy FeH at $P = 0$ and c_H = 1, 3, 5% calculated by the SMM.

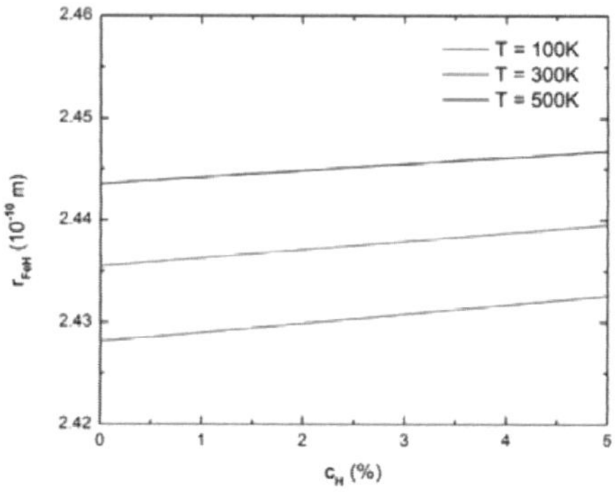

Figure 2. $r_{FeH}(c_H)$ of alloy FeH at $P = 0$ and T = 100, 300, 500 K calculated by the SMM.

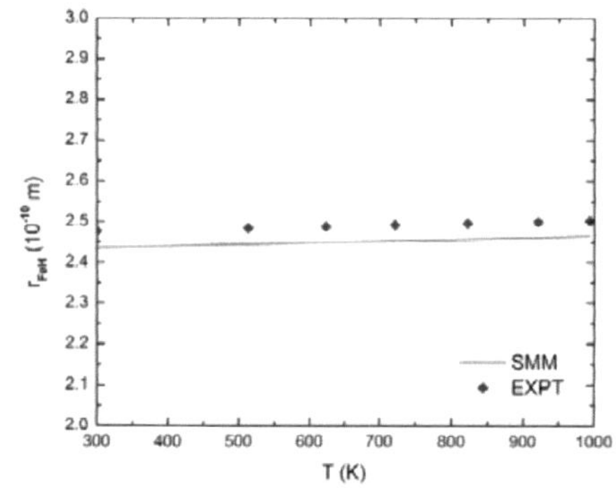

Figure 3. $r_{Fe}(T)$ of Fe at $P = 0$ calculated by the SMM and from EXPT [11].

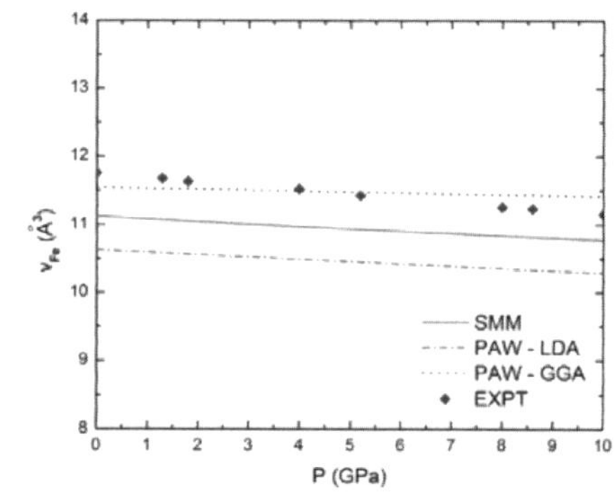

Figure 4. The volume of the unit cell $v_{Fe}(P)$ of Fe at 300K obtained from the SMM, PAW-LDA [12], PAW-GGA [12], EXPT [12].

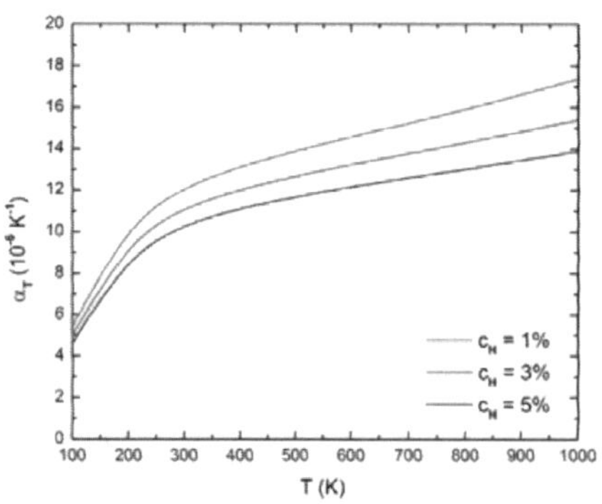

Figure 5. $\alpha_T(T)$ of alloy FeH at $P = 0$ and c_H = 1, 3, 5% calculated by the SMM.

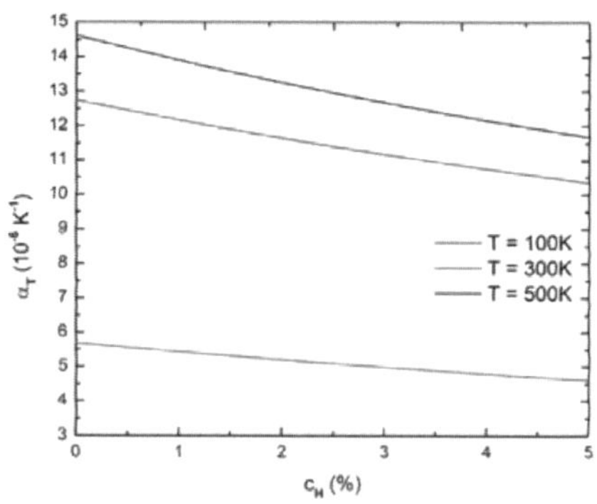

Figure 6. $\alpha_T(c_H)$ of alloy FeH at $P = 0$ and T = 100, 300, 500 K calculated by the SMM.

IOP Conf. Series: Materials Science and Engineering **485** (2019) 012017 doi:10.1088/1757-899X/485/1/012017

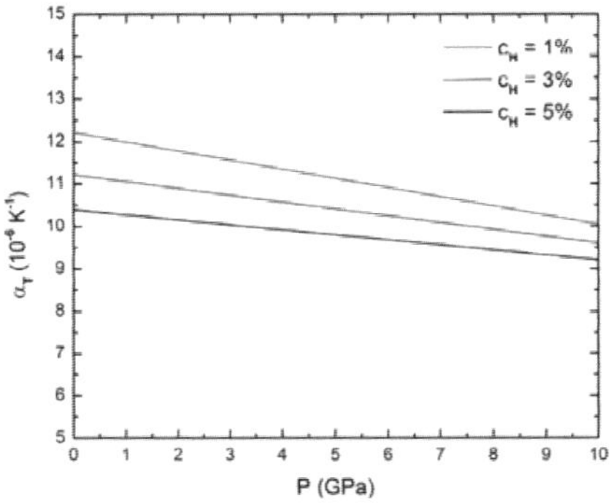

Figure 7. α_T (P) of alloy FeH at $T = 300$ K and $c_H = 1, 3, 5\%$ calculated by the SMM.

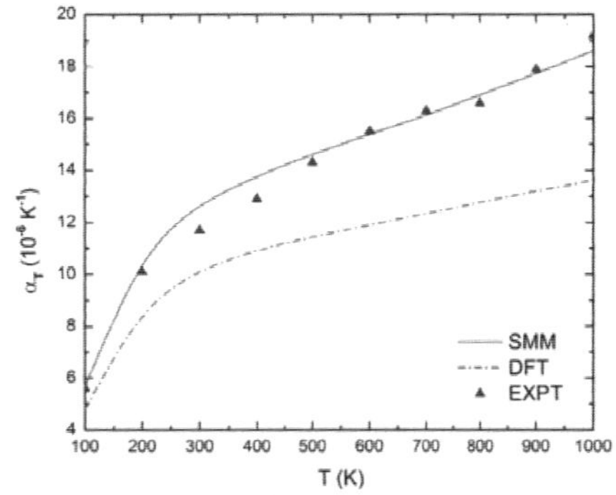

Figure 8. α_T (T) of metal Fe at $P = 0$ obtained from SMM, DFT [13] and EXPT [14].

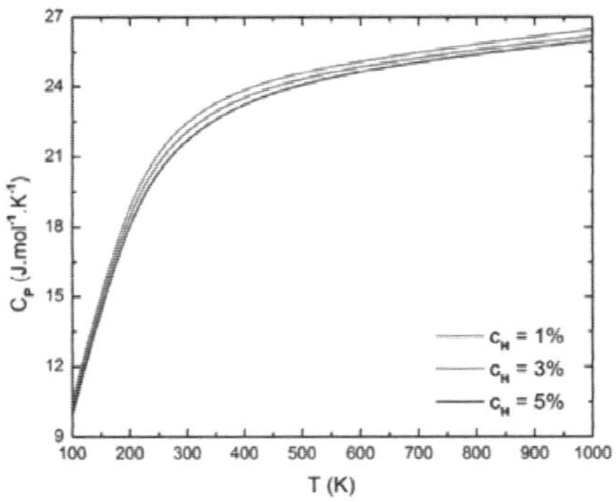

Figure 9. C_P (T) of alloy FeH at $P = 0$ and $c_H = 1, 3, 5\%$ calculated by the SMM.

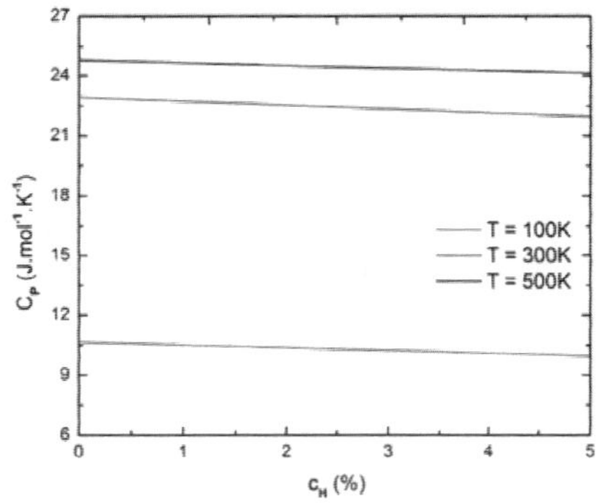

Figure 10. C_P (c_H) of alloy FeH at $P = 0$ and $T = 100, 300, 500$K calculated by the SMM.

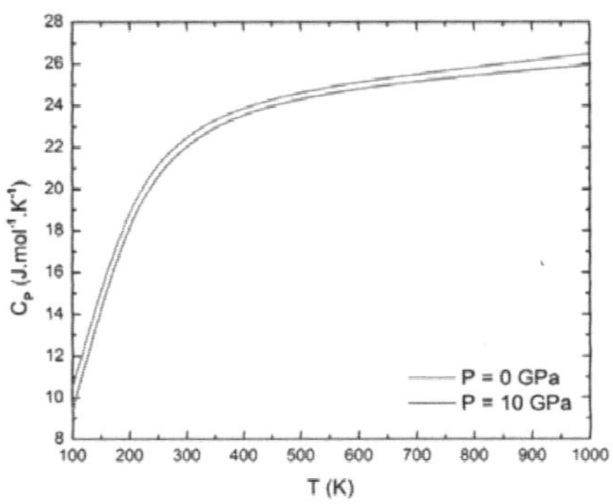

Figure 11. C_P (T) of alloy FeH at $c_H = 1\%$, $P = 0$ and 10 GPa calculated by the SMM.

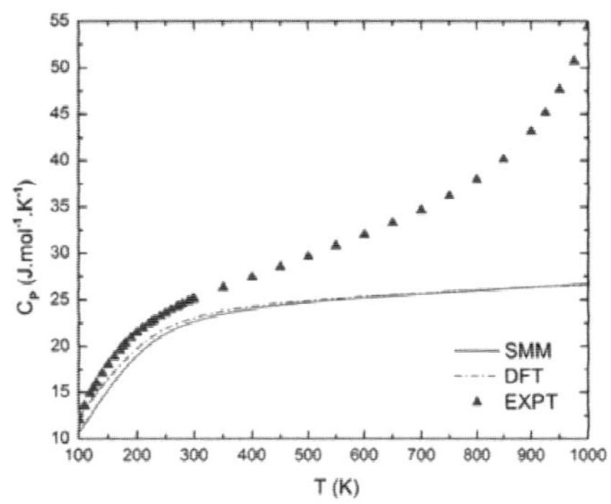

Figure 12. C_P (T) of metal Fe at $P = 0$ obtained from the SMM, DFT [13] and EXPT [15].

4. Conclusion

By using SMM, with the minimum condition of cohesive energies and the method of three coordination spheres, we found the mean nearest neighbor distance, the free energy, the isothermal compressibility, the isothermal and adiabatic elastic modulus, the thermal expansion coefficient, the heat capacities at constant volume and at constant pressure, the entropy of binary interstitial alloy with BCC structure with the concentration of interstitial atoms which is very small. The obtained expressions of these quantities depend on the temperature and concentration of interstitial atoms. At zero concentration of interstitial atoms, the thermodynamic quantities of the alloy become ones of main metal in alloy. The theoretical results are applied to interstitial alloy FeH. At zero concentration of H interstitial atoms, our calculated results for the nearest neighbor distance, the thermal expansion coefficient and the heat capacity at constant pressure are in rather good agreement with the experimental results.

References

[1] Sha X and Cohen R E 2006 *Phys. Rev. B* **73** 104303.
[2] Adams J J, Agosta D S, Leisure R G and Ledbetter H 2006 *J. Appl. Phys.* **100** 113530.
[3] Terasaki H, Ohtani E, Sakai T, Kamada S, Asanuma H, Shibazaki Y and Suzuki A 2012 *Phys. Earth Planet. Inter.* **194** 18.
[4] Pépin C M, Dewaele A, Geneste G, Loubeyre P and Mezouar M 2014 *Phys. Rev. Lett.* **113** 265504.
[5] Shibazaki Y, Ohtani E, Fukui H, Sakai T, Kamada S, Ishikawa D and Takemura K 2012 *Earth Planet. Sci. Lett.* **313** 79.
[6] Tang N and Hung V V 1988 *Phys. Stat. Sol. (b)* **149** 511.
[7] Hung V V and Masuda-Jindo K 2000 *J. Phys. Soc. Jpn.* **69** 2067.
[8] Tinh B D, Hoc N Q, Vinh D Q, Cuong T D and Hien N D 2018 *Adv. Mater. Sci. Eng.* **2018**.
[9] Magomedov M N 1987 *J. Fiz. Khimic* **61** 1003.
[10] Olander D R 1971 *J. Phys. Chem. Solids* **32** 2499.
[11] Basinski Z S, Hume-Rothery W and Sutton A L 1955 *Proc. R. Soc. Lond. A* **229** 459.
[12] Dewaele A, Torrent M, Loubeyre P and Mezouar M 2008 *Phys. Rev. B* **78** 104102.
[13] Dragoni D, Ceresoli D and Marzari N 2015 *Phys. Rev. B* **91** 104105.
[14] Gray D E 1972 *American Institute of Physics handbook* (New York, Toronto, London: McGraw-Hill Companies).
[15] Desai P D 1986 *J. Phys. Chem. Ref. Data* **15** 967.

IOP Conf. Series: Materials Science and Engineering **485** (2019) 012018 doi:10.1088/1757-899X/485/1/012018

Thermodynamic properties of FCC interstitial alloy AuSi with defects

T H Nguyen[1], Q H Nguyen[2], G Coman[3], C Tran Dinh[2] and V Le Hong[4]

[1]University of Transport and Communications, 3 Cau Giay, Dong Da, Hanoi, Vietnam
[2]Hanoi National University of Education, 136 Xuan Thuy, Cau Giay, Hanoi, Vietnam
[3]Dunarea de Jos University of Galati, 47 Domneasca, Romania
[4]Tran Quoc Tuan University, Co Dong, Son Tay, Hanoi, Vietnam

nthoaly@utc.edu.vn

Abstract. The analytic expressions for free energy, concentration of the equilibrium vacancies and thermodynamic quantities such as the isothermal compressibility, thermal expansion coefficient and heat capacities at constant volume and at constant pressure of FCC interstitial alloy AB with defects under pressure are derived by the statistical moment method. The theoretical results are applied to AuSi interstitial alloy. Our calculated results for main metal Au are compared with the experimental data.

1. Introduction

The point defects in crystals, including vacancies, is widely investigated and plays an important role in many properties of a material. Thus an investigation of the point defects in solid is of special interest [1, 2].

The melting curve of Au was studied by experiments [3, 4] and calculations [5, 6]. Thus, when we study the point defects in crystals, Au is one of the most important research topics [7, 8].

There are various methods to investigate the defective crystals, among which the statistical moment method (SMM) is a modern one. In this paper, using SMM [9-11] we derive the analytic expressions of the concentration of equilibrium vacancies, the free energy and the thermodynamic quantities for FCC interstitial alloy AB with defects, under pressure. We then applied these results to the AuSi interstitial alloy.

2. Method

Considering the interstitial alloy AB with FCC structure, where the main atoms A are in the corners and face centers, the interstitial atoms B are in body centers. The closest atoms to the atom B are called A_1. The second closest atoms to the atom B are called A_2. When the interstitial alloy AB has N atoms and n vacancies in lattice points, from the minimum condition of real Gibbs thermodynamic potential $\left(\frac{\partial G_{AB}^R}{\partial n_v}\right)_{P,T,N} = 0$, we can find the equilibrium vacancy concentration n_v as follows

$$n_v = \frac{n}{N} = \exp\left(-\frac{g_{vAB}^f}{k_{Bo}T}\right) = \exp\left(-\frac{g_{vAB}^f}{\theta}\right), \quad (1)$$

where k_{Bo} is the Boltzmann constant, g^f_{vAB} is the change of Gibbs thermodynamic potential when a vacancy is formed and is determined by the distribution of atomic concentrations c_X

$$g^f_{vAB} = \sum_X c_X g^f_{vX} \ (X = A, B, A_1, A_2),\ c_B = \frac{N_B}{N},\ c_{A_1} = 6c_B,\ c_{A_2} = 8c_B,\ c_A = 1 - 15c_B. \quad (2)$$

When P = 0, we have

$$n_v(0,T) = \exp\left(-\frac{\sum_X c_X g^f_{vX}(0,T)}{\theta} \right),$$

$$g^f_{vX}(0,T) = \frac{n_1(\psi^0_X - \psi^1_X) + \Delta\psi^0_X}{N} = \frac{n_1(\psi^0_X - \psi^1_X) + (B_X - 1)\psi^0_X}{N},\ B_X \approx 1 + \frac{Nu^0_{0X}}{4\psi^0_X}, \quad (3)$$

where ψ^1_X / N is the free energy of an atom X on the first coordination sphere with the vacancy as the center of sphere, ψ^0_X / N is the free energy of an atom X in perfect crystal, n_1 is the number of atoms on the first coordination sphere (for FCC lattice, n_1 = 12) and $\Delta\psi^0_X / N$ is the change of free energy when the atom X releases from the lattice point to form a vacancy. If we set $x^i_X = \frac{\hbar}{2\theta}\sqrt{\frac{k^i_X}{m_X}}$, $X^i_X = x^i_X \coth x^i_X$, $Y^i_X = x^i_X / \sinh x^i_X$ $(i = 0, 1)$, the free energy ψ^i_X has the form

$$\psi^i_X \approx \frac{N}{2}u^i_{0X} + 3N\theta\left[x^i_X + \ln(1 - e^{-2x^i_X}) \right] + 3N\left\{ \frac{\theta^2}{k^{i\,2}_X}\left[\gamma^i_{2X} X^{i\,2}_X - \frac{2\gamma^i_{1X}}{3}\left(1 + \frac{X^i_X}{2}\right)\right] + \right.$$

$$\left. + \frac{2\theta^3}{k^{i\,4}_X}\left[\frac{4}{3}\gamma^{i\,2}_{2X} X^i_X \left(1 + \frac{X^i_X}{2}\right) - 2\left(\gamma^{i\,2}_{1X} + 2\gamma^i_{1X}\gamma^i_{2X}\right)\left(1 + \frac{X^i_X}{2}\right)\left(1 + X^i_X\right)\right]\right\}, \quad (4)$$

where m_X is the mass of atom X, the cohesive energy u^i_{0X} and the crystal parameters $k^i_X, \gamma^i_{1X}, \gamma^i_{2X}$ for atom X in perfect alloy (i = 0) and defective alloy (i = 1) are determined in [11].

When $P \neq 0$, if v^*_{AB} is the activation volume [12] we have

$$n_v(P,T) = n_v(P,0)\exp\left(-\frac{Pv^*_{AB}}{\theta} \right), \quad (5)$$

By using the equation of state of the interstitial alloy AB with FCC structure at 0 K and pressure P

$$Pv^i = -r^i_1\left(\frac{1}{6}\frac{\partial u^i_{0X}}{\partial r^i_{1X}} + \frac{\hbar\omega^i_X}{4k^i_X}\frac{\partial k^i_X}{\partial r^i_{1X}} \right), \quad (6)$$

we can find the nearest neighbor distance of atom X from the following formula

$$y^i_X(P,T) = \sqrt{\frac{2\gamma^i_X(P,0)\theta^2 A^i_X(P,0)}{3k^{i\,3}_X(P,0)}}, \quad (7)$$

$$r^i_{1B}(P,T) = r^i_{1B}(P,0) + y^i_{A_1}(P,T), r^i_{1A}(P,T) = r^i_{1A}(P,0) + y^i_A(P,T),$$

$$r^i_{1A_1}(P,T) \approx r^i_{1B}(P,T), r^i_{1A_2}(P,T) = r^i_{1A_2}(P,0) + y^i_B(P,T). \quad (8)$$

The mean nearest neighbor distance between two atoms in perfect interstitial alloy AB with FCC structure is approximately determined by

$$r_{AB}^0(P,T) = r_{AB}^0(P,0) + y_{AB}^0(P,T),$$
$$r_{AB}^0(P,0) = (1-c_B)r_{1A}^0(P,0) + c_B\sqrt{2}r_{1B}^0(P,0),$$
$$y_{AB}^0(P,T) = (1-15c_B)\,y_A^0(P,T) + c_B y_B^0(P,T) + 6c_B y_{A_1}^0(P,T) + 8c_B y_{A_2}^0(P,T). \quad (9)$$

Approximately, the nearest neighbor distance between atoms, in an alloy with defects, is equal to that of an ideal alloy.

The thermal expansion coefficient of real alloy has the form [11]

$$\alpha_{AB}^R = \sum_X c_X(\alpha_X^0 + n_v n_1(\alpha_X^1 - \alpha_X^0) + n_v \Delta_X \alpha_X^0), \quad \alpha_X^i = \frac{y_X^i}{r_{01X}^i T}\left(1 + \frac{\theta}{2r_{1X}^i}\frac{dr_{1X}^i}{d\theta}\right). \quad (10)$$

The heat capacity at constant volume for the real alloy is determined by [11]

$$C_{VAB}^R = \sum_X c_X(C_{VX}^0 + n_v n_1(C_{VX}^1 - C_{VX}^0) + n_v \Delta_X C_{VX}^0),$$

$$C_{VX}^i = 3Nk_B\left\{Y_X^{i\,2} + \frac{2\theta}{k_X^{i\,2}}\left[\left(2\gamma_{2X}^i + \frac{\gamma_{1X}^i}{3}\right)X_X^i Y_X^{i\,2} + \frac{\gamma_{1X}^i}{3}\left(1+Y_X^{i\,2}\right) - \gamma_{2X}^i\left(Y_X^{i\,4} + 2X_X^{i\,2}Y_X^{i\,2}\right)\right]\right\}. \quad (11)$$

The heat capacity at constant pressure for the real alloy is determined by [11]

$$C_{PAB}^R = C_{VAB}^R + \frac{9TV_{AB}\alpha_{TAB}^R}{\chi_{TAB}^R}. \quad (12)$$

3. Numerical results

For AuSi alloy, we use the *n-m* pair potential where potential parameters are given in Table 1 [13]

$$\varphi(r) = \frac{D}{n-m}\left[m\left(\frac{r_0}{r}\right)^n - n\left(\frac{r_0}{r}\right)^m\right]. \quad (13)$$

Table 1. The parameters *m, n, D,* r_0 for Au and Si materials [13].

Material	m	n	D (10^{-16} erg)	r_0 (10^{-10} m)
Au	5.5	10.5	6462.540	2.8751
Si	6	12	45128.34	2.295

Approximately,

$$\varphi_{\text{Au-Si}} \approx \frac{1}{2}\left(\varphi_{\text{Au-Au}} + \varphi_{\text{Si-Si}}\right) \quad (14)$$

Our numerical results are illustrated in Fig.1 to Fig.10. When the concentrations $c_{Si} \to 0$, we obtain thermodynamic quantities of Au as shown in tables from Table 2 to Table 7 and figures from Figure 1 to Figure 4.

In Table 3, the EXPT1 is from the linear extrapolation of thermal expansion, the EXPT2 is from the differential dilatometry, the EXPT3 is from the specific heat, the EXPT4 is the volume and lattice parameter of quenched samples, the EXPT5 is the quenched-in enthalpy and the EXPT6 is from the microscopic observations of quenched samples.

Table 2. Temperature dependence of the equilibrium vacancy concentration $n_v(T)$ in Au at zero pressure determined by the SMM, the CAL [7] and the EXPT [14].

Method	T (K)	973	1023	1073	1123	1173	1223	1273	1323	1337
SMM	n_v (10^{-4})	0.50	0.91	1.58	2.62	4.16	6.38	9.48	13.71	15.14
CAL[7]		0.77	1.30	2.09	3.22	4.79	6.89	9.64	13.15	15.00
EXPT[14]		0.3	0.6	1.2	2.1	2.4	3.3	4.8	6.6	7.2

Table 3. The equilibrium vacancy concentration n_v in Au at the melting temperature (T_m = 1337K) from the SMM and the EXPT1-5 [8].

Method	SMM	EXPT1	EXPT2	EXPT3	EXPT4	EXPT5	EXPT6
n_v (10^{-4})	15.14	14	7	40	7	20	3

Table 4. The dependence of the volume ratio V/V_0 on pressure, for Au, with defects at 300K calculated by the SMM and the first-principles CAL [15].

Method	P (GPa)	2	4	6	8	10	12
SMM	V/V_0	0.9912	0.9805	0.9706	0.9614	0.9529	0.9450
CAL[15]		0.9885	0.9779	0.9679	0.9586	0.9499	0.9416

Table 5. The dependence of the nearest neighbor distance $r_{Au}(10^{-10}$ m) on the temperature, for Au, with defects at zero pressure calculated by the SMM and from the EXPT [17].

T (K)	293	574	676	777	875	971	1075	1179	1280	1324
SMM	2.8536	2.8655	2.8700	2.8745	2.8790	2.8835	2.8885	2.8936	2.8988	2.9011
EXPT [17]	2.8793	2.8928	2.8970	2.9013	2.9062	2.9112	2.9168	2.9232	2.9281	2.9310

Table 6. The dependence of thermal expansion $\alpha_T(10^{-6}$ K^{-1})and the heat capacity at constant pressure C_P (cal.mol^{-1}.K^{-1}) on temperature, for Au, with defects at zero pressure calculated by the SMM and from the EXPT [19].

Method	T (K)	300	400	500	600	700	800	1000	1200
SMM	α_T	14.44	15.00	15.53	16.06	16.62	17.20	18.47	19.94
EXPT[19]		14.1	14.5	15.0	15.4	15.9	16.5	17.9	19.5
SMM	C_p	6.00	6.10	6.18	6.25	6.32	6.40	6.56	6.76
EXPT[19]		6.07	6.18	6.28	6.40	6.52	6.65	6.90	7.15

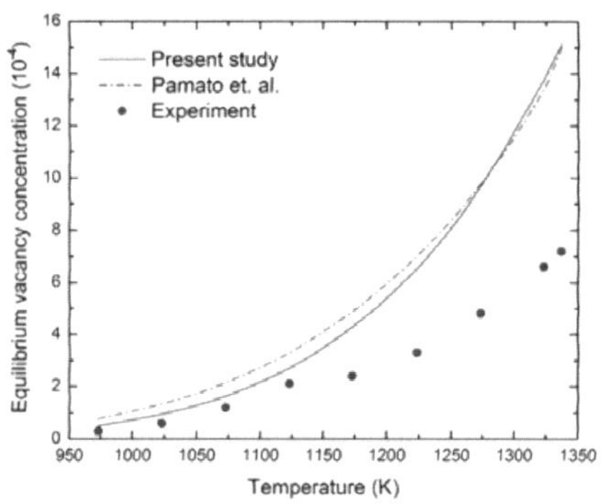

Figure 1. n_v (T) of Au at P = 0 from the SMM, the CAL [7] and the EXPT [14].

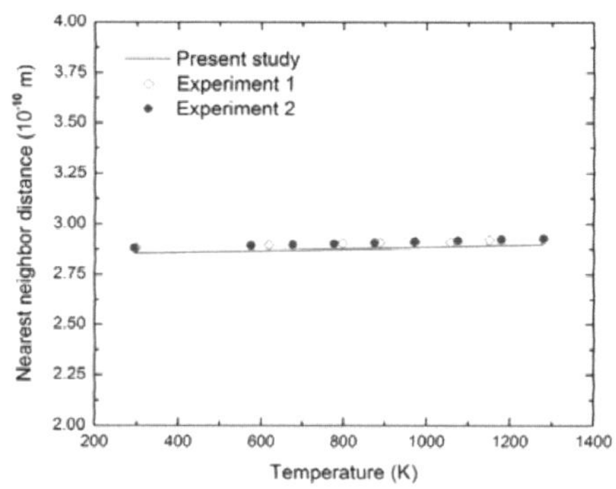

Figure 2. r_{Au} (T) of Au with defects at P = 0 from the SMM, EXPT 1 [16] and EXPT 2 [17].

IOP Conf. Series: Materials Science and Engineering **485** (2019) 012018 doi:10.1088/1757-899X/485/1/012018

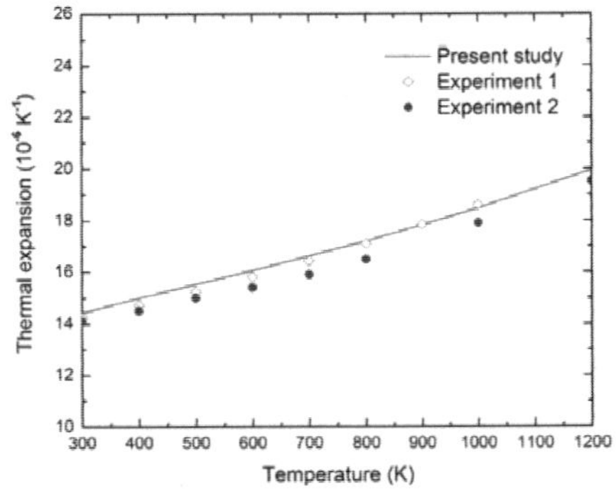

Figure 3. α_T (T) of Au with defects at $P = 0$ from the SMM, EXPT 1 [18] and EXPT 2 [19].

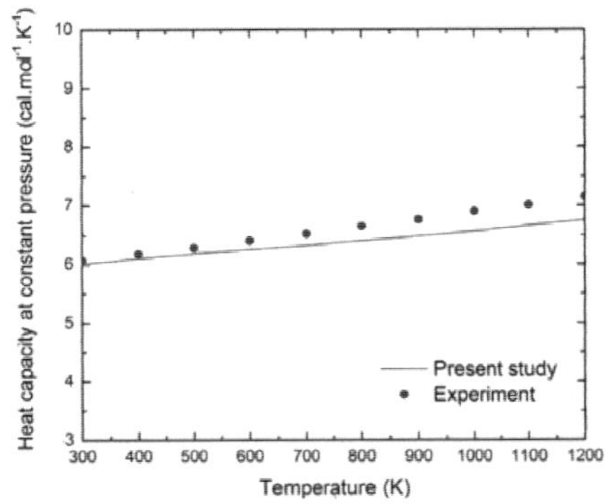

Figure 4. C_p (T) of Au with defects at $P = 0$ from the SMM and the EXPT [19].

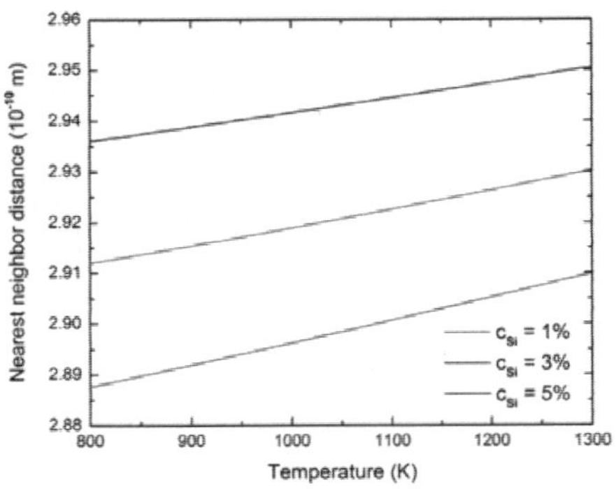

Figure 5. r_{AuSi} (T) of alloy AuSi with defects at $P = 0$, c_{Si}= 1%, 3%, 5% from the SMM.

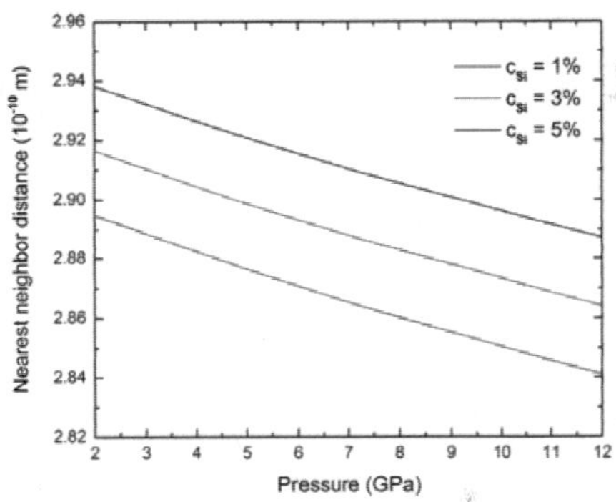

Figure 6. r_{AuSi} (P) of alloy AuSi with defects at $T = 1200$ K, $c_{Si} = 1\%, 3\%, 5\%$ from the SMM.

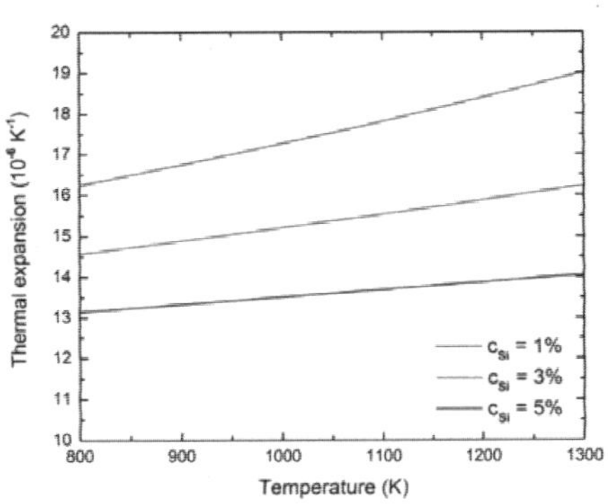

Figure 7. α_T (T) of alloy AuSi with defects at $P = 0$, c_{Si}= 1%, 3%, 5% from the SMM.

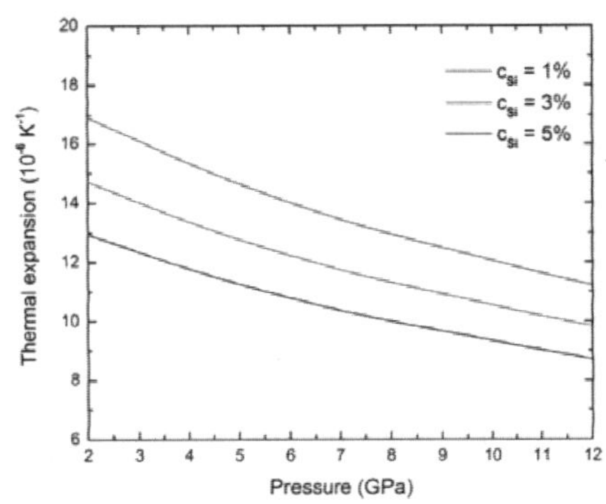

Figure 8. α_T (P) of alloy AuSi with defects at $T = 1200$ K, $c_{Si} = 1\%, 3\%, 5\%$ from the SMM.

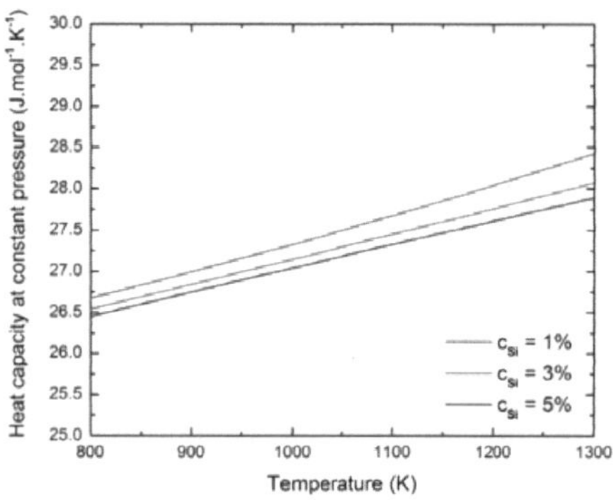

Figure 9. $C_P(T)$ of alloy AuSi with defects at $P = 0, c_{Si} = 1\%, 3\%, 5\%$ from the SMM.

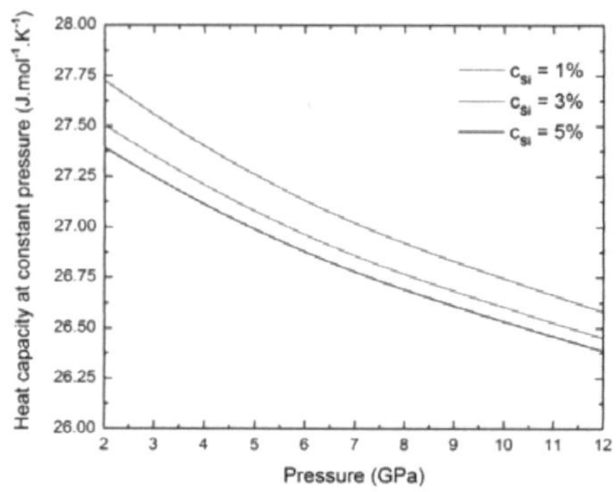

Figure 10. $C_P (P)$ of alloy AuSi with defects at $T = 1200$ K, $c_{Si} = 1\%, 3\%, 5\%$ from the SMM.

4. Conclusion

Our results of the thermodynamic theory for interstitial alloys with vacancy and FCC structure based on the SMM are applied to AuSi alloy in the 300 to 1337 K interval of temperature, in the interval of concentration of interstitial atoms from 0 to 5% and in the interval of pressure from 0 to 12 GPa. The calculated results for the nearest neighbor distance, the thermal expansion coefficient and the heat capacity at constant pressure for an alloy with vacancies are more similar to the experimental results than in the case of an ideal alloy. The calculated results for the equilibrium vacancy concentration, the nearest neighbor distance, the volume ratio, the thermal expansion coefficient, the heat capacity at constant pressure of the main metal Au are in good agreement with experiments and other calculations. That proves that the point defects has a significant contribution on the thermodynamic quantities of an interstitial alloy, at high temperatures.

References

[1] Simmons R O and Balluffi R W 1960 *Phys. Rev.* **117** 52.
[2] Siegel R W 1978 *J. Nucl. Mats.* **69** 117.
[3] Mirwald P W and Kennedy G C 1979 *J. Geophys. Res. B* **84** 6750.
[4] Sumita T, Kato M and Yoneda A 1998 *Rev. High. Pres. Sci. Tech.* **7** 254.
[5] Godwal B K, Ng A and Dasilva L 1990 *Phys. Lett. A* **144** 26.
[6] Hung V V and Hai N T 1997 *J. Phys. Soc. Jap.* **66** 3499.
[7] Pamato M G, Wood I G, Dobson D P, Hunt S A and Vočadlo L 2018 *J. Appl. Cryst.* **51** 470.
[8] Kraftmakher Y 1998 *Phys. Rep.* **299** 79.
[9] Hung V V and Hai N T 1998 *Int. J. Mod. Phys.* B **12** 191.
[10] Tang N and Hung V V 1998 *Phys. Stat. Sol. (b)* **149** 511.
[11] Hoc N Q, Hanh P T M and Viet L H 2017 *HPU No.2 J. Sci.* **50** 118.
[12] Aziz M J 1997 *Appl. Phys. Lett.* **70** 2810.
[13] Magomedov M N 1987 *J. Fiz. Khimic* **61** 1003.
[14] Simmons R O and Balluffi R W 1962 *Phys. Rev.* **125** 862.
[15] Yokoo M, Kawai N, Nakamura K G, Kondo K I, Tange Y and Tsuchiya T 2009 *Phys. Rev. B* **80** 104114.
[16] Dutta B N and Dayal B 1963 *Phys. Stat. Sol. (b)* **3** 473.
[17] Suh I K, Ohta H and Waseda Y 1988 *J. Mater. Sci.* **23** 757.
[18] Anderson O L, Isaak D G and Yamamoto S 1989 *J. Appl. Phys.* **65** 1534.
[19] Gray D E 1972 *American Institute of Physics handbook* (New York, Toronto, London: McGraw-Hill Companies).

IOP Conf. Series: Materials Science and Engineering **485** (2019) 012019 doi:10.1088/1757-899X/485/1/012019

Glass fibers for impact protection systems. A review

G G Ojoc[1], C Pirvu[2], S Sandu[3] and L Deleanu[1]

[1] Department of Mechanical Engineering, „Dunarea de Jos" University, Galati, Romania
[2] National Institute for Aerospace Research „Elie Carafoli" INCAS, Bucharest, Romania
[3]3Scientific Research Center for CBRN Defense and Ecology, Bucharest, Romania

lorena.deleanu@ugal.ro

Abstract. This paper is a review about glass fiber fabrics used for impact protection. The authors comment on types of fabrics, tests done for pointing out the impact resistance. Many of applications refer to ballistic protection but, as one may notice, experimental studies in open literature are disperse from the point of view of parameter ranges. Many articles on the subject do not have details on materials (fabrics and adhesives) used to manufacture the protection samples and on technology for obtaining the final system of protection. Few tests are done according to national, international or military standards because of difficulties in sample manufacturing at industrial scale or the costs related to sample producing. This review could be useful for the start of a research on ballistic panels based on or including glass fiber fabrics, for making the researchers familiar to their particular failure mechanisms.

1. Introduction

High-performance fibers have a history of only about 100 years, but they offer innovative solutions for protection against impact [1, 2]. Nylon and silk fibers had been used to make body armors, but with limited efficiency. Nylon, patented by DuPont got on the market in 1939. In the 1960s, the same firm developed polyparaphenylene terephthalamide (PPTA), now known as Kevlar, a much stiffer semirigid rod molecule and produced fibers of very high crystallinity. Today, these ones and similar fibers like Twaron are essential in manufacturing individual protection systems. Other polymeric fibers are particularly designated for ballistic protection, based on ultrahigh density polyethylene ultrahigh-molecular-weight polyethylene (UHMWPE) (trade names as Dyneema, Spectra), polybenzobisoxale (PBO), polybenzobisthiazole (PBT), polybenzenimidazole (PBI) [3]. Even if glass fibers have some shortcommings, they are used for vehicle and other systems protection against ballistic impact.

Fiber tensile strength increases with decreasing fiber diameter and is limited by defects, residual stresses, and structural inhomogeneities. The risk of finding defects decreases with decreasing fiber diameter for polymeric fibers, carbon fibers, ceramic and glass fibers. Nowadays, the commercial carbon fibers range from 4 μm to 10 μm in diameter; for polymeric and most ceramic and glass fibers, diameters are in the range of 10 μm to 15 μm. Fibers processed by chemical vapor deposition, such as boron fibers, tend to have much larger diameters, typically 100-150 μm [4].

2. A short presentation of glass fibers

The manufacturing of glass fibers and fabrics is still an energy consuming sector as the glass is melt-extruded and drawn into fibers typically at 1000°C to 1200°C.

IOP Conf. Series: Materials Science and Engineering **485** (2019) 012019 doi:10.1088/1757-899X/485/1/012019

Figure 1 presents the chemical composition of different grade of glass (symbols are explained under the figure). If one analyzes the graphs of their mechanical characteristics in Figure 2, it is obvious that grade S, including S2 is recommended for ballistic applications. One may notice that glass is based on SiO_2 and Al_2O_3, the highest sum of these two ceramics being of 90% for the grade S, a glass fiber that is especially used for its mechanical resistance.

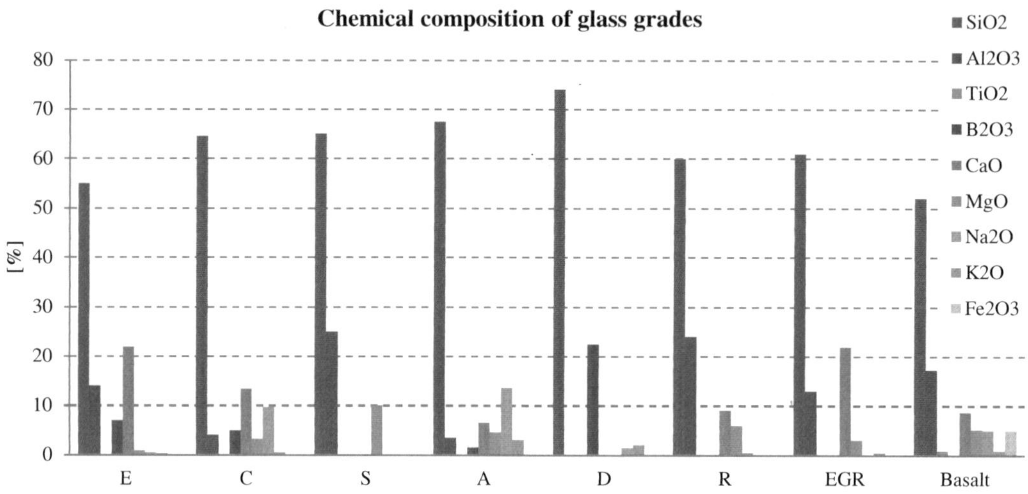

Figure 1. Chemical composition of glass grades: A - durability, high mechanical strength and high electric resistivity, C - high corrosion resistance, D - low dielectric constant, E - high mechanical strength and high electric resistivity, AR - alkali corrosion resistance, R- high strength and acid corrosion resistance S - highest tensile strength, S2- high strength, elasticity modulus and stability [5]

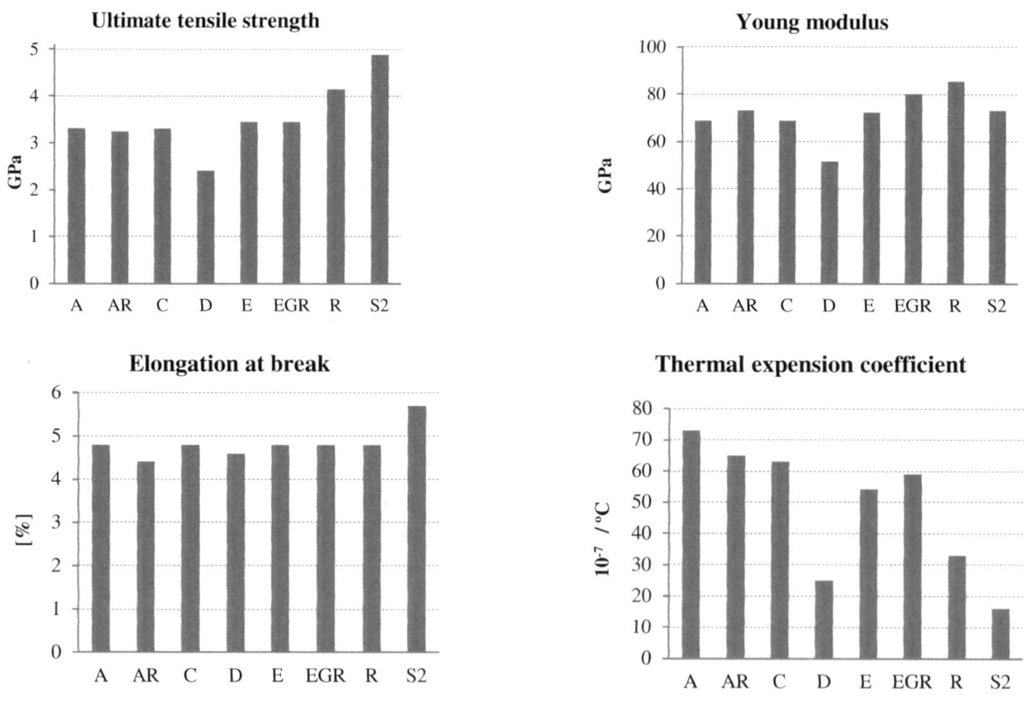

Figure 2. Characteristics of glass fibers (after [5])

Glass fibers fall into two categories: low-cost general-purpose fibers and special-purpose fibers. Over 90 % of glass fibers are general- purpose products. These fibers are known by the designation E-glass. Special-purpose fibers, which are of commercial significance in the market today, include glass fibers with high corrosion resistance (ECR-glass), high strength (S-, R-, and T E-glass), with low dielectric constants (D-glass), high-strength fibers, and pure silica or quartz fibers, which can be used at very high temperatures.

Characteristics of glass fiber fabrics (woven or not) are: higher mechanical properties in flexural strengths, a specific density between polymeric and steel fabrics, impact resistance, good mold conformability, very good wet out, low consumption of resins when producing layered composites.

3. Fabrics for protection systems

A criterion that is of high importance for impact protection structure including glass fibers is the glass fiber architecture. Studies on systems including stratified composites are done at different level: from micro level for the fibers and yarns, meso level that deal with fabrics and layers, to macro level when the simulation and experimental results are given for the entire protection structure. Recent advances in computer hardware make possible the simulation of panels under ballistic impact considering yarns and several layers [6].

Properties of composites reinforced by long fibers significantly depend on fiber nature, yarn architecture and matrix and the technology to put these together, fiber volume fraction, fiber orientation, laminate thickness. Glass fibers have high strength at low costs; carbon fibers have very high strength, stiffness and low density, aramid fibers have high strength and low density, preventing the fire spread and they are penetrable by radio waves. Polyesters are the most often matrices because they offer good properties at relatively low costs. Epoxies and polyamides have better properties, but they are expensive. Strength of composites increases by higher fiber volume fraction and fiber orientation parallel to load direction. Although glass and carbon fibers are relatively common, polymer, metal and ceramic fibers are used in specific applications, including military ones.

Rao and Farris [7] investigated the influence of twist on strength and modulus and found that fiber yarns exhibit the best tensile strength at an optimum twist angle of about 7°. In ballistics, the most common weave patterns are plain and basket weaves. Fabrics are also produced with unidirectional yarns, with a single orientation or more (two, three and four), for the latest, the angle between plies varying from [0°, 90°], [-45°, 45°] to [0°, 90°, -45°, 45°]. Cunniff [8] observed that loosely woven fabric or unbalanced weave led to poor ballistic performance.

Faur-Csukat [9] manufactured fabric composites with 55%wt fibers by manual lay-up, followed by compression. The ballistic performance of carbon-, glass-, aramid-, and polyethylene- fiber fabrics used to manufacture protection panels showed that their efficiency was as following: glass is better than aramid, which is better than or equal to ultrahigh-molecular-weight polyethylene (UHMWPE), which is better than carbon fibers.

Generally, fibers with large strain at high strain rate are better energy absorbers than those with low strain at break. [10], [11]. The fiber-matrix interface plays a critical role in managing the impact resistance. It was observed that weaker interfacial interaction resulted in higher energy absorption [12], [13] [14], but each matrix has to pass the experimental proof for being accepted, Composites with fabrics exhibit particular failure mechanisms as fiber-matrix debonding, delamination (between layers), slippage, cracking net, but also friction, favor for energy absorption.

Weave architecture also influences the ballistic performance of composites. Under the conditions investigated, the performance of basket-weave fabrics was better by about 10% than that of plain-weave fabrics. [9]. Satin and twill weaves also tended to absorb more energy than the plain weaves [15], possibly because the fiber are not loosing their strength to face the characteristic crimping in weaving. The architecture of the fabric is more important in thicker composites than in thinner composites, as the small crimp angle decreases stress concentration.

Improved ballistic performance may be obtained by using 3D woven fabrics instead of 2D woven fabrics [16] but this new architecture by 3D stitching is prone to loosen the yarns. Walter et al. [17]

analyzed results from 3D woven glass-fiber composites and observed that delamination along the weak layer is a severe shortcoming for these composites under impact at high strain rates. In general, Z-stitching increased the resistance to failure and it restricted damage to a smaller total area than that in unstitched samples. Hosur et al. [15] reported a decrease in ballistic limit in Z-stitched targets, although no explanation of this decrease was provided but it could be the fact that yarns seems to be have more space among them when stitching in such way.

Shockey et al. [18] studied single-ply Zylon fabrics and observed that absorbed energy was proportional to fabric areal density, but that ballistic effectiveness was not strongly dependent on mesh density or weave tightness. Chitrangad [19] recommended a range of 0.60 to 0.95 for the cover factor (the ratio of the area covered by the yarns to the whole area of the fabric) of fabrics for ballistic applications. Lower value characterizes fabrics too loose, but higher value induces too much bending stress in yarns, during weaving. The V_{50} of composite fabrics with higher elongation in weft yarns and lower elongation-to-break in warp yarns was greater than that of fabrics made from a single material, which may be due to the lesser influence of yarn crimp. By considering yarn crimp in modeling, Tan et al. [20] obtained more accurate results. The number of fabric plies also affects the ballistic performance (typically, there may be 18-56 plies). Shockey et al. [18] noticed an increase of the specific energy absorbed by stratified targets due to friction between layers. Also, a model designed by Ionescu et al. [41] estimated that friction reduce the residual velocity. The influence of interply distance on ballistic performance has also been investigated [8], [21]. The influence of projectile geometry becomes less important with the increased number of plies. [22] A 3D woven structure was studied in a fabric composite [23] designed to provide greater through-thickness direction reinforcement than in conventional 2D woven fabrics; this structure showed higher ballistic performance and led to fewer penetrated layers under impact.

Kolopp et al. [24] did an experimental impact study has been conducted on sandwich structures to identify and improve armor solutions for aeronautical applications: a non-perforated panel with minimal weight and back deformations. Medium-velocity impacts (120 m/s) have been conducted using a 127 g spherical projectile. Two potential choices of front skin have been identified for the sandwich structure: 3 mm thick AA5086-H111 aluminium plates and aramid stitched fabrics (8 to 18 plies). Impact tests indicate that aluminium honeycomb core associated with aluminium skins show mitigated results. However, the combination of dry fabric front skin and aluminium honeycomb show better performances than aluminium sandwiches, with a global weight decrease.

Applications of glasș fiber fabrics and their composites include automotive, shipbuilding, energy generation, construction, chemical/petrochemical, nuclear, manufacturing and ballistic protection [4].

In automotive industry, the glass fiber ratio by weight is between 31% for pan floor (with polyesther matrix, firm Das, 2011) to 55% for body and door (with epoxy resin, Rocky Mountain Institute) and 69% for chassis (with epoxy resin Suzuki & Takahashi, 2005). For impact protection panels, this percentage could be higher as the end-users want a low surface density and a small thickness [25].

4. Issues related to protection system design

Figure 3 presents a flow chart in designing a new protection system, pointing out a closer coupling of researches for both material development, modeling and simulating community, resulting in significantly reduced time for development of new armor. Restricted information issues make the solutions to be punctual or only partially effective [26]. The elements of an impact protection system are not themselves new (Figure 4 presents a theoretical and general structure. [27], a similar one being also detailed in Figure 5 [16]), but the emphasis shifts from design–produce–shoot test–re-design to simulation iterations, and from designing with materials on the market to designing with materials proved to have impact resistance. The feedback loop between protection system design and material design contrasts with current practice, in which a one-way flow puts new materials on the shelf to be tried in the produce-shoot test-analysis process [26] (Figure 3).

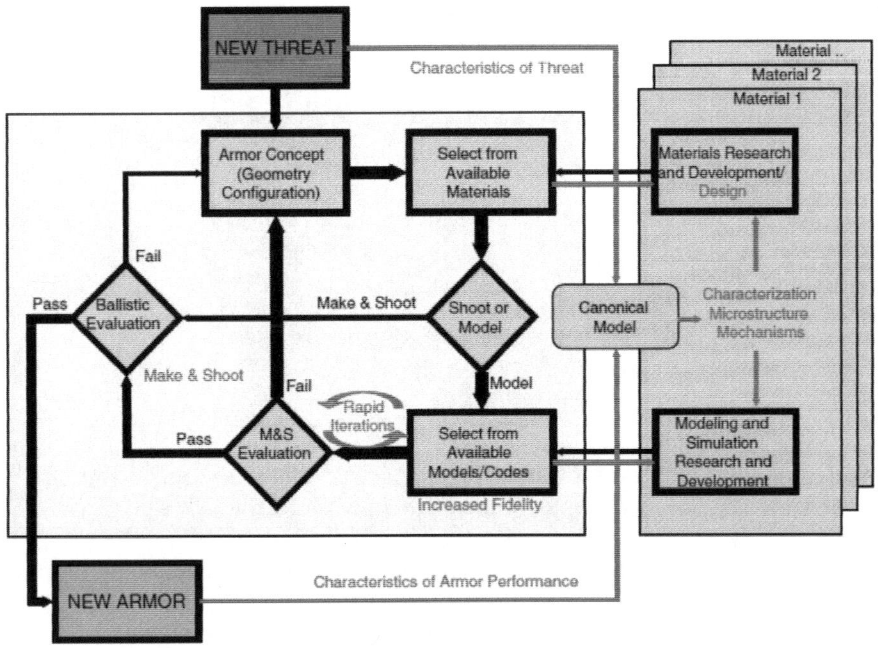

Figure 3. Flow chart for (re-)designing an armor [26]

Figure 4 presents a typical impact protection system. Medvedovski [28] pointed out that a ballistic protecting system consists of a monolithic ceramic or composite ceramic-metal plate, covered by ballistic polymeric fabrics with a high tensile strength (such as Kevlar, Twaron, Spectra, Dyneema), lining or laminated in polyethylene, placed on the back of ceramic or ceramic–metal composite, aluminum thin sheets may be as backing material. In some cases, a spall shield is attached on the front of armor.

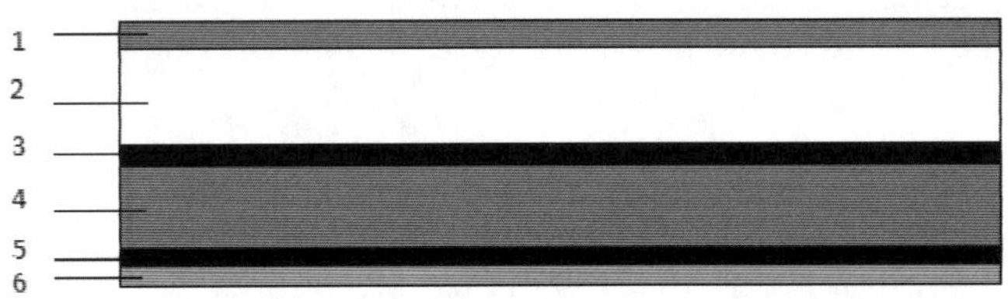

	Material	Functions
1	Composite	Low velocity, impact abrasion protection
2	Ceramic tile	
3	Elastomer	Ballistic protection, structural integrity
4	Composite	
5	Metal mesh/Plate	Displacement attenuation
6	Fiber composite (glass, polymeric)	Fire, smoke and toxicity protection

Figure 4. An impact protection system (after [26])

Under the impact of the projectile (velocity of 700–1000 m s^{-1}), the hard ceramic pack is cracked and the residual energy is absorbed by the soft reinforced backing material that must support the post-impact fracturing of the ceramic pack and the defeated bullet. Soft covering also protects the system against possible damage associated with vibrations. Figure 5 presents a system with several of the elements from Figure 4, tested by [16].

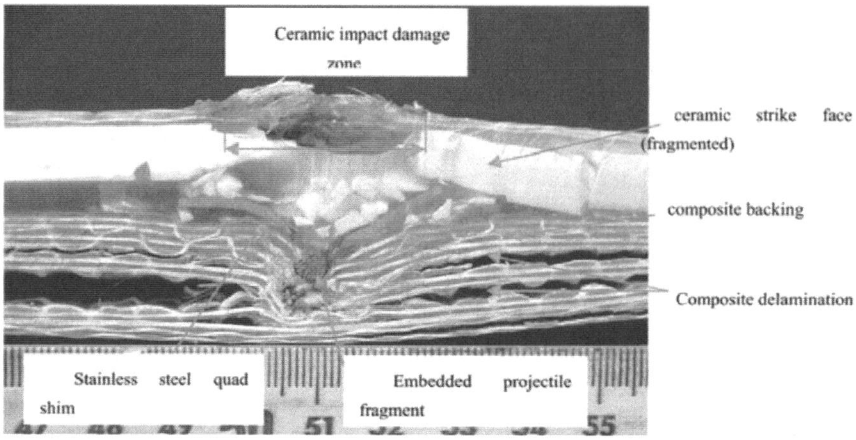

Figure 5. Water jet cross section of a vehicle armor panel, with 3D woven backing. [16] Fabric type, thickness, impact velocity projectile

The complexities of armor systems make the assessment of weight depending on the event they are designed for. What is lightweight for vehicles is extremely heavy for personnel. Thus, in assessing whether an armor system is sufficiently lightweight, one cannot look at the absolute weight of the system. Because protection systems are designated to protect a particular area [26], their weight is better described by its areal density, $\mathbf{A_d}$:

$$\mathbf{A_d} = \frac{\textbf{system weight}}{\textbf{area to be protected}} \quad \left[\mathbf{kg/m^2}\right] \tag{1}$$

But this characteristic of the system does not indicate its effectiveness. The effectiveness of two protecting systems is only assessed by comparing their performance against the same threat. The effectiveness of a given protection system is called its mass effectiveness, $\mathbf{E_m}$, the ratio of the areal density of rolled homogeneous armor (RHA), a common steel for tank armor that will stop a particular threat, to the areal density of the given protection system that will stop that same threat:

$$\mathbf{E_{m(protection\ system)}} = \frac{\mathbf{A_{d(RHA)}}}{\mathbf{A_{d(protection\ system)}}} \tag{2}$$

The mass effectiveness of the protection system indicates how effective it is against a specific threat and generally suggests whether the system may be considered lightweight— that is, the higher $\mathbf{E_m}$ value, the lighter is the weight of the system. One of the issues of any system is that $\mathbf{E_m}$ does not translate from one threat to another and only tests done.

5. Tests and failure mechanisms in glass fiber composites for ballistic protection

The densities of B_4C (2.52 g/cm^3) and SiC (3.29 g/cm^3) are less than that of Al_2O_3 (3.98 g/cm^3), but due to its easy sinterability and lower cost of raw powders, alumina is still preferred in vehicle protection, where the extra weight can be tolerated, while the lighter ceramics are now used in body armor [26].

When a projectile hits the individual fiber or yarn longitudinal and transverse waves propagate from the impact point. Most of the kinetic energy transfers from the projectile to the principal yarns (those coming directly in contact with the projectile); the orthogonal yarns, which intersect the principal yarns, absorb less energy [29] [30] [31]. The transverse deflection continuously increases until it reaches the breaking strain of fibers and causes failure. Glass fiber composites have specific failure mechanisms that could be investigated by simulation [32], [33] and post-mortem investigations: breakage of fiber bonds and yarns, yarn pull-out, remote yarn failure, wedge-through

IOP Conf. Series: Materials Science and Engineering **485** (2019) 012019 doi:10.1088/1757-899X/485/1/012019

effects (hole smaller than the diameter of projectile), fibrillation and splitting of the fiber, and effects of friction between the projectile and the fabric, resin, yarns and fibers (Figure 6).

In Figure 7, the panel made of glass fibers and epoxy resign arrested the bullet well, the fiber ratio, calculated for the cross section of structure being about 20%, but technology parameters are quite severe [34].

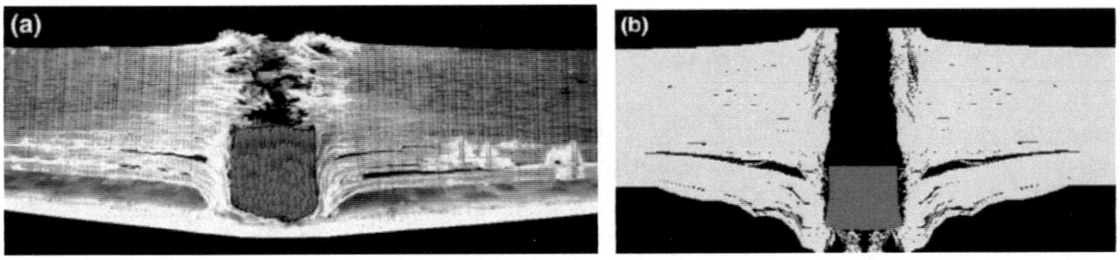

Figure 6. Experimental (a) and computational (b) results for an impact of 0.50 caliber FSP with a 25.4 mm, E-glass fiber reinforced poly-vinylester-epoxy matrix. Initial projectile velocity, 605 m/s [32]

Figure 7. Composite glass fiber + resin, thickness of 13 mm, after the impact with 9 mm FMJ at 420 m/s [34]

Ansari and Chakrabarti [35] presented experimental and numerical analysis of perforation behaviour of laminated composite reinforced with unidirectional glass fiber have been presented due to impact by 52 g blunt projectile. The influence of oblique impact on the ballistic performance of laminated target has been studied by considering four impact angles (0°, 30°, 45° and 60°) and impact velocities in between 50-500 m/s. The experimental impact tests on laminated target with fiber orientation (0°/90°/90°/0°) is performed with pneumatic gun. In general, delamination is the major cause of damage in the target plate struck by blunt projectile, normal to the target the amount of damage decreases as the impact angle increases. The delamination in the target plate occurs mostly due to matrix failure in tension or excessive stress along the thickness direction in case of normal impact whereas inter laminar stress causes maximum delamination in the target under oblique impact.

Another research of these authors remarks that for different sandwich composite plates made of glass fiber and Kevlar/epoxy laminate, sandwich composite having glass fiber laminate sandwiched in between Kevlar/epoxy offer good penetration resistance even better than that offered by single Kevlar/epoxy composite plate. The energy absorption in glass fiber composite plate is found to be more than the Kevlar/epoxy composite plate for all the incidence velocities. The energy absorption in sandwich composite plate KGK is found to be more than the other two combinations GKK and KKG and even more than that in case of Kevlar/epoxy plate of same thickness for all impact velocities. The differences in the energy absorption by sandwich composite plates KGK and KKG are 20.64 J and 138.33 J for incidence velocity of 100 m/s and 500 m/s, respectively, showing that the KGK sandwich composite serves better than the Kevlar/ epoxy as an energy absorber (here, K - Kevlar, G - glass fiber) [36].

Table 1 presents several experimental works in order to point out the diversity of glass fabrics and matrixes, but also the variety of projectiles.

Table 1. Protection plate based on glass fibers.

Author(s) **Years**	Plate / pannel / Basic materials Resin	test
Herbert et al. 2007 [37]	E-glass: E_3LTi 10800, E_2LTi 7200, E_2LTi3600 vinyl ester resins and urethane	testing range of 3.08–7.53 MPa peak incident pressure
Sabet 2011 [38]	E-glass fiber reinforcement • chopped strand mat of 400 g/m^2, • plain weave, 400 g/m^2, • satin weave, 300 g/m^2, • unidirectional 300 g/m^2, • cross-ply unidirectional fiber [0°, 90°]. specimen areal density 0.4...0.9 g/cm^2	80...160 m/s 3 and 6 mm thickness smooth barrel gas gun plate 15 cm x 15 cm **projectile**: sharp tipped (30° conical head, total length of 30 mm, shank length of 15 mm, weight of 9.74 g)
Ansari 2017 [35]	unidirectional glass fiber target plate of size 140 mm × 140 mm × 3.3 mm and (0°/90°/90°/0°) by hand layup	52 g blunt projectile of diameter 19 mm fully clamped boundary condition 150-300 m/s
Armenakas 1973 [39]	glass-fiber-reinforced epoxy plates 316 S-glass fibers with 0.177 m as length, Φ123 μm; matrix epoxy was then prepared by mixing 10 g of Epon 828, 10 g of Epon 871 and 2.6 g of Epon curing agent D.	high rates of strain (30,000 in/in/min) and low rates of strain (0.0265 in./in./min to 26.5 in/in/min) The stress-strain relation of composites is linear up to failure.

Yuan [40] presented results of impact experiments for studying strength against spall and delamination in glass–fiber + epoxy resin composites. Two architectures are investigated—S2 glass woven fabrics in Cycom 4102 polyester resin matrix and a balanced 5-harness satin weave E-glass in a Ciba epoxy (LY564) matrix. The samples were impacted using an 82.5 mm bore gas-gun. The delamination strength of the plates was given as a function of the normal component of impact stress and applied shear-strain by subjecting samples to normal impact compression and combined shock compression and shear loading, respectively. The spall strengths of the two composites decreased with increasing levels of normal impact compression. Superposition of shear-strain on the normal impact compression was found to be detrimental to the spall strength. The E-glass reinforced composite was found to have higher spall strength under both normal impact compression and combined compression and shear loading as compared to the S2-glass composite. These relatively low spall strength levels of the S2-glass and the E-glass fiber reinforced composites have important implications in designing glass fiber light-weight integral armor.

Herbert et al. [37] evaluated the response of E-glass reinforced vinyl ester and urethane panels of varying structures subjected to shock loading and drop weight impact. Shock waves are created using a shock tube with a testing range of 3.08–7.53 MPa peak incident pressure (Figure 8). Drop weight impact performance was measured by energy absorbed by the samples, depth of penetration, and extent of internal damage. Glass preforms having total areal weights (4.88 and 7.32 kg/m^2) were infused with one of three types of vinyl ester and one urethane resin. Urethane panels having areal weight of 7:32 kg/m^2 performed better than similar vinyl ester resin panels. It was also found that of two materials with identical vinyl ester resins having an areal weight of 88 kg/m^2, the one with a finer glass structure consistently performed better under shock wave and drop weight impact testing.

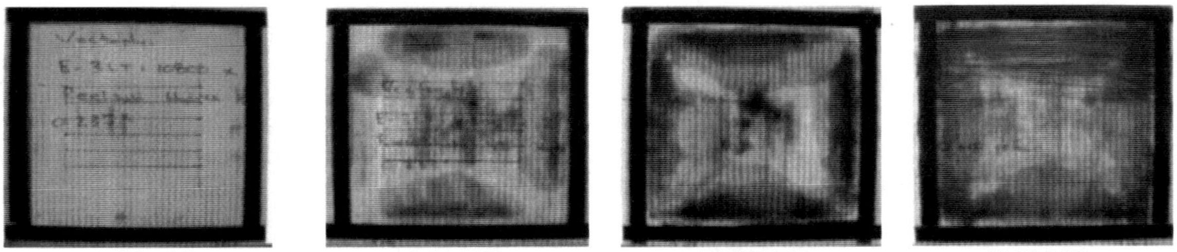

3.08 MPa 4.54 MPa 5.72 MPa 7.12 MPa

Figure 8. Delamination of panels – 2 x 108 vinyl ester, 5.59 mm thickness [37]

Sabet et al. [38] investigated composites based on glass reinforced polyester under impact velocity of 80...160 m/s. Five different types of E-glass fiber reinforcement were used, including chopped strand mat, plain weave, satin weave, unidirectional and cross-ply unidirectional fiber reinforcements. The projectile used was a sharp tipped (30°) conical head with total length of 30 mm and shank length of 15 mm, with weight of 9.74 g. Composite plates of 150 mm x 150 mm were prepared with 3 and 6 mm thickness. Results showed higher ballistic limit velocity for 3 mm plates with cross-ply chopped strand mat reinforcement followed by unidirectional reinforcement and plain weave. Plates with satin weave and bi-directional fibers were almost at same level. The thicker specimens (6 mm), with plain weave fabrics showed better ballistic performance, followed by cross-ply unidirectional, satin weave, unidirectional and CSM reinforced plates. Dominant failure modes were: fiber tension, fiber shear failure for thin-walled and severe delamination for thick-walled plates. Plates with plain weave and cross-ply unidirectional reinforcements showed relatively higher ballistic limit velocity as compared to the other type of reinforcement. Energy absorption associated with plain weave and cross-ply unidirectional reinforcement plates has higher values (Figure 9). The overall damage area in all specimens increased towards the exit side, in the shape of a cone, some specimens showing four petals on the distal side together with sever delamination (Figure 10).

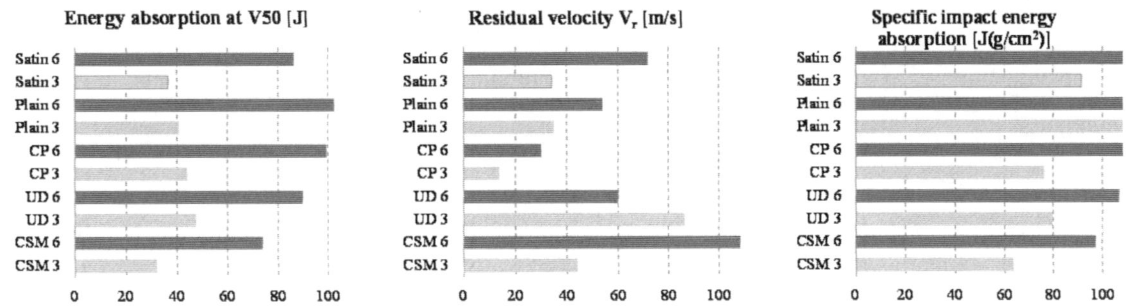

Figure 9. Characteristics of several panels obtained with different types of glass fiber and the same matrix: CSM - randomly oriented glass chopped strand of 400 g/m^2, UD - unidirectional fabric of 300 g/m^2, CP - cross ply laminate, [0°, 90°] with unidirectional fibers, Plain - a plain-woven roving cloth of 400 g/m^2, Satin - a satin cloth of 300 g/m^2, 3 - 3 mm thickness, 6 - 6 mm thickness [38]

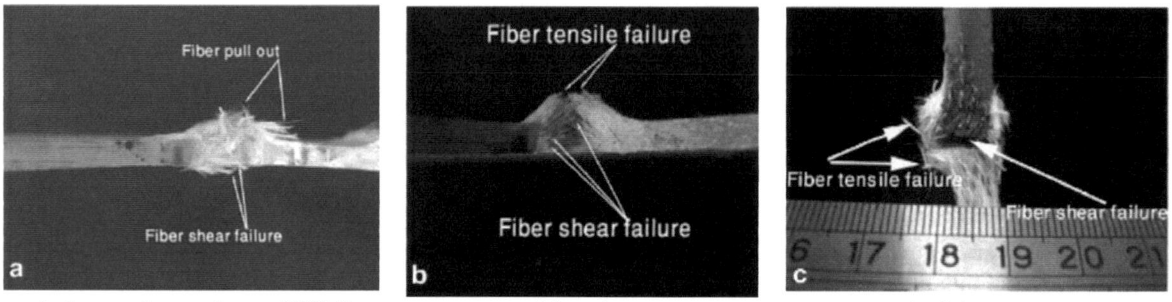

a) chopped strand mat (CSM) b) satin c) plain weave

Figure 10. Quarter section of 3 mm thin-walled reinforced samples (95 m/s) [38]

The differences between simulation and experimental data could be qualitatively explained by the energy dissipation by friction between layers as it is pointing out in [41].

Shukla et al. [16] tested vehicle armor panels with 2D and 3D woven composite backings. The armor panels with 3D woven backing had a higher ballistic efficiency than the 2D baseline panels, with controlled delamination and fewer complete penetrations. However, due to higher complete penetrations, it is believed that the projectile was able to penetrate the backing through direct tearing of the in-plane crimped fibers, without energy absorption through delamination. The majority of the delamination within the 3D composite panels transversed the full length of the panel. Because of the uncrimped nature of the warp and weft fibers, the individual lamina effectively act as a woven net to stop penetration, and allow kinetic energy to be absorbed mainly through intralaminary delamination. The delamination severity (opening) in the 3D composite panels decreased with increasing areal density, with the most controlled delamination seen in the vehicle armor panels with 2.77 kg/3D and 5.38 kg/3D composite backings.

Conclusions

Even if the glass fiber composites are challenged by polymeric, other ceramic or metal fiber, they still have been of interest for ballistic application, either as part of or as the entire of a protection system or as they exhibit good impact resistance at lower costs. Researches are conducted to increase the ratio of fibers, but also in finding more suitable matrix for a specific application.

This review could be useful for the start of a research on ballistic panels based on or including glass fiber fabrics, for making the researchers familiar to their particular failure mechanisms.

References

[1] Eichhorn S J, Hearle J W S, Jaffe M, Kikutani T (eds) 2009 *Handbook of textile fibre structure, Vol. 1: Fundamentals and manufactured polymer fibres*, (Woodhead publishing limited)

[2] Boccacinni A R 2005 Continuous fibre reinforced glass and glass-ceramic matrix composites (Spinger)

[3] Cavallaro P V, 2011 Soft Body Armor: An overview of materials, manufacturing, testing, and ballistic impact dynamics, Naval Undersea Warfare Center Division Newport, Rhode Island

[4] Mallick P K 2007 *Fiber reinforced composites. Materials, Manufacturing and Design* (3rd ed), (CRC Press, Francis and Taylor Group)

[5] Sathishkumar T P, Satheeshkumar S, Naveen J, 2014 Glass fiber-reinforced polymer composites – a review, *J of Reinforced Plastics and Composites*, **33**(13) 1258–1275

[6] Nilakantan G, Nutt S 2017 Effects of ply orientation and material on the ballistic impact behavior of multilayer plain-weave aramid fabric targets, *Defence Technology* 1-14

[7] Rao Y, Farris R J 2000 Modeling and experimental study of the influence of twist on the mechanical properties of high-performance fiber yarns *J Appl Polym Sci* **77**(9) 1938-1949

[8] Cunniff P M 1992 An analysis of the system effects in woven fabrics under ballistic impact *Textile Research J* **62**(9) 495-509
[9] Faur-Csukat G 2006 A study on the ballistic performance of composites *Macromolecular Symposia* **239**(1) 217-226
[10] Naik N 2004 Composite structures under ballistic impact *Composite Structures* **66**(1-4) 579-590
[11] Barre S, Chotard T, Benzeggagh M L 1996. Comparative study of strain rate effects on mechanical properties of glass fibre-reinforced thermoset matrix composites *Composites Part A - Applied Science and Manufacturing* **27**(12) 1169–1181
[12] Park R, Jang J 1998 A study of the impact properties of composites consisting of surface-modified glass fibers in vinyl ester resin. *Composites Science and Technology* **58**(6) 979-985
[13] Tanoglu M., McKnight S, Palmese G, Gillespie Jr J 2001 The effects of glass-fiber sizings on the strength and energy absorption of the fiber/matrix interphase under high loading rates *Composites Science and Technology* **61**(2): 205-220
[14] Agbossou, A., Cohen, I., Muller, D., 1995 Effects of interphase and impact strain rates on tensile off-axis behavior of unidirectional glass fiber composite—Experimental results *Engineering Fracture Mechanics* **52**(5) 923
[15] Hosur M, Vaidya U, Ulven C, Jeelani S 2004 Performance of stitched/unstitched woven carbon/epoxy composites under high velocity impact loading *Composite Structures* **64**(3-4), 455-466
[16] Shukla A, Grogann J, S. Tekalur, A. Bogdanovich, R. Coffelt. 2005 Ballistic resistance of 2D & 3D woven sandwich composites. 625-634 in *Sandwich Structures 7: Advancing with Sandwich Structures and Materials: Proc of the 7th Intern Conf on Sandwich Structures* Thomsen O, Bozhevolnaya E, Lyckegaard A (eds) (Springer)
[17] Walter T, Subhash G, Sankar B, Yen C 2009 Damage modes in 3D glass fiber epoxy woven composites under high rate of impact loading *Composites Part B: Engineering* **40**(6) 584-589
[18] Shockey D A, Elrich D C, Simmons J W 2001 *Improved barriers to turbine engine fragments* Interim report III, DOT/FAA/AR-99/8, III. Menlo Park, Calififornia SRI International
[19] Chitrangad I 1993 *Hybrid Ballistic Fabric,* Patent no. 5,187,003. http://www.freepatentsonline.com/5187003.pdf
[20] Tan V B C., Shim V P W, Zeng X 2005 Modelling crimp in woven fabrics subjected to ballistic impact *Intern J of Impact Engineering* **32**(1-4) 561-574
[21] Lim C T, Tan V B C, Cheong C H 2002 Perforation of high strength double-ply fabric system by varying shaped *Intern J of Impact Engineering* **27**(6) 577-591
[22] Prosser R A, Cohen S H, Segars R A 2000. Heat as a factor in the penetration of cloth ballistic panels by 0.22 caliber projectiles *Textile Research J* **70**(8) 709-722
[23] Grogan J, Tekalur S A, Shukla A, Bogdanovich A, Coffelt R A 2007 Ballistic resistance of 2D and 3D woven sandwich composites *J of Sandwich Structures & Materials* **9**(3) 283-302
[24] Kolopp A, Rivallant S, Bouvet C 2013 Experimental study of sandwich structures as armour against medium-velocity impacts *Intern J of Impact Engineering* **61** 24-35
[25] *** 2016 *Bandwidth study on energy use and potential energy saving opportunities in the manufacturing of lightweight materials: glass fiber reinforced polymer composites,* Energetic Incorporated, DOE/EE-1344
[26] *** 2011 *Opportunities in protection materials science and technology for future army applications,* Committee on Opportunities in Protection Materials Science and Technology for Future Army Applications National Research Council, ISBN 978-0-309-21285-4
[27] Gooch W Jr, 2006 Overview of the development of ceramic armor technology—Past, present and the future, *30th Intern Conf on Advanced Ceramics and Composites*, Cocoa Beach SUA
[28] Medvedovski E 2010 Ballistic performance of armour ceramics: Influence of design and structure. Part 1, *Ceramics International* **36** 2103–2115
[29] Bhatnagar A 2006 *Lightweight ballistic composites* (CRC Press, New York)

[30] Abrate S 2016 *Damage in laminates from low-velocity impact in Dynamic deformation, damage and fracture in composite materials and structures* (Woodhead Publishing) 35-70
[31] Duodu E. A., J. N. Gu, Z. Shang, W. Ding, S. Tang, 2017 Damage Induced by High-Velocity Impact, *Iran J Sci Technol Trans Mech Eng* **41** 97–107
[32] Grujicic M, Pandurangan B, Angstadt D C, Koudela K L, Cheeseman B A, 2007 Ballistic-performance optimization of a hybrid carbon-nanotube/E-glass reinforced poly-vinyl-ester-epoxy-matrix composite armor *J Mater Sci* **42** 5347–5359
[33] Basaran B, 2007 *Computational analysis of advanced composite armor systems*, PhD, Middle East Technical University
[34] Deleanu L, Georgescu C, Pirvu C, Sandu S 2017 *Development of protection system,* Research contract 725/2017, "Dunarea de Jos" University of Galati, Romania
[35] Ansari M M, Chakrabarti A, 2017 Impact behaviour of GFRP and Kevlar/epoxy sandwich composite plate: Experimental and FE analyses, *J of Mechanical Science and Technology* **31**(2) 771~776
[36] Ansari M M, Chakrabarti A, Iqbal M A 2017 An experimental and finite element investigation of the ballistic performance of laminated GFRP composite target *Composites Part B* **125** 211-226
[37] Hebert M, Rousseau C-E, Shukla A 2008 Shock loading and drop weight impact response of glass reinforced polymer composites, *Composite Structures* **84** 199–208
[38] Sabet A, Fagih N, Beheshty M B 2011 Effect of reinforcement type on high velocity impact response of GRP plates using a sharp tip projectile *Intern J of Impact Engineering* **38** 715-722
[39] Armenakas A E, Sciammarella C A 1973 Response of Glass-fiber-reinforced Epoxy Specimens to High Rates of Tensile Loading *Experimental Mechanics* 433-440
[40] Yuan F, Tsai L, Prakash V, Rajendran A M, Dandekar D P 2007 Spall strength of glass fiber reinforced polymer composites, *Intern J of Solids and Structures* **44** 7731–7747
[41] IonescuT F, Pirvu C, Badea S, Georgescu C, Deleanu L 2017 The influence of friction characteristics in simulating the impact bullet – stratified materials *15th International Conference on Tribology, SerbiaTrib'17* Kragujevac Serbia

IOP Conf. Series: Materials Science and Engineering **485** (2019) 012020 doi:10.1088/1757-899X/485/1/012020

Effects of the application of a non-conventional treatment in magnetic field on a steel for industrial gearings

C P Papadatu, M Bordei and S Ciortan

Dunarea de Jos University of Galati, Faculty of Engineering, 47 Domneasca street, Galati, Romania

papadatu.carmen@yahoo.com

Abstract. An alloyed steel grade for machine parts construction used in industry have been considered and this material was subjected to a non-conventional treatment. The samples have been tested using an Amsler stand for dry friction tests and the diffractometric analysis completed this study. The plasma treatment plant used was the INI 150 made by the Institute of Radiation Physics and Technology in collaboration with the "Electrotechnics" Enterprise and the Nuclear Apparatus (I.C.E.FIZ.) from Romania and was destined for technological research. The samples considered for tests were type rollers, with 10 mm width, with different diameters to obtain different sliding degrees {ξ}, considering different values for the normal load (Q). This paper is a short review of the researches from last years and realizes a synthesis and a correlation of the results to formulate conclusions on the efficiency of the application of non-conventional treatments in magnetic field on special pieces from industry.

1. Introduction

The gearings, as a part of the machine construction used in industry, are subjected to friction. Physical phenomenon known as friction process represents the resistance to motion of two bodies brought into contact. The contact can be formed between the surface layers. If the distance between the surfaces is sufficiently small, the forces between the atoms and molecules bond those surfaces, forming junctions which determine the adhesion level. In the case of the gearings in contact directly the forces are very large and it is important to have superficial layers with characteristics of resistance in accordance with the solicitation of exploitation. That's why the researchers try to find more opportunities to increase the mechanical properties of the superficial layers.

An important quantity of the aluminum in the structure of the steel increases the thermo-magnetic treatment power and the results are the best. At the same time, the existence of aluminum content in the structure of the steel causes some hardening problems which are countered by the Chromium existence [1, 2].

It was considered an unconventional treatment in magnetic field. The magnetic field applied during the improvement treatment of the steel determines local oscillation through the magnetostriction which provokes local plastic deformations of the steel.

In literature, studies regarding the researches on the influence of the magnetic field applied during the heat treatments on the hardness of the steels for bearings and on the quantity of the rezidual Austenite (A_{rez}), have been presented. Analyzing the influence of the improvement treatments in magnetic field on the mechanical characteristics of the steel RUL 1 (100 Cr6) and on the wear behavior on tribo-models or on real pieces, resulted the following observations:

a).Through the improvement treatment in magnetic field realized in alternative current (A.C.) the improvement of the mechanical characteristics have been obtained, excepting the resilience (KV/KCU), comparing to the classic improvement treatment of the same steel (without magnetic field).
b). The tests on tribomodels for different thermal treatment variants applied to RUL1 steel have led to the conclusion that the thermo-magnetic treatment applied to this steel increased durability, implicitly, reliability. The influence is higher as it is applied at a lower quenching temperature [1].

For example, at the hardening temperature 810-820°C –for RUL1 steel- the heating in magnetic field - Alternative Current (A.C.) - with the intensity of H=31Oe conducts to the increasing of the reliability with 22%, the cooling in magnetic field - direct current (D.C.) at the magnetic field intensity H=975Oe conducts to the increasing of the durability with 32%, and the tempering in magnetic field (D.C.) with the magnetic field intensity H=1200Oe, conducts to the increasing of the durability with 53%.

From a structural point of view, X-ray diffractometric analyzes showed that residual austenite (Arez) content decreased and the degree of tetragonality of Martensite (c/a) decreased too, resulting in more stable homogeneous structures, thus achieving a better behavior in the wear of the bearings [1, 2].

In literature, effects of the thermo-magnetic treatments applied to quick tool steels (Rp5, AISI/SAE M2) have been studied [3]. This steel – reach alloyed – supported a classic thermal treatment (the improvement) and a thermo-magnetic treatment. It were obtained the same conclusions as in the case of the RUL1(100Cr6) steel for bearings: the hardness of the steel have been increased, the residual Austenite quantity decreased and other mechanical characteristics have been improved too.

It was obtained a positive influence of the magnetic field applied during (total or partial) the improvement treatment of the steels.

In the chemical composition of these types of steels, there are both: a large proportion of carbon and the carburizing alloying elements (W, Mo) contents capable to give birth to carbides of the M6C type - most important for hardening by dispersing precipitation and for heat stability. Before the hardening treatment, the structure of the steels for tools is a structure similarly with the structure obtained after the globular annealing treatment (as in the case of the steel which will be analysed in this paper).

The initial structure of the Rp5 steel was formed by a base of Ferrite (F) alloyed with the following elements: W, Mo, Cr,V, at which were added primarly and secondary carburs as: M_6C, $M_{23}C$, MC, M_3C (M represents the metallic components). In the case of Rp5 steel, Ferrite has a proportion by 71% and the carburs have 28%. Through hardening treatment, 60% from metallic carburs have been disolved in Austenite (A) in the mass of this steel. Austenite (A) was enriched in Carbon and alloying elements. If the temperature increased suddenly- more than 1230°C for Rp5 steel grade, the Austenite was enriched in Carbon until 0,5%, but the hardening temperature have been limited by the increasing of the Austenite boundery [2, 3].

In the presence of the magnetic field, the ferromagnetic germs of martensite generate pressures on the surrounding austenite by magnetostriction. There is an acceleration of the martensitic transformation velocity. If continues to increase the ferromagnetic domains, the magnetostriction effects increase. The internal forces generates limits of separation between Austenite (A) and Martensite (M), permiting the evolution of the transformation A $\rightarrow$M, until a specific label, resulting a smaller content of Austenite (A) [2, 4].

Therefore, the action of the magnetic field increases the number of martensitic germs in the same volume, resulting a finer martensite than in the classical thermal treatment conditions.

Because of the magnetostrictive internal forces, the diffusion conditions of the atoms of Carbon inside the elementar cells have been created. Results, the tetragonality grade of Martensite decreases. Also, the fragility of the steel decreases [1- 4].

A higher quantity of Martensite was obtained during the tempering of a 13Cr6Ni2Mo supermartensitic stainless steel (X2CrNiMoV13-5-2) using magnetic field [5]. It can be observed that each phase of heat treatment in the magnetic field influences the durability in a different way: hardening treatment in the magnetic field leads to an increase in durability by 49% compared to classical treatment. The high tempering treatment with cooling in magnetic field determines an increasing of the durability by 40%, comparing to the classic treatment.

On the other hand, if Aluminum and Chromium contents increase in the structure of the steel the residual austenite quantity decreases more rapidly. The Martensite quantity and implicit, the hardness of the steel increase significantly, more than in the case of the steels with approx. the same content of Carbon but with lower quantity of Aluminum. As a consequence, the magnetic field intensity, the content of the Carbon and the content of the Aluminum from the steels have an important influence. Because of these aspects, the tendency of breaking decreases and the probability of the fragile breakage no longer exists [2, 6].

Magnetostriction determines local oscillations and results local plastic deformations [2-5].

The variation of the number of germens of Martensite depends by the content of the Carbon and by the intensity of the magnetic field applied during the heat treatments [2].

In table 1 were presented these values variations.

Table 1. The variation of the number of germens of Martensite [2, 3]

Germens number (M)	0.03%C	0.4%C	0.6%C	0.8%C	1%C (100Cr6)	1.6%C
H = 0.8 MA/m	1.11	1.34	1.18	1.18	1.20	1.35
H = 1.6 MA/m	1.23	1.30	1.34	1.39	1.44	1.81
H = 4MA/m	1.68	1.90	2.06	2.26	2.46	4.35

2. Experimental procedure

For the experimental procedure, the samples have been realized as rollers from a steel grade for the improvement treatment, with the following chemical composition principal content: Fe 72.89%, 0.38 % C, 1.02% Al, **1.38** % Cr, 0.17 % Mo, 0.50 % Mn, 0.25 % Si, 0.26 % Ni, 0.02 % P, 0.02 % S. The existence of the Molybdenum content in the composition of the steel decreases the stiffening phenomenon.

The first stage from the complex program of treatments consisted of thermo-magnetic treatments. The treatment t1 represents a hardening treatment (at 920 °C) followed by a high tempering (at 620°C), being a classic improvement treatment.

The treatment t2 represents a complete martensitic hardening treatment in weak alternative magnetic field (with cooling in water) and high tempering treatment (with cooling in water, in strong magnetic field - Alternative Current (A.C.). In this case, the sample was introduced in the centre of the electrical coil located in the walls of a cylindrical oven. All treatment have been made completely in magnetic field (A.C.) and the results obtained through the treatment t2 have been presented in figure 1.

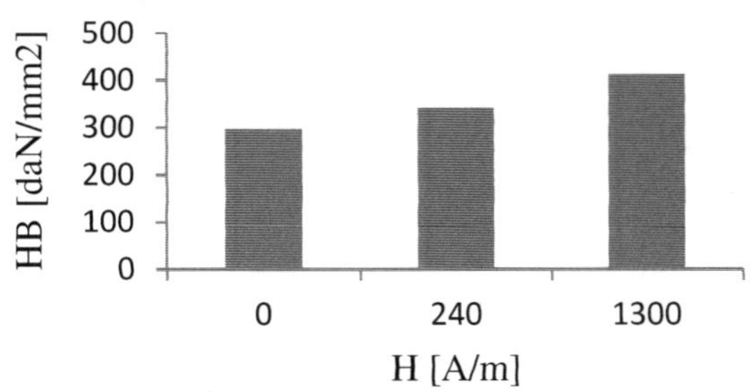

Figure 1. The hardness of the steel (HB) depending on the intensity of the magnetic field (H), in the case of the thermo-magnetic treatment T2

In figure 1 was presented the effect of the intensity of magnetic field (H) on the hardness of the steel (HB) in the case of a complete martensitic hardening process in weak alternative magnetic field (with cooling in water) followed by a high tempering treatment with cooling in water, in a strong magnetic field - Alternative Current (A.C.).

It must be mentioned that all these treatments have been made completely in magnetic field (A.C.).

There were considered different values for the intensity of the magnetic field. A higher intensity of the magnetic field (H) applied during treatment implies a better hardness of the steel (HB), in the case of the treatment t2.

The treatment t3 represents a thermo-magnetic treatment just with cooling in water in magnetic field-Alternative Current (A.C.).

Treatment t4 represents a thermo-magnetic treatment corresponding to magnetic field- Direct Current (D.C.) applied just during the cooling in water.

The second stage of the treatments presumes the existence of the ionic nitriding treatment (plasma nitriding) applied at 530 ° C. This treatment was applied after the improvement treatments, a classic one (T1) or in magnetic field: Alternative Current (T3) and Direct Current (T4) treatments. T2 = t2 and plasma nitriding. The temperature of the thermo-chemical treatment was considered at 530 ° C, being specific for this kind of the improvement steel [7 - 9]. In case of a low temperature (100 ° C – 510 ° C), plasma nitriding produces the expanded austenite (the S-phase) with good behavior at friction process. The phase γ (Fe_4N) appears at higher temperatures than 500 ° C and reduce the thickness of the S-phase. In this paper it was considered an improvement steel grade alloyed with Aluminum and Chromium and the treatments temperatures have been chosen for this case [6].

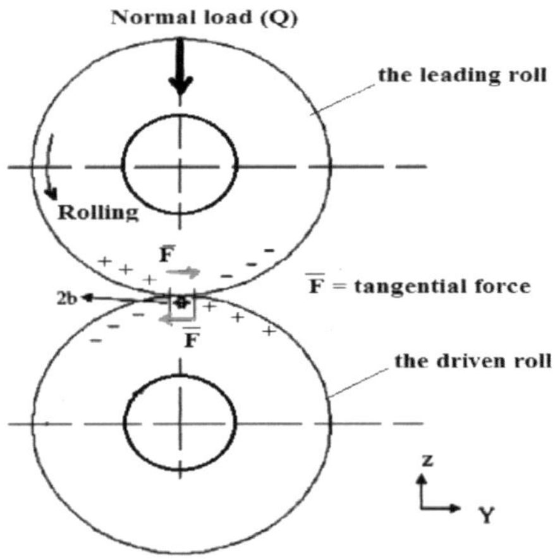

Figure 2. Rolling-sliding configuration: "+" represents tensile stresses; "-" represents compressive stresses, in the case of the wear tests

The increasing of the depth of the superficial layers in the case of unconventional treatment applied has been reported in accordance with the depth of the superficial layer for the same steel which suffered a classic improvement treatment before a thermo-chemical treatment.

The wear tests have been made on Amsler machine, type roller on roller, with different diameters to obtain different sliding degrees (for example, for ξ = 20%, the outer diameter of the conducting roller had 43.9 mm and the second roller had 40 mm. Rolls width was 10 mm. The wear moment Mf = 40 daN·cm, the normal load (Q) was 150 daN, the friction coefficient (μ) was 0.121, the lenght of friction zone of the rollers (Lf) was 74.606 m. Figure 2 presents the scheme of the wear tests.

3. Results and Discussion

After each hour of wear tests, during three hours, for each sample, the calculations have been made depending on diffractometric analysis. The results [2] presented in Table 2 have been obtained.

Table 2. Results obtained during three hours of wear tests [2]

Code samples	**~ % I_{Fe3N}**	**~ σ_{II} B_{Fe3N}**	**~ % I_{Fe4N}**	**~ σ_{II} B_{Fe4N}**	**~ c/a B_{211}**	**~ σ_I Δθ**	**t [h]**	**Q [daN]**	**ξ [%]**
322(T1)	4	4	20	1.81	2.526	-	0	-	-
322(T1)	6	4.40	26	2.40	2.526	-	1	150	20
322(T1)	4	4	20	1.81	2.260	-	2	150	20
322(T1)	20	5	26	3.33	2.260	-0.10	3	150	20
CA1(T3)	16	1.50	16	3.00	1.750	-0.10	0	-	-
CA1(T3)	15	2	17	2.98	1.750	-	1	150	20
CA1(T3)	18	4	18	3	2	-	2	150	20
CA1(T3)	18	4	18	2.70	2.160	-0.15	3	150	20
CA0(T4)	10	3.20	12	3	-	-	0	-	-
CA0(T4)	6	2.94	10	3.20	-	-	1	150	20
CA0(T4)	5	3	8	2.70	-	-	2	150	20
CA0(T4)	5	3	8	2.70	-	-	3	150	20

IOP Conf. Series: Materials Science and Engineering **485** (2019) 012020 doi:10.1088/1757-899X/485/1/012020

The contact geometry of the friction couple (roller on roller), the technological parameters (surface quality, thermo-chemical treatment) and the exploitation conditions (for example, the thermal solicitation) are the factors which influence this kind of wear process [6, 7, 8].

Phases from superficial layer versus the wear tests duration corresponding to the non-conventional treatment in magnetic field – Direct Current T4, have been presented in figure 3.

In figures 4 and 5, the evolution of phases Fe_3N and Fe_4N from superficial nitride layer and the degree of tetragonality of Martensite (c/a) have been presented, for case of classical treatment without the magnetic field influence (T1) and respectively, for non-conventional treatment in magnetic field – Alternative Current (T3).

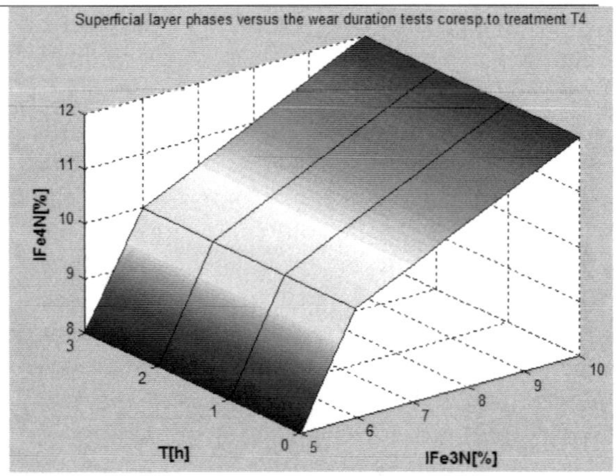

Figure 3. Phases distribution depending on the wear tests duration corresponding to the treatment T4

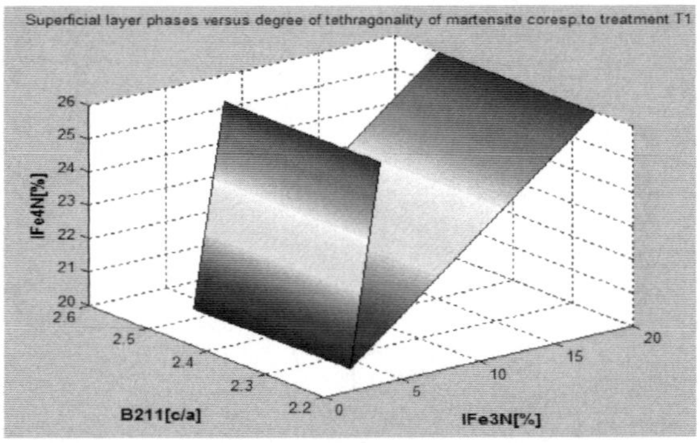

Figure 4. Phases distribution and degree of tetragonality of Martensite (B_{211}), corresponding to the treatment T1

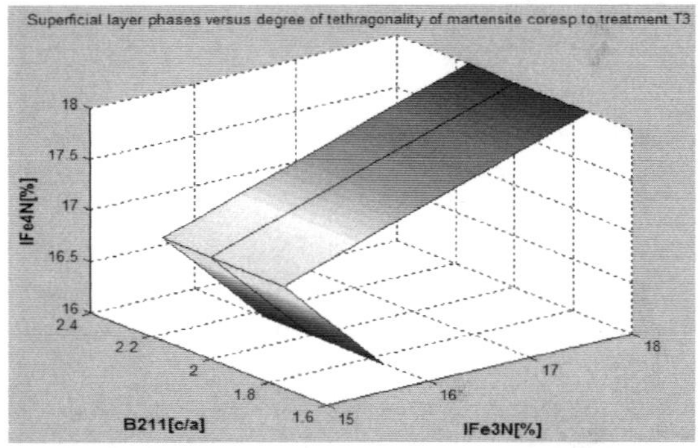

Figure 5. Phases distribution in superficial layer and the degree of tetragonality of Martensite corresponding to the treatment T3

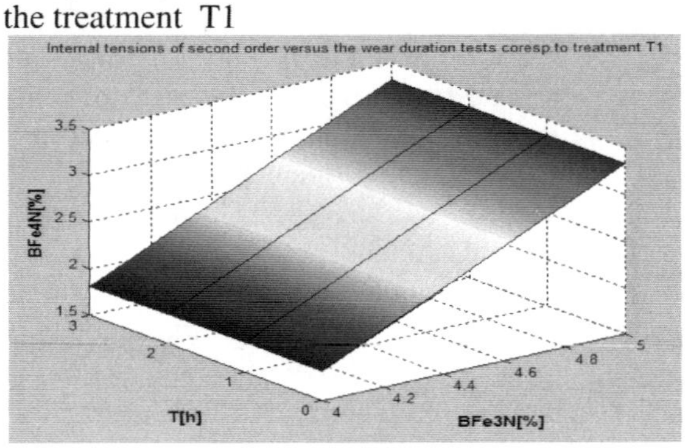

Figure 6. Internal tensions of second order depending on the wear tests duration corresponding to the treatment T1

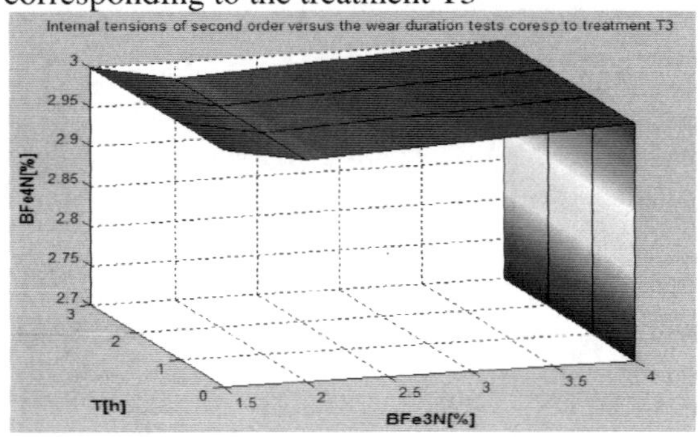

Figure 7. Internal tensions of second order depending on the wear tests duration corresponding to the treatment T3

The hardness of Martensite (M) leads to high mechanical strength and low plasticity which are the result of the complexity of its structure, a solid solution supersaturated with interstitial carbon and tetragonal network formed by processes of sliding. It's important to study the degree of the tetragonality of Martensite which influences the hardness of the superficial layers.

IOP Conf. Series: Materials Science and Engineering **485** (2019) 012020 doi:10.1088/1757-899X/485/1/012020

Figures 6 and 7 presents the evolution of the internal tensions of second order depending on the duration of the wear tests, in both cases: in classical treatment regime T1 and in the case of the non-conventional treatment T3. The internal tensions of the phases determine a lower resistance of the Fe_3N phase during the cyclical fatigue.This fact was determined by mechanical oscilations created by the magnetic field (Alternative Current) through the permanently changes of the field lines directions, because of the magnetostriction existence. The magnetostriction - caused by the magnetic field - changes the re-crystallization conditions and the speed of germination. In this sitation, the hardness of the superficial nitrided layers increased. The degree of tetragonality of Martensite (c/a) was higher in the case of γ'-Fe_4N (see figure 5).

After plasma (ion) nitriding, the nitrided superficial layer had a higher depth in the case of applying the magnetic field. Figure 8 present the evolution of the micro-hardness values in superficial layer, depending by the treatment, following the thickness of the white layer [6].

The following notes have been used:
HVa = micro-hardness in the case of the treatment T3;
HVc = micro-hardness in the case of the treatment T4.

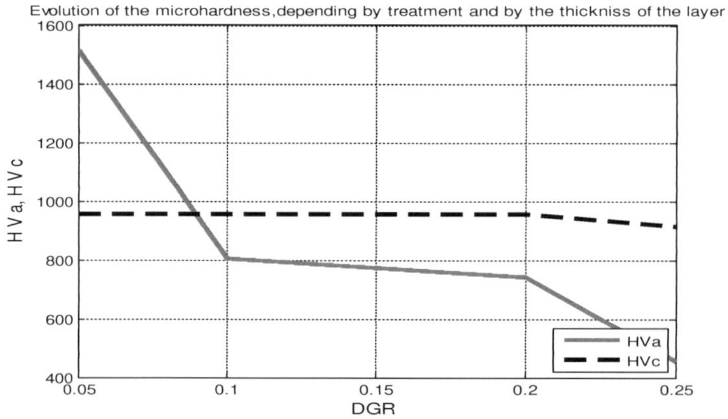

Figure 8. Micro-hardness of the nitrided layers in the case of the non-conventional treatments T3 and T4

It can be observed an improvement of the mechanical properties, for example the micro-hardness of the superficial layer increased in the case of the non-conventional treatment (T3) because of the distribution of the γ'-Fe_4N phase, especially in the case of the incresing of the normal load (Q) and for the increasing of the sliding degrees (ξ). The structural and magnetic properties of epitaxial γ'-Fe_4N iron nitrides films have been investigated by Costa – Krämer J L, Borsa D M, and others, in [14] and they explain why this phase is so important. According with [7, 10], the magnetization properties described are so far consistent with a single phase of epitaxial film, having a cubic structure and a positive anisotropy constant, i.e., the [100] directions are easy magnetization axes. All the magnetic characteristics presented so far are then dictated by the value of the anisotropy constant. This value can be estimated from the hysteresis loops obtained applying the field along a hard magnetization axis, if the value of the saturation magnetization is known [6, 8, 10-13].

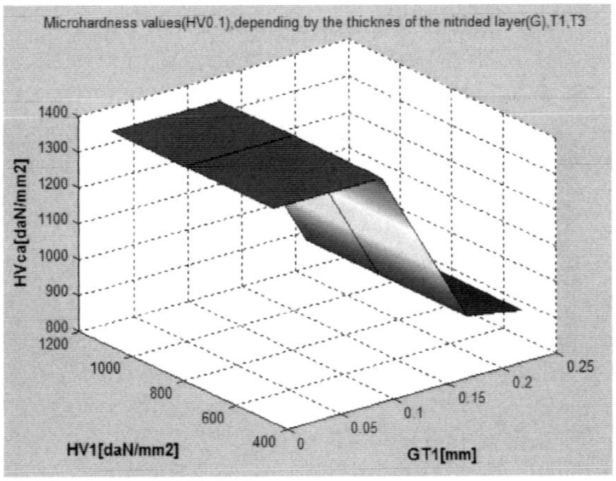

Figure 9. Microhardness values of the white layer after the treatments T1 (HV1) and T3 (HVca) depending on the thickness of the layer (GT1), before the wear tests.

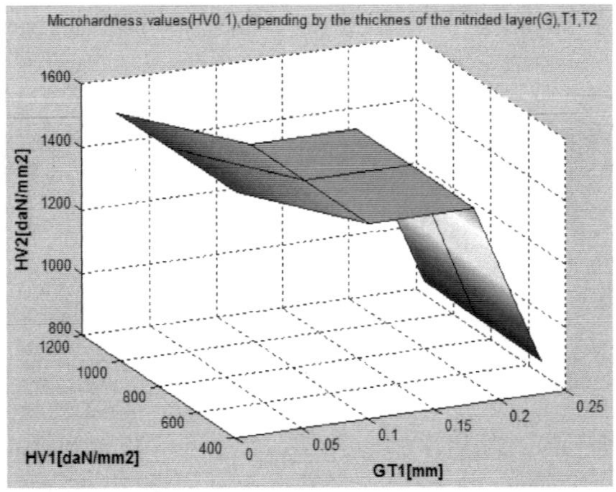

Figure 10. Microhardness of the white layer after the treatments T1 and T2 depending on the thickness of the layer (GT1), before the wear tests [11].

Figures 9 and 10 presents the evolution of the the microhardness values after the treatments T1 and T3 respectively T1 and T2 depending on the thickness of the nitrided layer (GT1) obtained through plasma (ion) nitriding, before the wear tests.

4. Conclusions

After the first stage of the treatment the hardness of the steel increased with 38% in the case of the non-conventional treatments (alternative current-magnetic field) because the Martensite amount increased, comparing to classic treatment case. The micro-hardness of the superficial layer of steel increased with approx. 20 - 30% in the case of plasma nitriding applied after thermo-magnetic treatment, depending on the treatment regime applied, comparing to the classic treatment applied on the witness samples.

From the diffractometric point of view, the samples present changes regarding the content of Fe_3N and Fe_4N phases in the thermo-chemically treated superficial layer, during the wear process, considering the non-conventional thermo-magnetic treatment applied as a basic treatment, before plama nitriding and comparring to classic treatment. This situation implies changes in structure and implies the increasing of the hardness of the material, including in superficial layers thermo-chemical treated after thermo-magnetic treatments. For example, Fe_4N amount increasing implies a higher micro-hardness in superficial layers of the steel and a higher mechanical resistance at friction wear tests. Magnetic field influences positively the amount of Fe_4N and mechanical properties of the steel.

References

[1] Gheorghies C and Stefanescu I 2010 *Journal of Iron and Steel Research, International* 17 (9) 46-52, https://doi.org/10.1016/S1006-706X(10)60141-0]

[2] Papadatu C P 2005 Researches on improving the properties and reliability of some steel grades used for manufacturing metallurgical equipment *PhD. Thesis* Dunarea de Jos University Galati, Romania

[3] Baltes L 1998 Researches regarding the thermo-magnetic treatments of the tools steel *PhD. Thesis* Brasov, Romania

[4] Dulucheanu C, Bancescu N and Severin T 2011 The influence of quenching medium on the structure and mechanical properties of the dual-phase steel with 0.99% C and 1.9% Mn *Metalurgia international* **XVI** (Special Issue)

[5] Bojack A, Lie Zhao, Morris P and Sietsma J 2014 In Situ Thermo-magnetic Investigation of the Austenitic Phase During Tempering of a 13Cr6Ni2Mo Supermartensitic Stainless Steel, *Metallurgical and Materials Transactions A* **45** (13) DOI: 10.1007/s11661-014-2551-3

[6] Papadatu C P 2017 Experimental studies regarding wear processes through dry friction of the superficial layer for an unconventional treated steel, *IOP Conference Series: Materials Science and Engineering* **174** (1) 012014

[7] Papadatu C P and Vasilescu E 2009 *Some aspects regarding the influence of non-conventional treatments on the superficial layers nitro-carburized characteristics,* 16th International Metallurgical and Materials Conference, Metal 2009, 22-25 May 2009, *Proc.Int.Conf. on Materials and Metallurgy* (Ostrava) Czech Republic pp. 563-568, ISBN 978-80-87294-03-01

[8] Papadatu C P 2007 *Posibilities to improve the quality of some steel grades from metallurgical industry (Posibilitati de crestere a calitatii unor oteluri utilizate in industria metalurgica)*, Dunarea de Jos Foundation Publishing House Galati, Romania, ISBN 978-973-627-371-1

[9] Levcovici S, Vasilescu E, Gheroghies L a. al. 2004 *Surface Engineering (Ingineria suprafetelor)*, Didactic and Pedagogic Publishing House, Bucharest, Romania, ISBN 974-30-2969-6

[10] Cavaliere A, Perrone A and Silvello 2016 *Engineering Science and Technology*, an International Journal **19** 292-312 http://www.elsevier.com/locate/jestch, Science Direct

[11] Papadatu C P, Sandu A V, Bordei M and Sandu G 2017 *Revista de Chimie* (Bucharest) **68** (4) 675-679 http://www.revistadechimie.ro

[12] Gheorghies C, Papadatu C P and Stefanescu I 2004 *Study on the model of behaviour of some non-conventional treated steels, during friction process*, *Proc.Int.Conf. on Metallurgical and Materials, R.P.D.(Rapid Product Development Portugal)*, CENTIMFE Marinha Grande, vol.6, Pp.1/6-6/6, Portugal, ISBN972-98872-2-5

[13] Papadatu C P, Cotet A, Vasilescu E and Andrei G 2016 *Experimental Researches regarding the possibility to increase the durability of a CrMo Alloyed Steel subjected to some Non-conventional treatments applied in magnetic field, Proc.Int.Conf. Metal 2016*, 25th Anniversary International Conference on Metallurgy and Materials (Brno) Czech Republic pp. 1133-1138

[14] Kosta-Krämer, Borsa D M at al. 2004 Structure and magnetism of single-phase epitaxial γ'-Fe4N, *Physical Review* **B** 69, published by The American Physical Society pp.144402-1…144402-8

[15] Papadatu C P and Vasilescu E. 2012 Some Aspects regarding the Influence of the Tribological on the Superficial Layers of Treated Steels with Plasma Nitriding, *Annals of Dunarea de Jos University* Fasc.IX **4** (1) 310-317

[16] Vasilescu E and Doniga A. 2012, *Modern Approaches in Design Optimization of Parts and the Development Trend of Surface Treatments, Annals of Dunarea de Jos University* Fasc.IX, **1** (1) 30-36.

IOP Conf. Series: Materials Science and Engineering **485** (2019) 012021 doi:10.1088/1757-899X/485/1/012021

Aspects regarding the plasticity of the steel subjected to non - conventional treatment

C P Papadatu and S Ciortan

Dunarea de Jos University of Galati, Faculty of Engineering, 47 Domneasca street,Galati Romania

papadatu.carmen@yahoo.com

Abstract. Plasticity is an important property to determine the capability of the steel to support plastic deformations, to obtain important pieces for industry. It is important to know the optimal cooling regime for this material because the steel have been rolling at high temperature or, the steel have been heated at high temperature after lamination process. It was considered a steel grade with a low carbon content, which can be find often in machine industry and in metallurgical operations. Three different mediums for directing of the cooling process have been considered and the properties of the steel have been modified. This study tried to determine a way to improve the properties of this kind of steel to obtain a higher durability corresponding to the exploitation conditions.

1 Introduction

The plasticity of the steel depends on its deformability. According to the literature, the deformability of the metallic materials characterizes their ability to be deformed permanently without breaking the inner linkages. Deformable metals and alloys are elasto-plastic-viscous bodies.

Temperature is a factor that determines the physico-chemical state of the alloys. With increasing temperature, the speed of the modifying of the properties through deformation increases and the possibility to appear micro-cracks in the deformation decreases.

Studying Hanning 's and Boulger's deformability variations depending by temperature, it can be seen that the temperature of deformability around (250-300) °C is due to the formation of precipitates that favor the fragility of "the tempering to blue". The sudden increasing in deformability between 600°C and 800 °C is due to inter-crystalline displacements and re-crystallization of steel. At the same time, the deformability of the steels depends on the carbon content [1, 2].

In the case of the steels with 0.1% C, 0.48% Mn, 0.24% Si, the micro-structural parameters which determine the mechanical properties of the steels are the sizes and the shapes of pearlitic (P) and ferritic (F) grains which form its microstructure. Ferrite has a low mechanical strength but it has good plasticity properties. Cementite (Cem) in steels has a good hardness but it is fragile [3, 4]. Cementite can cause cracks. Therefore, with the increasing of the carbon content, the hardness and the strength of the steels increase and the plasticity properties decrease [3, 5, 6-8].

IOP Conf. Series: Materials Science and Engineering **485** (2019) 012021 doi:10.1088/1757-899X/485/1/012021

The steels with low carbon content have high plasticity properties below 550℃ and the thermal stresses will not leads to the formation of cracks [1]. Damage in the case of materials represents the progressive or sudden deterioration of their mechanical strength which conducts to dangerous effects for products [1-4]. It is important to know how it is possible to improve the properties of the steels [3].
The paper focuses on directed fast cooled steels strips for automotive application after hot laminated process [3].

2. Technology of rolling of large thickness sheets before non-conventional treatment

In order to collect the samples, it was considered a scheme of hot rolling of steel sheet. The samples had the following dimensions: 10 x 150 x 200 mm and came from a slab with the dimensions: 150 x 1500 x 2500 mm, in the case of a thick sheet rolling consisting of one system type quarto (two pairs of rolls - four Rollers Cylinders), with a sheet length maximum of 3000 mm and a vertical rolling system, placed in front of the first system. The hot rolling scheme is shown in Table 1 [1].

The mill from Mittal Steel Galati [1] consists of a reverse Roll Mill type "Quarto" and the equipment for straightening, cutting and marking of finished products.

Table 1. Hot rolling scheme of the steel [1]

Nb. of passes	Type of rolling/ Passes/ crossing	Thick - ness [mm]	Width [mm]	Length [mm]	Absolute reduction ratio [mm]	Relative reduction ratio [%]	Reduction ratio coefficient γ	Obs.
0	-	150	1500	2500	-	-	-	-
1	Pressing/ Forge rolling	150	1450	2600	-	-	-	Rotation 90°
2	Rolling on width	185	2600	1750	23	16.9	1.20	-
3	Rolling on width	105	2070	2090	20	16	1.19	-
4	Pressing/repressing Forge rolling	105	2050	2620	-	-	-	Rotation 90°
5	Pressing/ Forge rolling	105	2050	2650	-	-	-	-
6	Rolling on length / flatting	85	2050	3250	20	19	1.24	-
7	Rolling on length / flatting	65	2050	4200	20	23.5	1.30	-
8	Rolling on length / flatting	45	2050	6100	20	30.8	1.44	-
9	Rolling on length / flatting	30	2050	9200	15	33.5	1.50	-
10	Rolling on length / flatting	20	2050	13700	10	33.3	1.50	-
11	Rolling on length / flatting	14	2050	19600	6	30	1.43	-
12	Rolling on length	11	2050	25000	3	21.4	1.27	-
13	Rolling on length	10	2050	27500	1	9.1	1.10	-

After the technology to obtain Hot-Rolled Sheet's Steel corresponding to the scheme presented in the table 1, the samples have been debited. The samples supported non-conventional treatments, watching the conduction of the cooling, modifying the speed of cooling after the heating of the steel samples.

The chemical composition of the steel was presented in Table 2.

Table 2. Chemical composition of the steel [1]

Element	C	Mn	Si	Al	P	S	Cu	Mo
Content [%]	0.14	1.51	0.31	0.041	0.01	0.01	0.18	0.009

3. Experimental procedure

After rolling process, according to EN 10025-2:2004, 24 experiments have been carried out – considering three groups of samples and each group supported three types of non-conventional treatments, taking in consideration four values of the initial temperatures.

There were considered three groups of samples of the steel grade and each group had eight samples. The Chemical composition of the steel has been presented in table 2 [1].

The samples were heating at the different temperatures: T1 = 850°C (for the first batch of samples), T2 = 900°C (for the second batch of samples), T3 = 950° C (for the third batch of samples) and T4 = 1000° C (for the last batch of samples). Three cooling regimes [1, 3] have been used: (1) cooling regime in normal conditions; (2) cooling regime in metallic box; (3) cooling in air flow (using a jet of cold air). If the cooling mediums differ, the speeds of the cooling differ too.

Traction tests at the room temperature were carried out according to ASTM: E8M-11 standard [1, 9, 10], while those related to elevated temperatures were carried out according to ASTM: E21-09 [1, 3, 11]. Charpy impact tests for resilience determination were carried out according to ASTM: E23-07ae1 standard [12].The preparation of metallographic specimens was conducted according to ASTM E3-11 standard [13]. All of mentioned standards can be found in Annual Book of ASTM Standards (2012) [1].

4. Results and discussion

In table 3, the cooling speed values for each different cooling conditions and the evolution of some mechanical properties of the steel [1], obtained after the experimental scheme, have been presented.

Table 3. Evolution of some mechanical properties of the steel depending on the cooling speed [1]

Cooling Conditions	Initial temperature [°C]	Z [%]	KV_{-20} (- 20°C)	Cooling Speed values (VR) - Average [°C /min]	KV_{+20} (20°C)	A5 [%]	Rm [N/mm^2]	σ_c (RC) [N/mm^2]
(1)	1000	78	73	5.03	93.5	35	549	385
(1)	950	78.1	83	6.20	91	34.2	547	386.5
(1)	900	73.7	80	6.82	87.5	43	551	396
(1)	850	77	61	7.35	81	32.5	562	406
(2)	1000	77.8	76	3.69	94.5	35.3	558	377
(2)	950	77.5	69	3.88	85.5	34	548	396
(2)	900	77.8	65	4.15	88	36.8	534	392
(2)	850	78.5	75	4.61	89	35.8	562	393
(3)	1000	76.8	56.5	19.62	98	33.8	542	367
(3)	950	76.6	66	20.21	96	34.8	549	440
(3)	900	72.6	85	20.95	98	33.8	555	390
(3)	850	77.8	77	21.84	90	33.2	551	391

Cooling Speed value (VR) – Average [°C/min] was calculated with the following relation:

IOP Conf. Series: Materials Science and Engineering **485** (2019) 012021 doi:10.1088/1757-899X/485/1/012021

$$Vr = \frac{t - t0}{\Delta\tau}, \quad [^{\circ}C/min.] \tag{1}$$

Where:
t = the initial temperature corresponding to each cooling regime, [^{o} C];
t_0 = the final temperature which was approximate to 20 ^{o}C;
$\Delta\tau$ = the cooling duration, [min.].

In some cases, the difference between the initial temperature and the final one dues that the behavior of the evolution of the cooling speed values was special, unexpected.

In figures 1 to 4 have been presented the evolutions of the mechanical properties depending on the cooling speed from the initial temperature, in case of the cooling in normal conditions (1), corresponding to the steel analyzed.

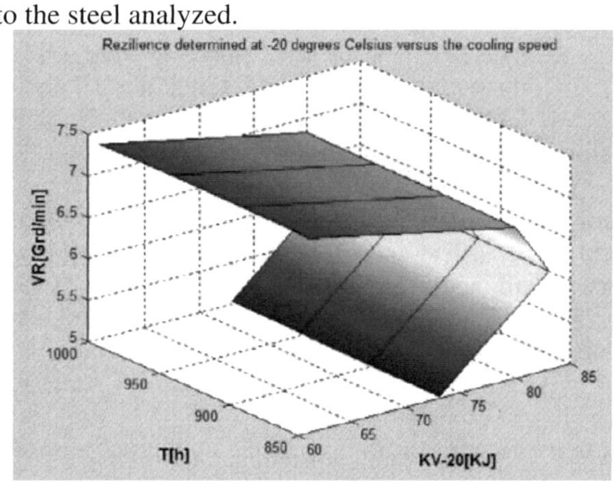

Figure 1. Resilience determined at (-20)$^{\circ}$C depending on the cooling speed (VR) in the case of the cooling regime (1)-normal conditions

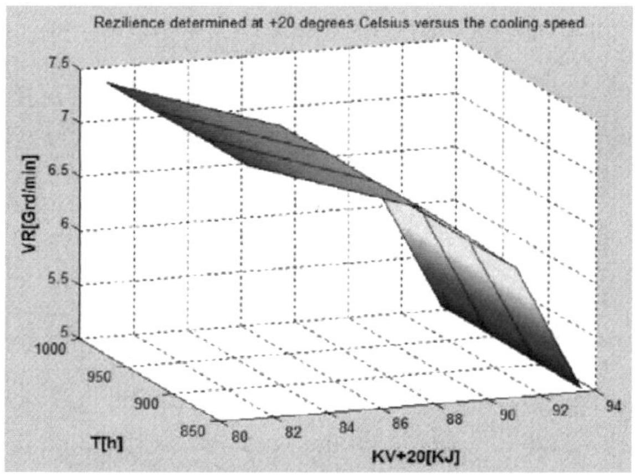

Figure 2. Resilience determined at 20°C depending on the cooling speed (VR) in the case of the cooling regime (1)-normal conditions

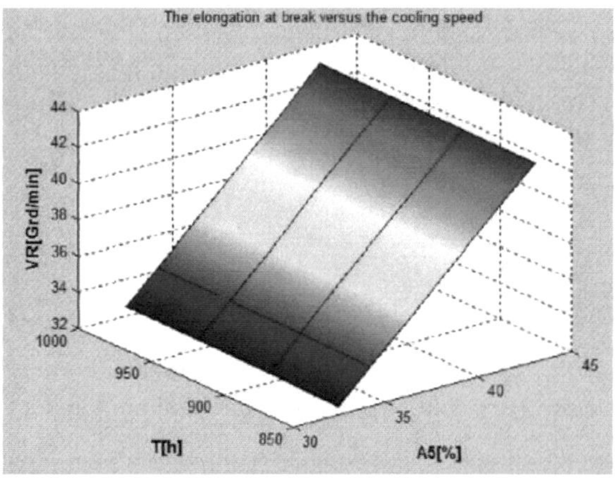

Figure 3. The elongation at break (A_5) depending on the speed of cooling (VR), corresponding to cooling regime (1)

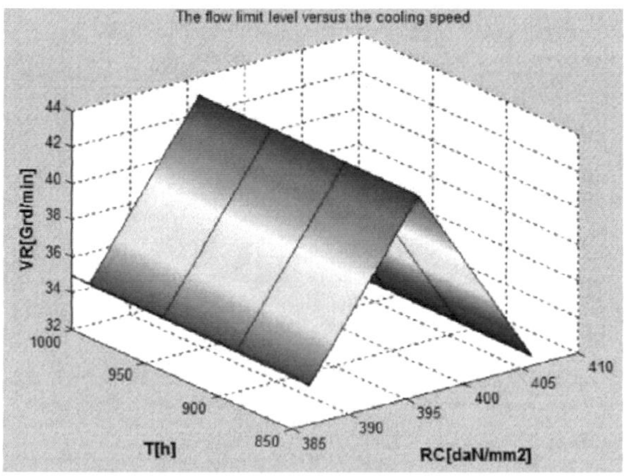

Figure 4. The flowing limit (RC) level depending on the speed of cooling (VR) in normal conditions (1)

In figures 5-8 were represented the evolution of the mechanical properties versus the cooling speed from the initial temperature, in case of the cooling in metallic recipient (2).

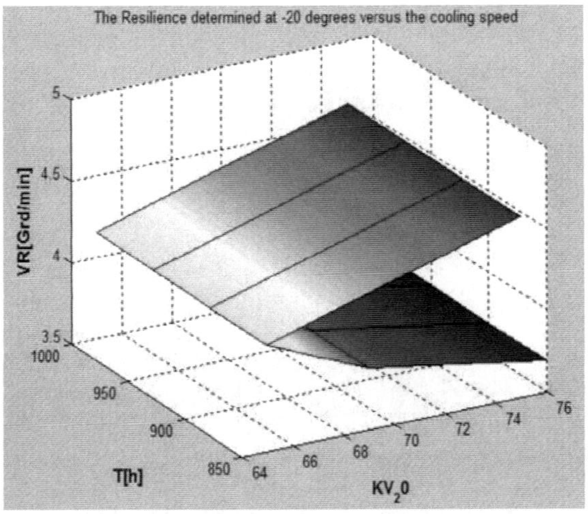

Figure 5. Resilience values determined at (-20)°C depending on the cooling speed (VR) in the case of the cooling regime (2) – in metallic box

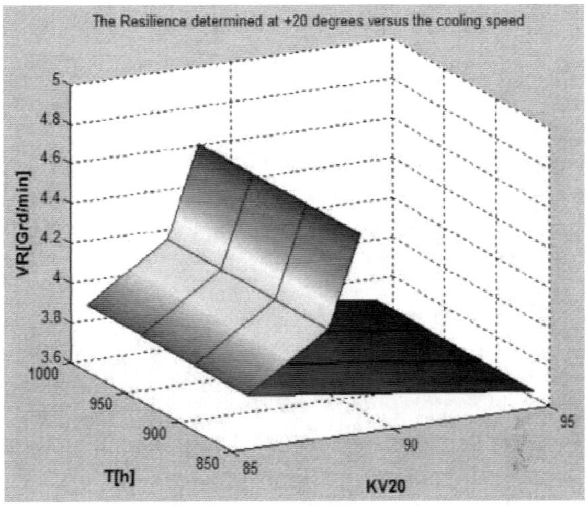

Figure 6. Resilience values determined at +20°C depending on the cooling speed (VR) in the case of the cooling regime (2) - in metallic box

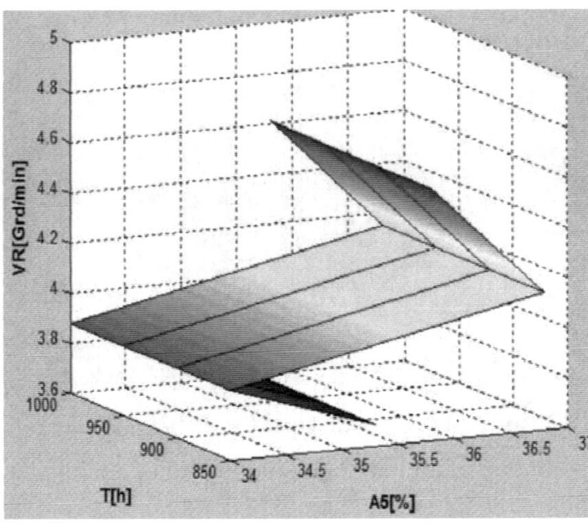

Figure 7. The elongation values (A_5) depending on the speed of cooling (VR), in case of the cooling regime (2), in metallic box

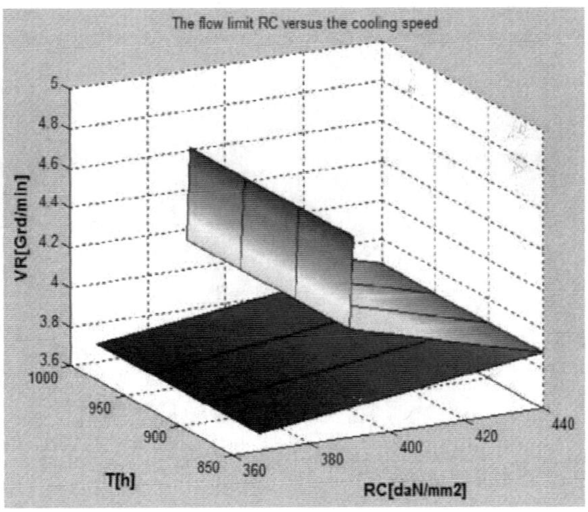

Figure 8. The flowing limit (RC) level depending on the speed of cooling (VR), in case of metallic box (2)

It is noticed that when the samples are cooling in a metal box from an initial temperature of 1000 ° C, a good resilience is achieved at ambient temperature (+ 20 ° C).

In figures 9-12 were represented the evolution of the mechanical properties depending on the cooling speed from the initial temperature, in case of the cooling in jet of air (3), corresponding to the steel analyzed.

IOP Conf. Series: Materials Science and Engineering **485** (2019) 012021 doi:10.1088/1757-899X/485/1/012021

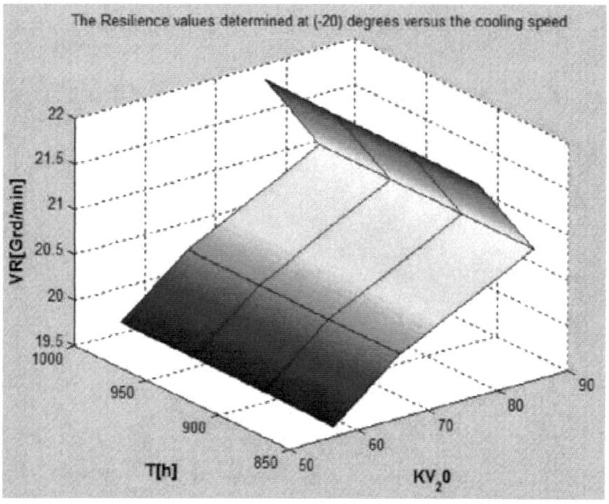

Figure 9. Resilience values determined at (-20)° C depending on the cooling speed (VR) in case of the cooling regime (3) - in cold jet of air

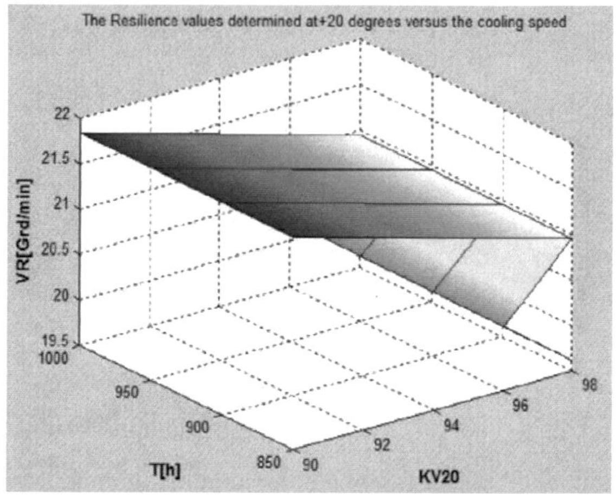

Figure 10. Resilience values determined at +20°C depending on the cooling speed (VR) in the case of the cooling regime (3)- in cold jet of air

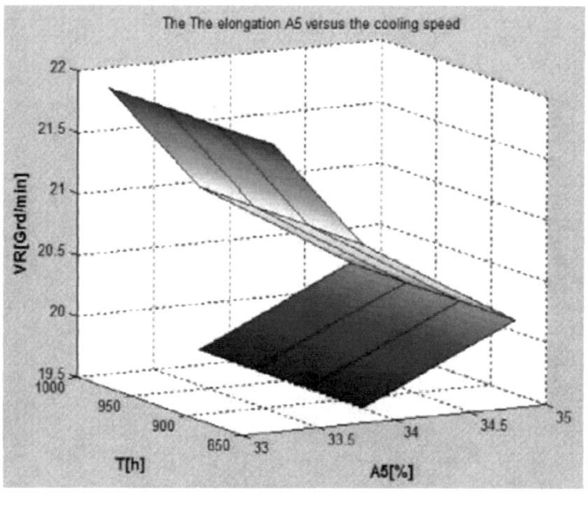

Figure 11. The elongation values (A_5) depending on the speed of cooling (VR), in case of the cooling regime (3) - in cold jet of air [10]

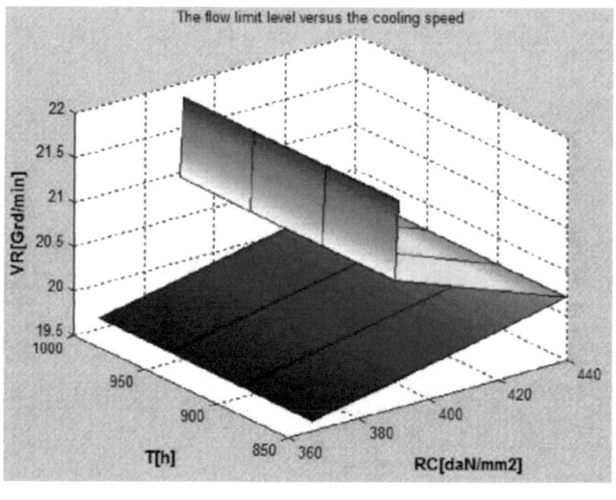

Figure 12. The flowing limit (RC) level depending on the speed of cooling (VR), in case of cold jet of air (3)

The temperature of the rolling of steel influences the characteristics of the steel. If it is applied a heating after rolling and it is studied the influence of the speed of cooling on some characteristics of that steel, it can be obtained a changes of the properties, implicit changes of the plasticity evolution.

It is noticed that a higher degree of thinning (Z) than in the case of the cooling in cold jet of air, have been obtained.

The flowing limit - the highest value - was obtained in the case of jet air-cooling from 950 ° C. In all cases, high flow limits were recorded at the cooling in jet of air from the temperatures of 950 ° C, 900 ° C and 850 ° F°C. After non-conventional heating process and cooling in different mediums, the steel flowing limits have exceeded the values of the same steel after rolling (σ_c = 360 daN/mm^2).

High resilience values determined at +20 ° C were obtained by cooling in an air jet. It can be said that a rapid cooling from high temperatures produces better resilience of the material analyzed, instead, in this case (cooling in the air jet), the elongation (A5) at breakage decreases. The plasticity properties decrease in this case.

The lowest resilience value determined at negative temperatures (-20 °C) was recorded for air-jet cooling from 1000 ° C. The resilience represents the tenacity of the steel. A good tenacity represents a good plasticity of the steel.

Tenacity is the ability of a material to absorb energy and to have plastic deformation before it breaks; the amount of energy absorbed during both deformation and fracture is a measure of the tenacity of the material.

The size of the deformation preceding the tear is a measure of ductility, and the force required to break is a measure of mechanical strength.

Figure 13 presents micro-structural aspects after non-conventional treatments. Microstructure in strings is appropriate to the structure of Hot-rolled steel/

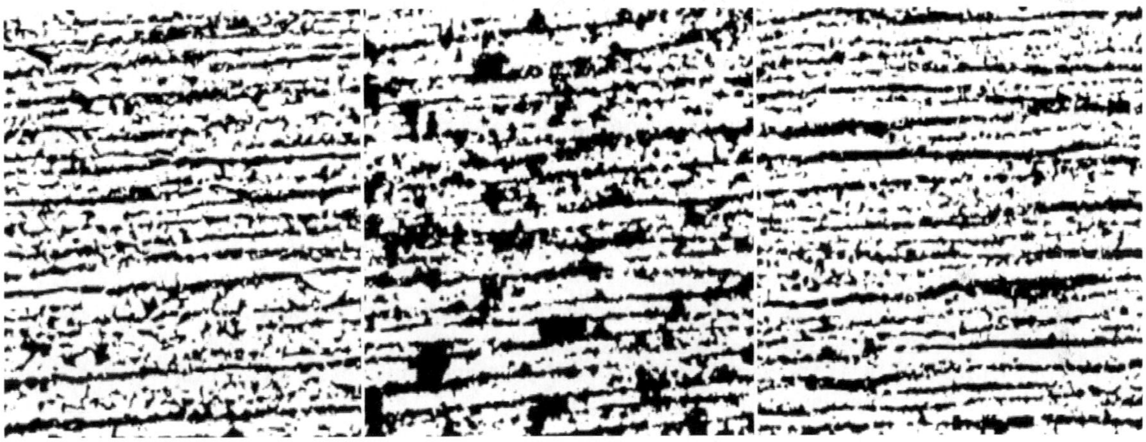

a). b). c).

Figure 13. a) Cooling in jet of air; Ti = 900°C, Nital Attack 2%, x100; b) Cooling in normal conditions; Ti = 900°C, Nital Attack 2%, x100; c) Cooling in metallic box; Ti = 900°C, Nital Attack 2%, x100.

A fine structure of the hot-rolled steel is observed in the case of the cooling in metallic recipient (box) and in the case of the cooling in jet of air. The basic structure is compound by Ferrite (F) and lamellae of Pearlite (P). The ratio is 25/75 % (P/F) and corresponds to figure 13c. The ratio 40/60 % (P/F) corresponds to the microstructure of figure 13b. The ratio 35/65 % (P/F) corresponds to the microstructure of figure 13a.The micro-structures in strings are specific to laminated structures. Manganese (Mn) content more than 2.75% determines the increasing of the tenacity and plasticity of the steel, comparing to the simple casted steel.

5. Conclusions

The experimental program shown that the plasticity characteristics differ appreciably according to process parameters as the cooling speed (VR) during non-conventional treatments and it is in accordance with non-conventional treatment applied. It was shown that the cooling conditions can modify the

IOP Conf. Series: Materials Science and Engineering **485** (2019) 012021 doi:10.1088/1757-899X/485/1/012021

mechanical characteristics of the steel with a low content in Carbon. For example, the highest resilience had been obtained in the case of the cooling in jet of air comparing with the normal conditions of cooling.

References

[1] Papadatu C P 1994 *Reserches and studies regarding the influence of the cooling regime on the structures and properties of the steels*, Research Project, Dunarea de Jos University Galati, Romania
[2] Dziedzic M and Turczyn S 2015 *Designing of Hot Rolling and Cooling Conditions for Dual Phase and Complex Phase Steel Strips, Proc. Int. Conf. on Metallurgy and Metals* METAL (Brno) Czech Republic pp. 182-188
[3] Papadatu C P, Sandu I G, Bordei M and Sandu A V 2017 *Rev. Materiale Plastice* (Bucharest), Romania 54 (4) pp.759-763.
[4] Liu Y G, Li M Q and Luo J 2013 Mater. Sci. Eng. A **574** p 1
[5] Papadatu C P, Sandu I G, Bordei M, Nabialek M, Sandu A V 2016 *Rev. Materiale Plastice* (Bucharest) Romania **53** (4) p. 771
[6] Vasilescu E. and Vasilescu I. 2006 *Nonconventional thermical processes of the rolling flat steel product, Proc. Int. Conf.* on Technology and Quality for Sustained Development TQSD (Bucharest) Romania p310/L1
[7] Angelova D. et al. Frattura ed Integrità Strutturale, 2016 **37** 265-271; DOI: 10.3221/IGF-ESIS.37.35
[8] Geru N 1991 *Analysis of the Structure of the Metallic Materials*, Technical Publishing House, Bucharest, Romania
[9] Papadatu C P, Sandu I G, Bordei M and Sandu A V 2016 Revista de Chimie (Bucharest) Romania **67** (11) p. 2306
[10] ***ASTM E8/E8M, Standard Test Methods for Tension Testing of Metallic Materials: https://www.astm.org/Standards/E8.htm
[11] ***ASTM E9-21, Standard Test Methods for Elevated Temperature Tension Tests of Metallic Materials: https://www.astm.org/Standards/ E21.htm
[12] ***ASTM: E23-07 ae1, Standard Test Methods for Notched Bar Impact Testing of Metallic Materials: https://www.astm.org/Standards/E23.htm
[13] ***ASTM E3-11, Standard Guide for Preparation of Metallographic Specimens: https://www.astm.org/Standards/E3.htm.

IOP Conf. Series: Materials Science and Engineering **485** (2019) 012022 doi:10.1088/1757-899X/485/1/012022

Comparative study of vibration-absorbing materials to improve the comfort of the crew on a river ship

C L Picu[1] **and M E Picu**[1]

[1] Faculty of Engineering and Agronomy in Braila, "Dunarea de Jos" University of Galati, Romania

constantin.picu@ugal.ro

Abstract. The purpose of this paper is to find a damping material for the vibrations generated by the equipment of a ship. That will improve the working and resting conditions of seafaring personnel on a pusher boat that travels over long distances. The experiments were made upstream the Danube. The convoy was made up of a self-propelled vessel and four barges. The water flow rate was 0.76 m/s and there were good visibility conditions and light wind. The accelerations transmitted to man were measured with Seat PAD 01dB triaxial accelerometers mounted on the floor of the ship and connected to the NetdB-Complex system for analysis and measurement. The working frequencies were 0÷100 Hz. Subjects were placed directly on the ship's deck, then on neoprene and rigid foam. It was found that for longitudinal vibrations, the r.m.s. accelerations are 9 times higher than standard if there is no attenuator, 4.8 times higher than standard for neoprene and 2.4 times higher in the case of rigid foam. For transverse vibrations, r.m.s. accelerations are 10.8 times higher than standard when there is no attenuator, 6.4 times higher in the neoprene case and 2.7 times higher for rigid foam. For vertical vibrations, accelerations r.m.s is 5.9 times higher than standard if there is no attenuator, 3.4 times higher for neoprene and 1.7 times higher for rigid foam. The results are similar in the case of transmissibility determination. It was shown the importance of a vibration attenuator for the health, comfort and implicitly for labour productivity of river sailing personnel.

1. Introduction

The mechanical vibrations generated by equipment during work lead to the occurrence of occupational diseases. The admissible vibrations limits depend on their physical characteristics: frequency, amplitude, acceleration, speed. Vibration generation can be: desirable (vibratory tools and machines) and undesirable (technical failures, means of transport).

Vibrations are transmitted directly to the worker's body in two main ways:

a) Action on the whole body (Whole Body Vibrations) through the lower limbs (if the worker stands in orthostatic position on the surface that vibrates: ground, floor, platform, etc.) or through the part of the body that is sitting on a chair that vibrates and through the lower limbs of pedals, etc. (if the worker is in the sitting position) with a frequency of 0÷20 Hz.

b) The action on the hand-arm system (Hand-Arm Vibrations), a system that supports, pushes, guides the vibratory (pneumatic) tools with a frequency of 20÷200 Hz.

Vibrations act on: vibration-sensitive receptors; internal organs in the abdominal and pelvic cavity; the osteoarticular system (mainly on the spine); directly on the muscles, tendons, etc. [1-6]:

a) Occupational diseases due to professional vibration with a frequency of 0÷2 Hz (motion sickness) occur in the personnel who work in aeronautical or maritime transport, etc. Linear movements are transformed into elliptical movements at head due to reflex reactions of the head, causing vestibular stimulation, responsible for clinical manifestations.
b) Occupational diseases due to professional vibrations with 2÷20 Hz frequency (general vibration illness) occur in the following professions: trucks, tractors, excavators, bulldozers drivers, etc. These vibrations lead to rhythmic movements of the organs in the abdominal and pelvic cavity, as well as to the displacements of the vertebra.
c) Occupational diseases due to professional vibrations with 20÷1000 Hz occur in all professions that use vibratory tools and machines that act on the hand-arm system: miners (due to pneumatic equipment), workers in the machine-building industry used for riveting, moulding casting etc.

For example, American sailors are exposed to a very high number of risks due to mechanical vibrations; pneumatic tools are one of the potential exposure hazards for HAV [7], but WBV is a negative factor [8].

Another example is given by Netudykhatka Oiu [9], which presents the connections between social, psychological, economic, physiological and hygienic factors, between age and occupation. Navigation for longer than 6 months, continuity, vibration, air temperature, noise, chemical factor, ship's diet, atmospheric humidity and motion air lead to deterioration of the seamen's physical and mental state: excessive fatigue, low working capacity, restless sleep, abnormal irritability, depressive state.

Chan et al., 2006 [10] shows that, in addition to all these presented factors, smoking and drink (outside the ship) contribute to the deterioration of the crew' health.

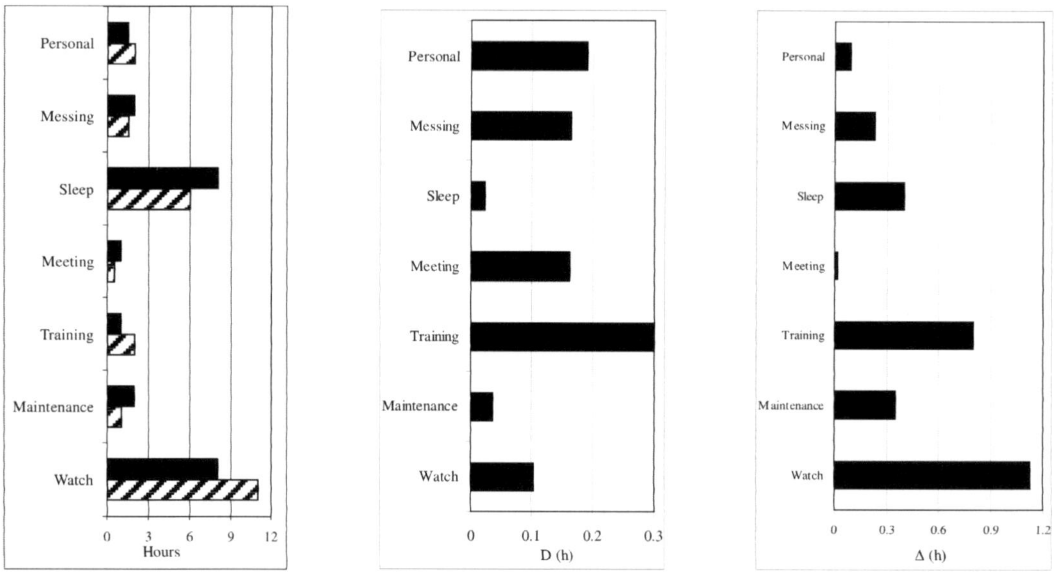

a) Real hours spent on ship **b)** Difference (D) **c)** Deviation (Δ)

Figure 1. Reported activities of sailor as compared to NSW
(■) – Real, (////) Navy standard

In this paper we will refer to the vibrations transmitted to the navigational staff of a riverboat (a pusher). In this case, things are different from on land workers because the sailing personnel besides working on the ship they are also living there for several weeks. In other words, their activities are divided into several categories: surveillance, maintenance, training, meeting, sleeping, free & personal time [11]. These categories do not comply with the periods indicated by the legislation in force [12], mostly to the detriment of sleep and personal time (Figures 1.a) and 1.b)). This is caused by the lack of personnel.

Figure 1.c) illustrates the deviation of sailor from the Navy Standard Workweek (NSW). The deviation was calculated using the following formula:

$$\text{Deviation} = \frac{(\text{Reported - Alloted})^2}{\text{Alloted}} \tag{1}$$

Sailor shows the greatest deviation from the Navy Standard Workweek (NSW) in the categories of standing watch and training [13].

The purpose of this paper is to find an attenuating material for the vibrations generated by a ship's equipment to improve the working and resting conditions of seafaring personnel on a long-distance Danube pushing ship.

2. Materials and methods

The experimental determinations took place in August 2018 on the upstream Danube, between Orşova (km 954) and Baziaş (km 1072), Romania (Figure 2), on certain segments of approximately 1 km long.

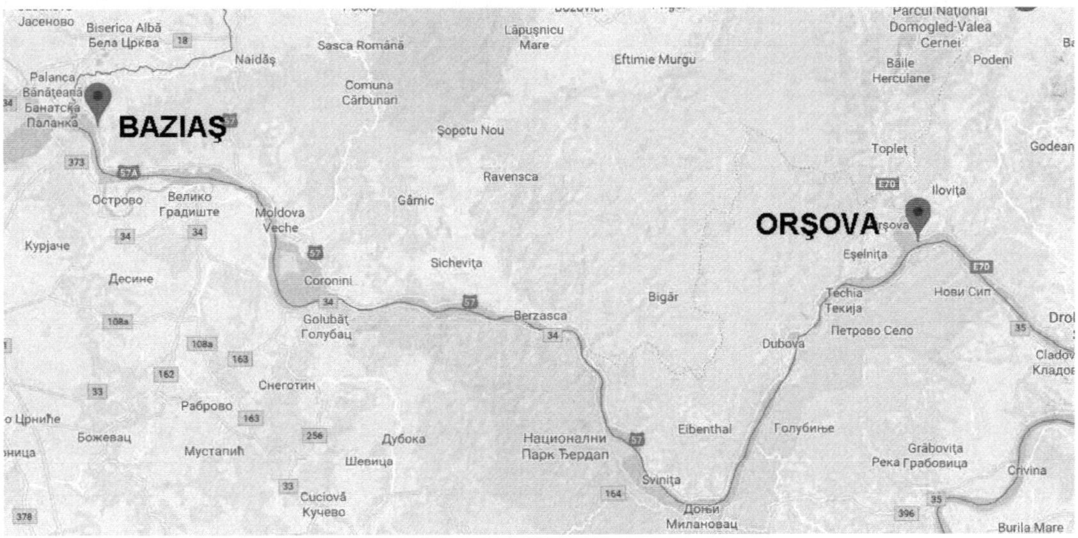

Figure 2. The distance over where measurements were made on the Danube: Orşova (km 954) and Baziaş (km 1072)

The convoy was made up of a propelled vessel and four barges coupled in front of the pusher (Figure 3). The characteristics of the vessel are: length 20.73 m, width 7.78 m, total height to the main deck is 3.30 mat the bow and 3.70 m at the stern; maximum draft 1.50 m; fresh water displacement of 152 tons and MP 2x500 hp. The propulsion system of the ship consists propellers in modern nozzles with rounded rear edges. Barges have a length of 70.05 m, width of 11.60 m, a useful weight of 1398 t.

The characteristics of the convoy are: length 160.83 m, width 23.20 m, draft 2.42 m, maximum load capacity 5512 t, real load 4614 t (83.71% - including pusher), maximum displacement maximum 6844 t, real displacement 5900 t. Speed of the water flow was 0.76 m/s, with good visibility and light wind (0÷5 m/s). Measurements were made over a distance of 118 km, over certain segments of approximately 1 km long. The time corresponding to each analyzed segment was 7÷8 min. The speed over the ground was approximately 7.92 km/h and over water approximately 10.66 km/h.

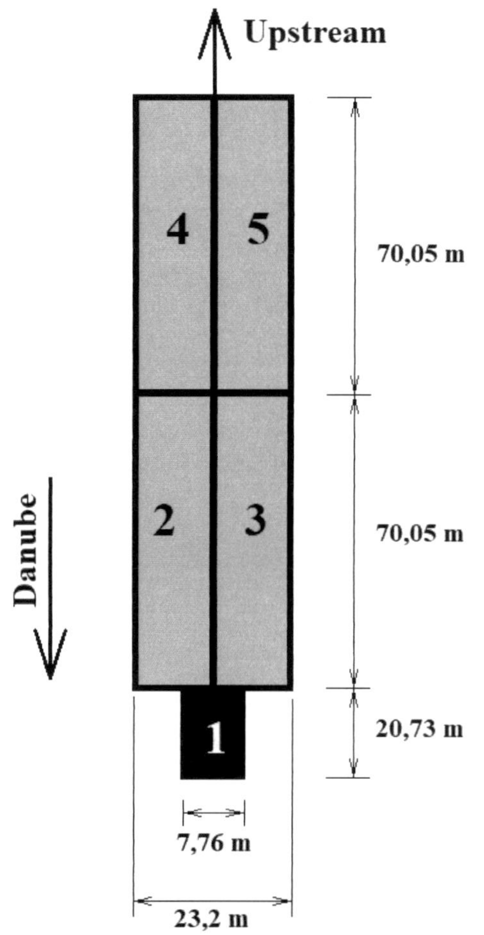

Figure 3. Convoy: (1) – Pusher; (2÷5) – barges (upstream)

The accelerations transmitted to man were measured with Seat PAD 01dB triaxial accelerometers mounted on the floor of the vessel and connected to the NetdB - Complex system for the analysis and measurement of vibration to the human body. Data was processed with the dBFA Suite - 01dB Control Software for data acquisition and post-processing. The working frequencies were 0÷100 Hz. Subjects were placed directly on the ship's main deck, then on neoprene and rigid foam (Figure 4), making comparisons according to the legislation in force. The elastic features of neoprene and rigid foam have been analyzed in a previous paper [14].

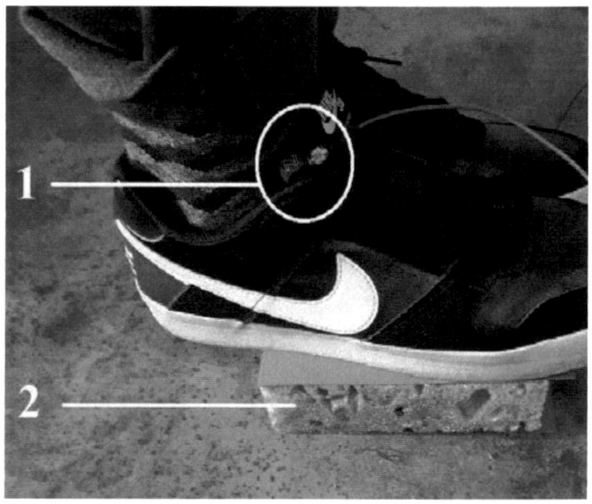

Figure 4. Subject placed on attenuating vibration materials: (1) – accelerometer, (2) – rigid foam

3. Results and discussions

The experiments took place on the Danube, upstream, during a 118 km trip, on certain sections of approximately 1 km. These sections were completed in 7 min and 40 s; the shore speed was 7.4 km/h, water speed 2.7 km/h and the engine revolution was 1800 rpm.

Longitudinal, transverse and vertical vibrations transmitted to the subjects by the ship's equipment were measured in 3 cases: the subject placed directly on the deck (Figure 4), the subject placed on neoprene and on rigid foam.

Were measured the amplitude of the output vibration frequency for the materials studied. Also, the transmissibility was determined according to the relationship [15]:

$$T = \frac{a_{w_{max}out}}{a_{w_{max}in}} \tag{2}$$

where $a_{w_{max}in}$ is the maximum acceleration input and $a_{w_{max}out}$ is the maximum acceleration output.

The working frequencies are shown in Table 1 (small frequencies of interest to the human body). 1/3 octave bands were used to provide a further in-depth outlook on vibration levels across the frequency composition.

IOP Conf. Series: Materials Science and Engineering **485** (2019) 012022 doi:10.1088/1757-899X/485/1/012022

Table 1. The frequency spectrum (Hz) which was used

Octave			The 1/3 octave		
Lower band limit	Middle of the band	Top band limit	Lower band limit	Middle of the band	Top band limit
			14.1	16	17.8
11	16	22	17.8	20	22.4
			22.4	25	28.2
			28.2	31.5	35.5
22	31.5	44	35.5	40	44.7
			44.7	50	56.2
			56.2	63	70.8
44	63	88	70.8	80	89.1
			89.1	100	112

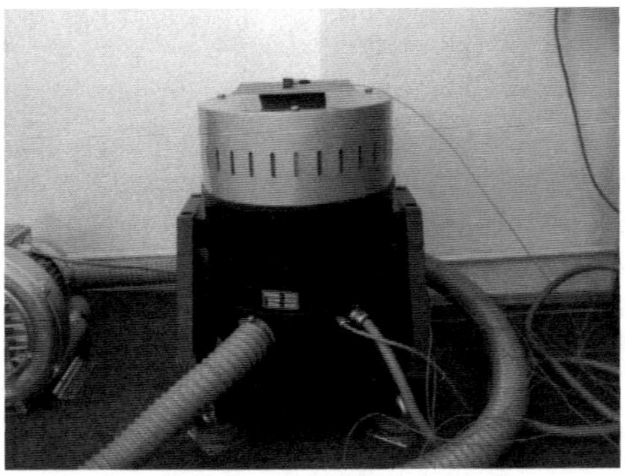

a) for neoprene

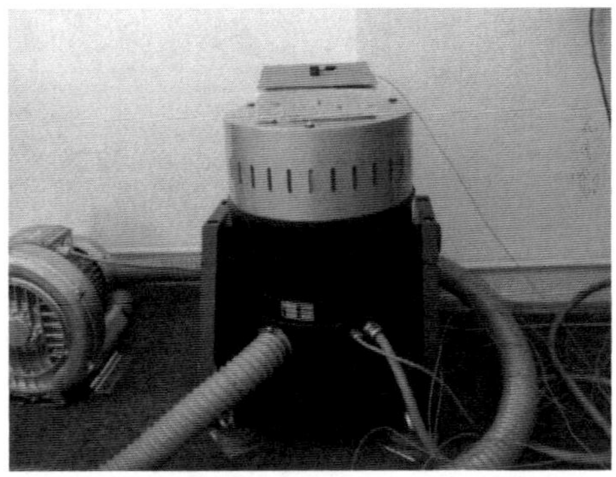

b) for rigid foam

Figure 5. Determination of the vibratory properties

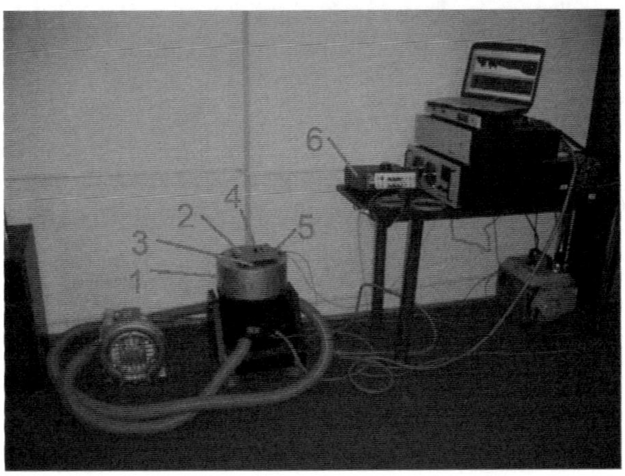

Figure 6. Experimental assembly to determine the vibration damping through different mediums (1) exciter, (2) contact sheets, (3) damper medium, (4) accelerometer for the final vibrations, (5) accelerometer for the initial excitation, (6) acquisition system NetdB

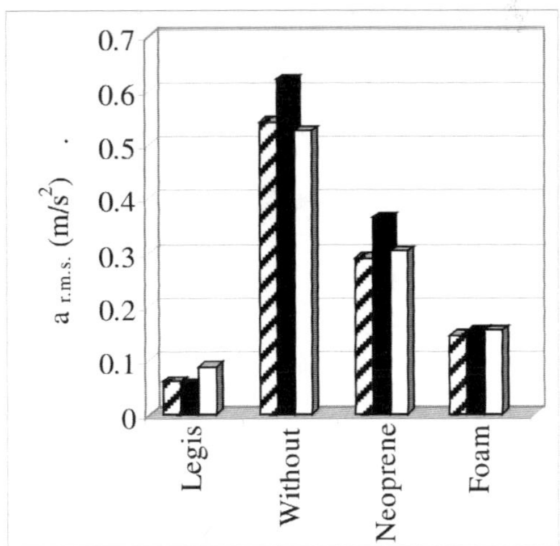

Figure 7. Variation $a_{r.m.s.}$ for the three situations: without damper, neoprene and rigid foam *versus* legislation (r.m.s. = root mean square)

The determination of the accelerations and transmissibility for the neoprene analysis (Figure 5.b)) and rigid foam (Figure 5.a)) was made with the equipment in Figure 6.

When determining transmissibility, the results are similar: for small frequencies (of interest to the human body), rigid foam proved to be better than neoprene (Figure 7).

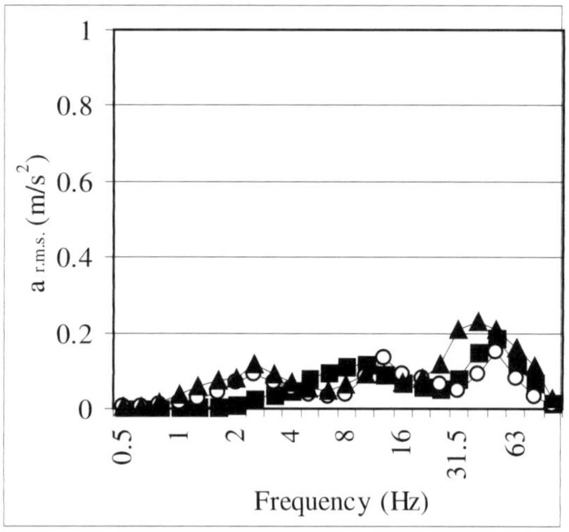

Figure 8. Variation of $a_{r.m.s.}$ according with legislation: (■) – Longitudinal; (o) – Transversal; (▲) – Vertical

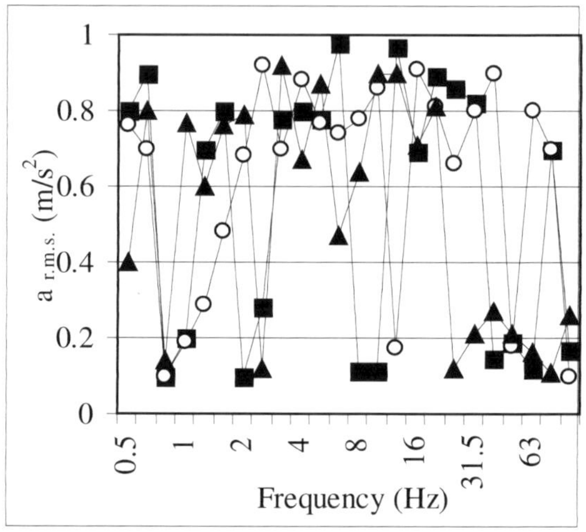

Figure 9. Variation of $a_{r.m.s.}$ without damper: (■) – Longitudinal; (o) – Transversal; (▲) – Vertical

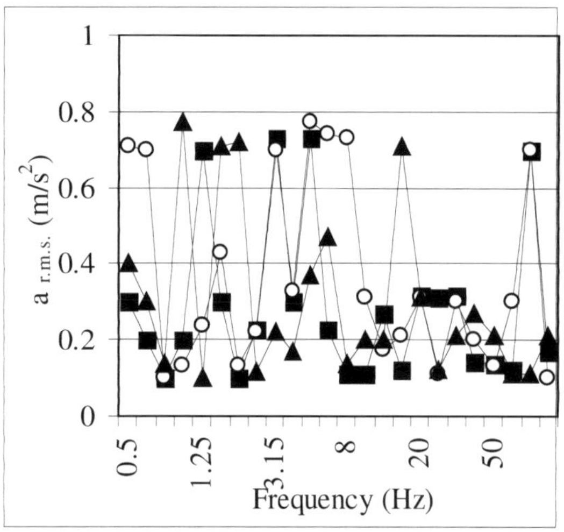

Figure 10. Variation of $a_{r.m.s.}$ with neoprene: (■) – Longitudinal; (o) – Transverse; (▲) – Vertical

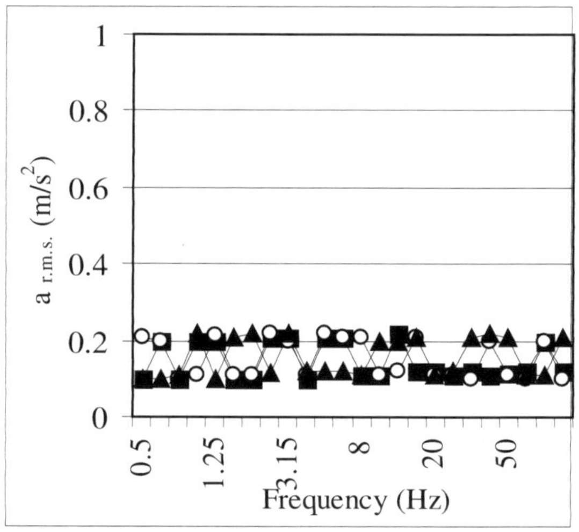

Figure 11. Variation of $a_{r.m.s.}$ with rigid foam: (■) – Longitudinal; (o) – Transverse; (▲) – Vertical

Following the determinations, it was found that:

a) for **longitudinal** vibrations, accelerations r.m.s. are 9 times higher than the standard (Fig. 8) if there is no attenuator (Fig. 9), 4.8 times higher when we use neoprene (Fig. 10) and 2.4 times higher than the standard for rigid foam (Fig. 11);

b) for **transverse** vibrations, accelerations r.m.s. are 10.8 times higher than standard when there is no attenuator (Fig. 9), 6.4 times higher for neoprene (Fig. 10) and 2.7 times higher for rigid foam (Fig. 11);

c) for **vertical** vibrations, accelerations r.m.s. are 5.9 times higher than standard if there is no attenuator (Fig. 9), 3.4 times higher for neoprene (Fig. 10) and 1.7 times higher for rigid foam (Fig. 11).

"Using the resonant method the natural frequency will be determined for an oscillating system with one degree of freedom (where the spring is made up from the material to be studied), also it will be determined the bandwidth of the frequency that corresponds to certain level decreases near the resonance area and amplitude oscillatory motion speed decrease." [14]

Abbreviation with f_0 the resonance frequency of the respective material and with f the working frequency, then transmissibility will be represented the as a function of the ratio $r = f/f_0$.

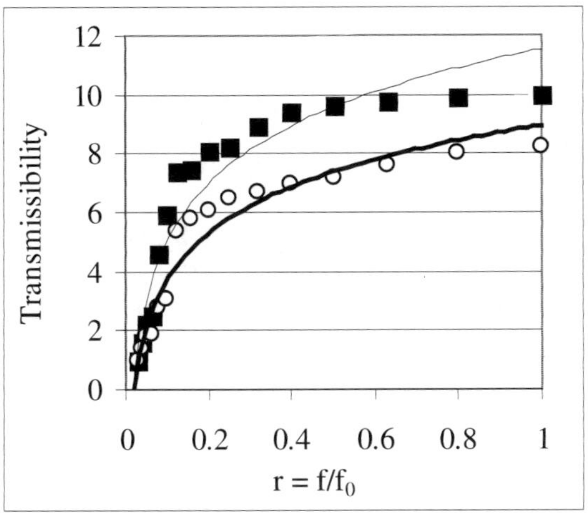

Figure 12. Transmissibility for r ∈ (0÷1)
(■) – neoprene (——); (o) – rigid foam (▬▬)

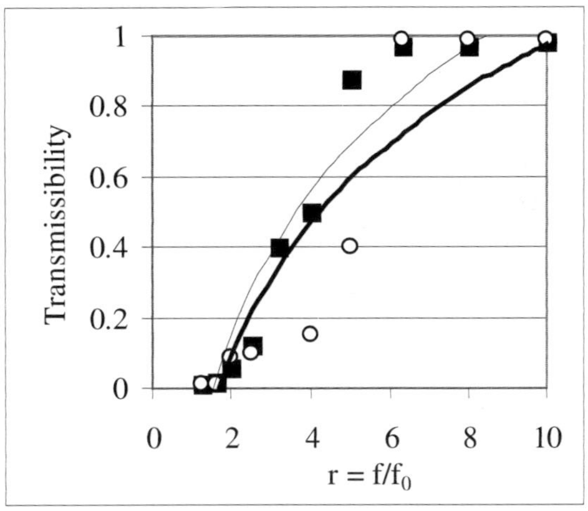

Figure 13. Transmissibility if r ∈ (1÷10)
(■) – neoprene (——); (o) – rigid foam (▬▬)

For r ∈ (0÷1):

in the case of neoprene: $T = 2.8244 \cdot \ln(r) + 11.533$ $(R^2 = 0.9087)$ (3)

in the case of rigid foam: $T = 2.877 \cdot \ln(r) + 8.9578$ $(R^2 = 0.9406)$ (4)

If r ∈ (1÷10):

for neoprene: $T = 0.5888 \cdot \ln(r) - 0.2568$ $(R^2 = 0.9225)$ (5)

for rigid foam: $T = 0.5507 \cdot \ln(r) - 0.2913$ $(R^2 = 0.8263)$ (6)

Figures 12 and 13 show that the transmissibility is greater for neoprene than for rigid foam (so rigid foam is better insulator for vibrations transmitted by the ship than neoprene).

4. Conclusions

From the studies conducted, it was easy to see that the vibrations transmitted by the ship's working equipment far exceed the limits indicated by the legislation in force. These vibrations, along with long ship stays, air temperature, noise, atmospheric humidity, and diet on the ship, can lead to irritability, depression, etc. All these inevitably lead to fatigue, to the change in circadian rhythm and implicitly to sleep problems.

These results have shown the importance of a vibration attenuator (especially for low frequencies) for the health, comfort, and implicitly work productivity of inland waterway sailors working and living on the ship 3-4 weeks. As a result, it was observed that the use of rigid foam carpets (where possible) can lead to an improvement in the quality of life of this type of worker.

The difference between transmissibility through neoprene, or rigid foam is more pronounced for low frequencies [r∈(0÷1)], where the neoprene transmissibility reaches up to 10 and rigid foam up to 9. It also results from the difference in free terms from equations (3) and (4), where for neoprene is 11.533, and 8.9578 in the case of rigid foam.

As for the range r∈(1÷10), the situations are similar for neoprene and rigid foam; in this case, the difference comes from the logarithm coefficient which is 0.5888 for neoprene and 0.5507 in the case of rigid foam (see equations (5) and (6)).

References

[1] Picu C L 2018 *Proc. 11th Eur. Cong. and Exp. on Noise Contr. Eng. (Hersonissos, Crete)* 2663-68
[2] Picu M E and Picu C L 2018 *MATEC Web Conf.* **148** 09005
[3] Picu M E and Picu C L 2017 *Springer Proc. in Phys.* **198** 389-95
[4] Bezyukov O K and Afanaseva O V 2014 *Life Sci J* **11**(5) 483-86
[5] Praetorius G, Österman C and Hult Carl 2018 TransNav: Int J Mar Navig Saf Sea Transp **12**(3) 587-95
[6] Picu C L and Rusu EVC 2017 *Whole body vibrations of a pushtow boat crew on the Danube* Mechal Testing and Diagnosis **VII**(1) 28-35
[7] Dunn S E 2006 Vibration level characterization from a needle gun used on U.S. naval vessels *Graduate Thesis*
[8] Wilhite C R 2007 Pneumatic tool hand-arm vibration and posture characterization involving U.S. navy shipboard personnel *Graduate Thesis*
[9] Netudykhatka O 1999 The age-related characteristics of the psychophysiological indices of sailors Likars'ka Sprava (Medical report) **1** 131-34
[10] Chan G, Moochhala S M, Zhao B, Tan D and Wong J 2006 A comparison of motion sickness prevalence between seafarers and non-seafarers onboard naval platforms *Int Marit Health* **57**(1-4) 56-65.
[11] Hursh S R, Redmond D P, Johnson M L, Thorne D R, Belenky G, Balkin T J, Storm W F, Miller J C and Eddy D R 2004 Fatigue models for applied research in warfighting *Aviat Space Environ Med* 75 (3 suppl.) A44-53
[12] Ministry of Labour and Social Solidarity 2002 General Labour Protection Norms
[13] Haynes L E 2007 A comparison between the Navy standard workweek and the actual work and rest patterns of U.S. Navy Sailors *Graduate Thesis*
[14] Picu A 2011 A study on improving minibus travel comfort by using vibrations attenuator material *Carpath J Earth Env* **6** 119-28
[15] Picu M E 2015 From exploratory research and monitoring mechanical noise and vibrations to the development of models of the human response under the action of stress caused by these factors *Habilitation Thesis*

IOP Conf. Series: Materials Science and Engineering **485** (2019) 012023 doi:10.1088/1757-899X/485/1/012023

Thermomechanical processing effects on the structure and properties of Fe-based SMAs. II. Evolution of damping behavior

M Popa[1], B Pricop[1*], V Bulbuc[2], E Mihalache[1], B Özkal[3] and L G Bujoreanu[1]

[1] Faculty of Materials Science and Engineering, "Gheorghe Asachi" Technical University of Iasi, Blvd. Dimitrie Mangeron, 61 A, 700050 Iasi, Romania
[2] The National Railway Company "CFR" SA, Regional Branch of Passenger Rail Iasi, Str. Piața Gării 1, 700090 Iași, Romania
[3] Particulate Materials Laboratory, Metallurgical and Materials Engineering Department, Istanbul Technical University, 34469 Maslak, Istanbul, Turkey

* bogdan.pricop@tuiasi.ro

Abstract. The present paper analyses the effects of five different heat treatment temperatures, between 973, 1073, ... and 1373 K as well as two mechanical alloying (MA) fractions, 0 and 40 vol. %, on the evolution of damping behavior of an Fe-Mn-Si-Cr-Ni Shape Memory Alloy (SMA) obtained by powder metallurgy (PM). From the 10 types of samples, rectangular specimens were cut by wire-spark erosion for dynamic mechanical analysis (DMA) tests. DMA tests were performed by temperature scans (TS) and strain sweeps (SS). TS were applied between room temperature (RT) and 673 K, at constant amplitude, and thus enabled determining the temperatures of internal friction maxima associated with antiferromagnetic-paramagnetic transition (Néel temperature), A_{50}^{N}, and ε-martensite reversion to γ-austenite, A_{50}^{ε}, as a function of MA'ed fraction and HT temperature. SS were applied during three cycles with increasing strain amplitude and performed at three temperatures: (i) T_1=RT; (ii) $T_2 < A_{50}^{\varepsilon}$ and (iii) $T_3 > A_{50}^{\varepsilon}$. The effects of HT temperature and MA'ed fraction on storage modulus (E') values were emphasized for each of the three temperatures $T_{1,2,3}$. An E'-plateau was observed and associated with the formation of ε stress induced martensite, during DMA-SS.

1. Introduction

After the first paper, published in 1982, on the shape memory effect (SME) observed at Fe-Mn-Si single crystals [1], polycrystalline Fe-(28-34) Mn-(4-6.5) Si (mass. %, as all chemical compositions will be listed hereinafter) alloys with almost perfect SME were reported in 1987 [2]. With the additions of Cr [3] and Ni [4], two commercial shape memory alloy (SMA) grades were developed, as Fe-28Mn-6Si-5Cr and Fe-14Mn-6Si-9Cr-5Ni.

The first reports on practical applications of Fe-Mn-Si-based SMAs were published in 1991, introducing lock rings for bicycle frame pipes [5]. Subsequently, constrained recovery SME applications [6] were developed, comprising pipe couplings [7], fishplates for crane rail fastening [8], concrete pre-straining rods [9] and embedded stripes for beam curvature control [10]. Up to the present date, the largest SMA applications, ever reported, have been the sixteen 2-tons anti-seismic dampers made in

2014 from Fe-15Mn-4Si-10Cr-8Ni SMA and installed from the first to the fourth floors of a 196-metre skyscraper, “JP Tower Nagoya” that was completed in November 2015 [11]. These dampers exploit the general capacity of SMAs to damp oscillations [12] as well as the particular feature of Fe-Mn-Si based SMAs to retransform, from ε-hexagonal close packed (hcp) martensite to γ-face centered cubic (fcc) austenite, not only by heating but also by counter-directional loading [13].

Most of these applications use bulk materials obtained by ingot metallurgy (IM). However, powder metallurgy (PM) accompanied by mechanical alloying (MA) have the potential to overcome most of drawbacks of IM, basically by improving the control of chemical composition and grain size, as succeeded in the case of other SMAs, such as Ti-Ni, or Cu base alloys [14].

Nevertheless, PM Fe-Mn-Si based SMAs, were only scarcely reported in literature. On the other hand, some of present authors have studied quintenary PM-MA’ed Fe-Mn-Si-Cr-Ni SMAs, ever since 2009. Thus, it was proved that PM specimens experienced higher stress levels than IM ones, due to larger amounts of α’ thermally induced martensite [15], which was additionally stress-induced with the increase of the number of mechanical cycles [16], the amount of MA’ed powder fraction [17] and pre-straining degree [18]. In addition, specific damping capacity experienced an increasing trend with pre-straining degree and MA’ed fraction [19]. Yet, the dynamic behavior of PM-MA’ed Fe-Mn-Si-Cr-Ni SMAs during strain sweeps performed by dynamic mechanical analysis (DMA) was never analyzed and reported. The evolution of dynamic stiffness is particularly important in the case of vibrating pipes, due to the turbulence of transported fluids that can be connected by coupling rings made from PM-MA’ed Fe-Mn-Si-Cr-Ni SMAs, the constrained recovery of which can be controlled by the deformation and thermomechanical training processing technologies [20].

The present paper investigates the effects of heat treatment (HT) and mechanical alloying (MA) on the evolution of damping behavior of an Fe-Mn-Si-Cr-Ni Shape Memory Alloy (SMA) obtained by powder metallurgy (PM).

2. Experimental details

Two groups of specimens, with nominal chemical composition 66Fe-14Mn-6Si-9Cr-5Ni (mass %), were pressed and sintered under argon atmosphere, being designated as: (i) 0_MA, from as-blended elemental powders and (ii) 40_MA comprising 40 vol. % MA’ed powders, obtained after 4 hrs. high-energy ball milling under Ar atmosphere in order to inhibit the chemical reactivity of oxygen-sensitive alloying elements. After blending, pressing and sintering, further increase of compactness was achieved by six consecutive hot rolling passes at 1373K, until samples’ thickness decreased from 4 to 1 mm. By spark-erosion wire cutting, specimens with “dog bone” (gauge dimensions 1×4×20 mm) configuration were prepared for dynamic mechanical analysis (DMA). Heat treatment (HT) was applied with 5 min-holding at five temperatures, 973, 1073, … 1373K, followed by water quenching. In this way, 10 different sets of specimens, with different MA’ed powder fraction and HT temperature were obtained, designated by MA’ed powder fraction and HT temperature (in ^{0}C).

DMA tests were performed on a NETZSCH DMA 242 Artemis, equipped with “three-point bending” specimen holder, under two variants: (i) temperature scans (TS) and (ii) strain sweeps (SS).

TS were applied between RT and 673 K, with a heating rate of 5 K/min and a frequency of 1 Hz. The present authors pointed out that, in PM-MA’ed Fe-Mn-Si-Cr-Ni SMAs, a second internal friction maximum occurs on TS-DMA diagrams recorded on heating, compared to only one found in literature. This low-intensity tanδ maximum, located at a temperature designated as A_{50}^{N}, could be associated with antiferromagnetic-paramagnetic transition characterized by Néel temperature (TN) [22]. In order to confirm the antiferromagnetic-paramagnetic transition, thermomagnetic measurements were done, on a Lake Shore VSM 7410 Vibrating Sample Magnetometer. Temperature variation rate was 2 K/min, applied magnetic field was H = 20 Oe and temperature range was RT-723 K.

SS, up to 20 μm-amplitude, were applied at three temperatures: (i) T_1 =RT, (ii) $T_2 < A_{50}^{\varepsilon}$ and (iii) $T_3 > A_{50}^{\varepsilon}$, at constant frequency of 1 Hz. DMA-SS diagrams were recorded under the form of storage modulus vs. (strain) amplitude, at the frequency of 1 Hz.

3. Experimental results and discussion

3.1. Heat treatment and mechanical alloying effects on the response observed by DMA thermal scans

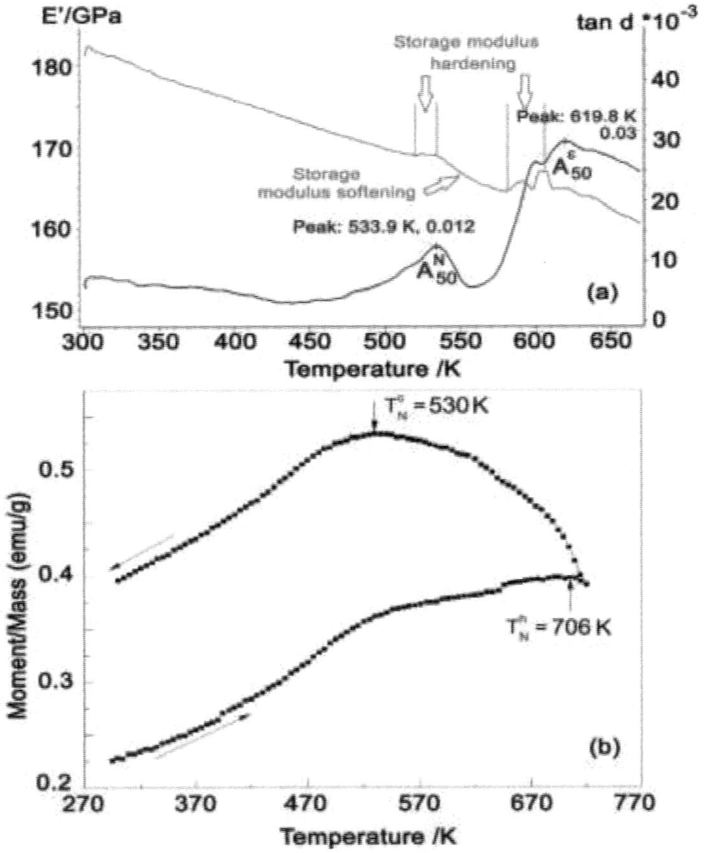

Figure 1. Temperature scan-DMA diagrams, recorded for specimen 0-MA_700, illustrating the variations with temperature of: (a) storage modulus (E') and internal friction (tan δ); (b) magnetization, during a heating-cooling cycle

One typical example for the variation of storage modulus and internal friction, during DMA-TS on heating, is illustrated in figure 1(a). Considering that during paramagnetic→antiferromagnetic transition the change in storage modulus is abrupt but the correlated internal friction peak height is small, because magnetic ordering relaxes much faster than the moving array of atoms associated with the martensitic transformation, it was assumed that the low intensity tan δ maximum could correspond to antiferromagnetic → paramagnetic transition. In addition, internal friction value is larger in austenite than in martensite because the former possesses a larger number of close packed planes. Therefore the second tan δ maximum, which has always larger internal friction values after the peak than before it, could fairly correspond to ε(hcp) →γ(fcc) reverse martensitic transformation. Figure 1(b) displays magnetization variation, during heating, for the same specimen.

Martensite reversion to austenite, during heating, is accompanied by a local drop in storage modulus (E'), since martensite is harder than austenite. Consequently, two modulus hardening segments were identified by arrows in figure 1(a). It is noticeable that the temperatures of modulus hardening segments coincide with the first halves of internal friction maxima. The beginning of second increase of E' (larger), during heating, can be associated with the ε(hcp) →γ(fcc) reverse martensitic transformation, in figure 1(a). In figure 1(b), above Néel temperature (TN) the subsequent magnetization decrease on heating corresponds to the paramagnetic state. One can argue that, due to the low amplitude of applied magnetic field, magnetization increase with temperature could be caused by thermal fluctuations that occur at high temperature or the formation of a new phase. Nevertheless, considering that a similar magnetization maximum was observed during cooling as well, we assume that the presence of the magnetization maxima can be related to antiferromagnetic-paramagnetic transition [23].

The variations of storage modulus and magnetization during heating, for the two fractions of MA'ed powders, corresponding to the lowest (973 K) and highest (1373 K) HT temperatures, were summarized in figure 2. It is noticeable that on most of the thermograms storage modulus increased before magnetization reached its maximum. This delay could be caused by the low value of applied magnetic field.

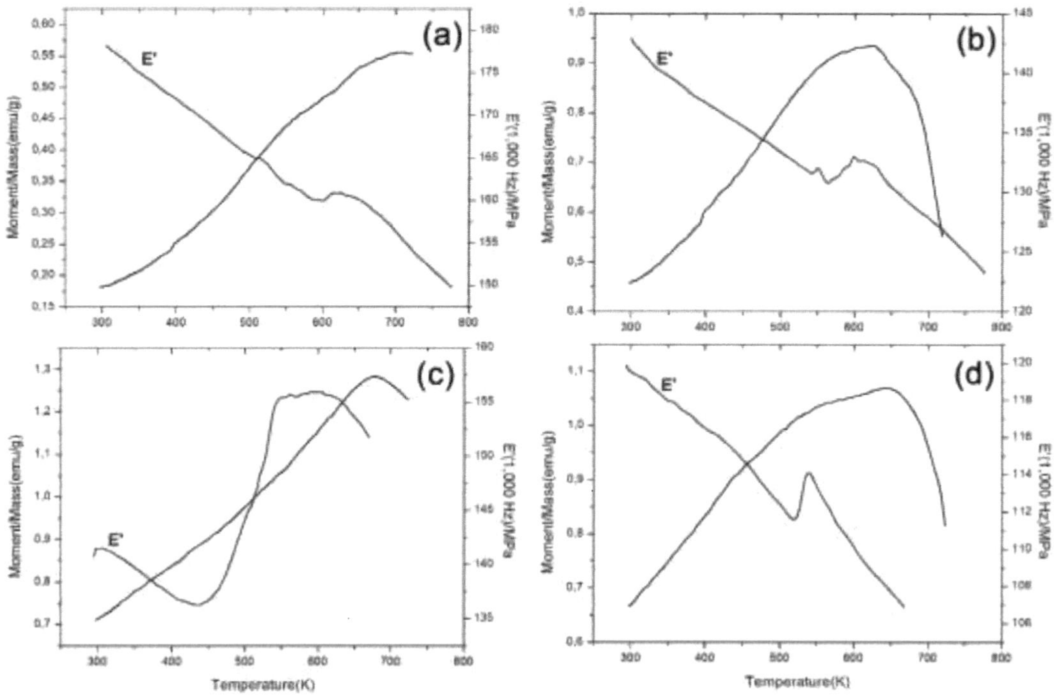

Figure 2. Heat treatment temperature and MA'ed powder fraction effects on the evolution of storage modulus and magnetization during heating: (a) 0_MA_700; (b) 0_MA_1100; (c) 40_MA_700; (d) 40_MA_1100.

The influence of HT temperature, on the positions of the two peaks, is not substantial because dynamic recrystallization was quite active during 1373 K-hot rolling. With increasing MA'ed powder fraction, both A_{50}^{N} and A_{50}^{ε} experience a global tendency to decrease.

3.2. Heat treatment and mechanical alloying effects on the response observed by DMA strain sweeps

As mentioned above, DMA-SS were performed at three different temperatures. In each case, three SS cycles were applied. A typical SS-DMA diagram comprising storage modulus (E') variation with strain amplitude, is shown in figure 3, corresponding to specimen 0_MA_700, tested at T_1 = RT.

In most of the DMA-SS diagrams recorded at RT, E' experienced a plateau between 0.02 and 0.06 % strain amplitudes. This plateau became longer in the second and third SS cycles. Since storage modulus remains constant, instead of continuously increasing with strain amplitude (as expectable in the case of work-hardenable materials), it can be assumed that the plateau can be associated with ε-hcp stress induced formation during dynamic loading. In order to have a better perception on storage modulus evolution during SS cycles, an insight of E' values at 0.02 % strain amplitude was shown in figure 3 comprising the three cycles. It appears that E' markedly increased from the first to the second cycle and almost stabilized in the third.

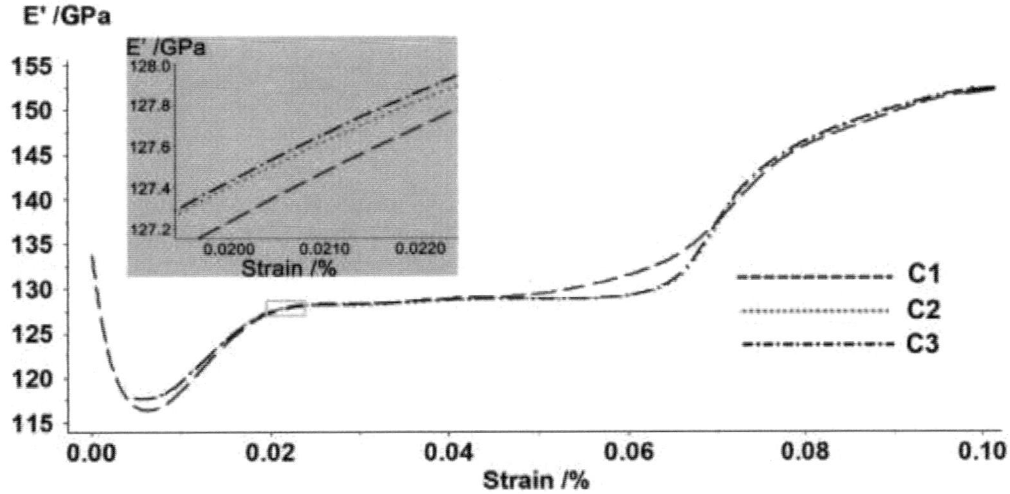

Figure 3. Variation of individual values of storage modulus during three cycles of DMA-strain sweep of 0_MA_700 specimen, tested at T1 = RT with inset at 0.02 % strain amplitude.

At the specimens tested at T_1: (i) the values of storage modulus tend to saturate with increasing strain amplitude, displaying a first E' plateau, at most of the specimens; (ii) further increase of strain amplitude caused additional E' saturation, yielding a second plateau (exceeding 200 GPa at specimen 0_MA_1000), which is noticeable only at a part of specimens.

At the specimens tested at higher temperatures (T_2 and T_3), most E' variations display a single plateau or two shorter plateaus at higher MA'ed powder fractions (40_MA).

4. Summary and conclusions

- PM-MA'ed Fe-Mn-Si-Cr-Ni SMAs had porosity degrees between approximately 17 % (at as blended powders) and 5 % at specimens with 40 vol. % mechanically alloyed powders, which contributed to the obtainment of mass magnetization values below 4.5 emu/ g at RT;
- DMA temperature scans revealed two tanδ maxima: a low-intensity one located between 463-572 K and a high-intensity one located at 532-650 K. In the opinion of the authors, the former was ascribed to antiferromagnetic ⇒ paramagnetic transition, at $\boldsymbol{A_{50}^{N}}$ and the latter to ε(hcp) ⇒ γ(fcc) reverse martensitic transformation, at $\boldsymbol{A_{50}^{\varepsilon}}$.
- With increasing MA'ed powder fraction, the temperatures of both $\boldsymbol{A_{50}^{N}}$ and $\boldsymbol{A_{50}^{\varepsilon}}$ tend to decrease, during DMA-temperature scan tests.
- Storage modulus (E') increased with strain amplitude and cycle number, experiencing a plateau at 0.02-0.06 % (associated with ε-hcp stress induced formation) during DMA-strain sweep tests. The presence of this modulus plateau recommends PM-MA'ed Fe-Mn-Si-Cr-Ni SMAs as potential candidates for RT-dampers, under the form of coupling rings of vibrating pipes.

Acknowledgements: This research work was supported by UEFISCDI by means of Project PN-III-P4-ID-PCE-2016-0468, Contract 76/ 2017.

References

[1] Sato A, Chishima E, Soma K and Mori T 1982 Shape memory effect in γ→ ε transformation in Fe-30Mn-1Si alloy single crystals *Acta Metall.* **30** 1177–83

[2] Murakami M, Suzuki H and Nakamura Y 1986 Effect of Si on the shape memory effect of polycrystalline Fe-Mn-Si alloys. *Trans. Iron Steel Inst. Japan* **27**

[3] Otsuka H, Yamada H, Maruyama T, Tanahashi H, Matsuda S and Murakami M 1990 Effects of

alloying additions on Fe-Mn-Si shape memory alloys. *ISIJ Int.* **30** 674–9
[4] Moriya Y, Kimura H, Ishizaki S, Hashizume S, Suzuki S, Suzuki H and Sampei T 1991 Properties of Fe-Cr-Ni-Mn-Si (-Co) shape memory alloys *Le J. Phys. IV* **1** C4--433
[5] Otsuka H 1991 Fe-Mn-Si based shape memory alloys *MRS Online Proc. Libr. Arch.* **246**
[6] Bujoreanu L G, Dia V, Stanciu S, Susan M and Baciu C 2008 Study of the tensile constrained recovery behavior of a Fe-Mn-Si shape memory alloy *Eur. Phys. J. Spec. Top.* **158** 15–20
[7] Druker A V, Perotti A, Esquivel I and Malarria J 2014 A manufacturing process for shaft and pipe couplings of Fe-Mn-Si-Ni-Cr shape memory alloys *Mater. Des.* **56**
[8] Maruyama T, Kurita T, Kozaki S, Andou K, Farjami S and Kubo H 2008 Innovation in producing crane rail fishplate using Fe-Mn-Si-Cr based shape memory alloy *Mater. Sci. Technol.* **24** 908–12
[9] Sawaguchi T, Kikuchi T, Ogawa K, Kajiwara S, Ikeo Y, Kojima M and Ogawa T 2006 Development of prestressed concrete using Fe-Mn-Si-based shape memory alloys containing NbC *Mater. Trans.* **47** 580–3
[10] Shahverdi M, Czaderski C and Motavalli M 2016 Iron-based shape memory alloys for prestressed near-surface mounted strengthening of reinforced concrete beams *Constr. Build. Mater.* **112** 28–38
[11] Sawaguchi T, Maruyama T, Otsuka H, Kushibe A, Inoue Y and Tsuzaki K 2016 Design concept and applications of Fe-Mn-Si-based alloys—from shape-memory to seismic response control *Mater. Trans.* **57** 283–93
[12] Torra V, Isalgue A, Lovey F C and Sade M 2015 Shape memory alloys as an effective tool to damp oscillations *J. Therm. Anal. Calorim.* **119** 1475–533
[13] Sawaguchi T, Bujoreanu L-G, Kikuchi T, Ogawa K, Koyama M and Murakami M 2008 Mechanism of reversible transformation-induced plasticity of Fe-Mn-Si shape memory alloys *Scr. Mater.* **59** 826–9
[14] Suzuki Y, Otsukaand K and Wayman C M 1998 Fabrication of shape memory alloys *Shape Mem. Mater.* 133–48
[15] Bujoreanu L G, Stanciu S, Özkal B, Comăneci R I and Meyer M 2009 Comparative study of the structures of Fe-Mn-Si-Cr-Ni shape memory alloys obtained by classical and by powder metallurgy, respectively *ESOMAT 2009 - 8th Eur. Symp. Martensitic Transform.* **05003**
[16] Pricop B, Söyler U, Comăneci R I, Özkal B and Bujoreanu L G 2010 Mechanical cycling effects at Fe-Mn-Si-Cr-Ni SMAs obtained by powder metallurgy *Physics Procedia* vol 10
[17] Pricop B, Söyler U, Özkal B, Lohan N M, Paraschiv A L, Suru M G and Bujoreanu L G 2013 *Influence of Mechanical Alloying on the Behavior of Fe-Mn-Si-Cr-Ni Shape Memory Alloys Made by Powder Metallurgy* vol 738–739
[18] Pricop B, Soyler U, Ozkal B, Suru M G, Lohan N M, Comaneci R I, Cimpoesu N, Musat V, Gurau G, Istrate B, Mihalache E and Bujoreanu L G 2015 A Study of Martensite Formation in Powder Metallurgy Fe-Mn-Si-Cr-Ni Shape Memory Alloys *Materials Today: Proceedings* vol 2 pp S789–92
[19] Pricop B, Mihalache E, Lohan M-N, Istrate B, Mocanu M, Ozkal B and Bujoreanu L-G 2015 Powder metallurgy and mechanical alloying effects on the formation of thermally induced martensite in an FeMnSiCrNi SMA *MATEC Web of Conferences* vol 33
[20] Pricop B, Söyler U, Lohan N M, Özkal B, Bujoreanu L G, Chicet D and Munteanu C 2012 Thermal behavior of mechanically alloyed powders used for producing an Fe-Mn-Si-Cr-Ni shape memory alloy *J. Mater. Eng. Perform.* **21**
[21] Sawaguchi T, Bujoreanu L-G, Kikuchi T, Ogawa K and Yin F 2008 Effects of Nb and C in solution and in NbC form on the transformation-related internal friction of Fe--17Mn (mass%) alloys *ISIJ Int.* **48** 99–106
[22] Mihalache E, Pricop B, Suru M-G, Lohan N M, Comăneci R I, Istrate B, Özkal B and Bujoreanu L-G 2015 Factors influencing martensite transitions in Fe-based shape memory alloys *MATEC Web of Conferences* vol 33 p 04002
[23] Wang L, Cui Y G, Wan J F, Zhang J H and Rong Y H 2013 Magnetic thermal hysteresis due to paramagnetic-antiferromagnetic transition in Fe-24.4 Mn-5.9 Si-5.1 Cr alloy *AIP Adv.* **3** 82126

The structural analysis of the material resulting from the destruction by explosion of a compressed air cylinder

F Potecaşu*, O Potecaşu

Faculty of Engineering"Dunărea de Jos" University of Galati, Domnească Street, 47, RO-800008, Galati, Romania

*fpotec@ugal.ro

Abstract: In this paper is presented the structural analysis carried out on samples from a compressed air cylinder subjected to explosion. The air cylinder was mounted on public transport bus. The risk that the incident would repeat with other similar cylinders carrying compressed air required a detailed study regarding the causes that led to the catastrophic degradation. The research by micro and macro structural analysis has highlighted several factors that have increased the damage. The optical micrographs are obtained using an Olympus metallographic microscope equipped with a digital camera, with the possibility to automatically assign the microstructure magnification and with automatic software for linear measurements. The research regarding the appearance of the inner surface of the compressed air cylinder container has highlighted an accentuated corrosion process, developed in the thickness of the cylinder wall with a non-uniform thickness. The break occurred near the welding, known as thermal influence zone (the analyzed cylinder being welded longitudinally). The research has shown that inter-crystalline corrosion associated with pitting (punctiform) corrosion developed between the dendrite branches leads to an intense metal degradation in the wall depth and, in limited cases, causes perforation of the metal wall, increasing the explosive fracture development. A factor that influenced the corrosion process in the ZIT area was the mounting of the compressed air cylinder with the welded area at the bottom of the component.

1. Introduction

A compressed air cylinder mounted on a public transport bus and subjected to an explosive rupture due to corrosion degradation was investigated in the Structural Materials Characterization Laboratory. The corrosion, as a result of the chemical and electrochemical reactions due to the material interaction with the environment, has as effect the partially or totally materials degradability, as in the present research.

The corrosion process may be defined as a heterogeneous oxidation-reduction reaction at the metal/medium interface, where the metal is oxidized and one or more components of the medium are reduced by the manufacturer's location the condensate drain hole near the longitudinal welding cord [1-3]. The component quality is given by the material nature, purity, and it depends on the elaboration and casting processes, respectively [4-7]. The mentioned component in this study, a compressed air cylinder was in the first phase visually analyzed, as a general condition, both on the outside and on the inside.

2. Materials and methods

2.1. Chemical analysis of steel sheet

Table 1 presents the results of the chemical composition on steel sheet. In accordance with the classification system EN 10020, the steel grade is classified as non-alloy steel.

Table 1. Chemical composition of the cylinder sheet determined by spectrophotometer.

	Analysis for sample I						Analysis for sample II					
No	Chemical element	%	±	Chemical element	%	±	Chemical element	%	±	Chemical element	%	±
1	Si	0,82	0,08	Si	0,88	0,08	Si	1,02	0,09	Si	1,11	0,09
2	W	trace	trace	W	trace	trace	Mo	trace	trace	Mo	trace	trace
3	Fe	98,99	0,39	Fe	98,94	0,37	Fe	98,74	0,43	Fe	98,63	0,39
4	Mn	0,19	0,03	Mn	0,17	0,03	Mn	0,17	0,03	Mn	0,21	0,03
5	Cr	trace	trace	Cr	trace	trace	Cr	0,06	0,02	Cr	0,05	0,02

2.2. Macrostructural analysis of the damaged compressed air cylinder

After the damage by explosion, the degraded compressed air cylinder component was brought to the laboratory and analyzed both as a modification of the shape and dimensions but also as regards the corrosion process (figure 1). The breaking of compressed air cylinder was occurred near the welding area (the analyzed compressed air cylinder being longitudinally welded). The compressed air cylinder was mounted with the welded area at the bottom. The drain valve was placed at the bottom near the weld area. Checking the appearance of the inner surface of the compressed air cylinder container has highlighted a severe and massive corrosion process.

In the macrostructures shown in figure 2 at 100x magnification, a thick film of relatively continuous, heterogeneous corrosion products, consisting of mixed oxides, is observed on the inner surface of samples taken from the compressed air cylinder body. The visible effect of corrosion on the steel wall of compressed air cylinder is the appearance of rust (hydrated iron oxides) as in figures 2a and 2b and the change in the thickness of the cylinder walls.

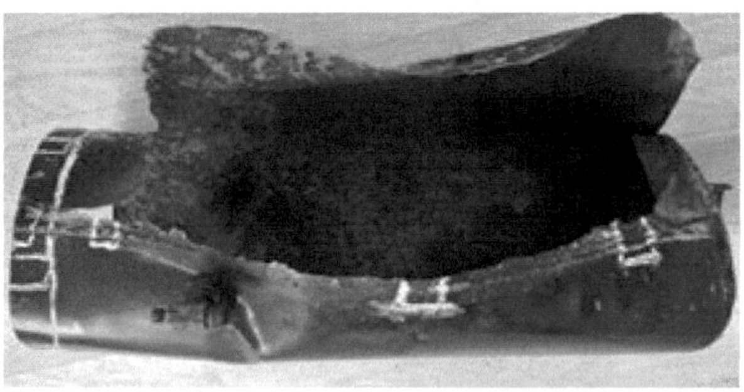

Figure 1. The surface aspect of compressed air cylinder subjected to explosive rupture.

The samples from the damaged body of compressed air cylinder for microstructural analysis in the cross section through the cylinder wall were performed. The samples included the weld joint and were cut perpendicular to the welding cord, (P_1) and seamless zone (P_2). In figure 3 are presented some representative metal fields in the samples transversally through the wall and including the welding cord. The general conclusion is that, in all cases, on the analyzed fields (in the material of welding – fusion zone-FZ, the heat-affected zone - HAZ, the base material – BM, in the immediate proximity of welding) the amount of oxides is very high. According to the standards, the samples are impure, with rating over 5. The largest amount of oxides (black points) being in FZ, then in HAZ and then in BM.

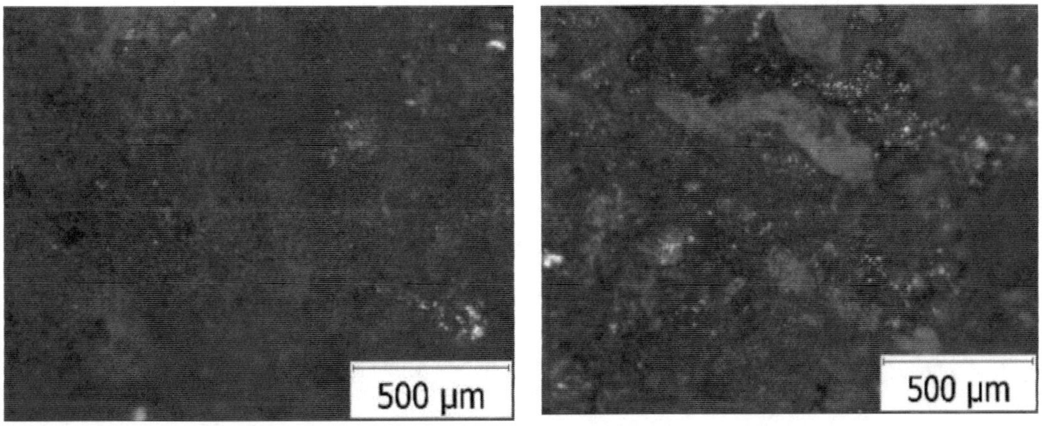

(a) (b)

Figure 2. The macrostructural analysis of the inside surface of the compressed air cylinder material: (a) the area without explosive rupture, (b) the area with explosive rupture.

2.3. Microstructural analysis

The qualitative and quantitative analysis of non-metallic inclusions of polished and unetched metallographic samples, under standardized conditions, using the chart comparison method with standard charts, at 100X magnification of optical microscope was performed. The expression of the steel purity was done by inclusions rating. Non-metallic inclusions are particles of impurities trapped mechanically in the solidified steel. They are not part of the steel structure but might come from the reactions that occur in liquid steel, especially from deoxidation and solidification.

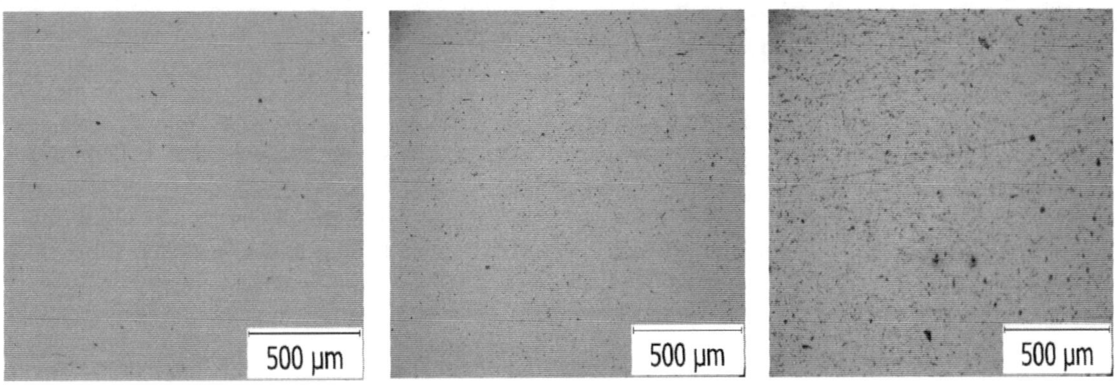

(a) (b) (c)

Figure 3. Analysis of non-metallic inclusions on the polished samples (100X): (a) in the cross-section of the base material (BM), the steel sheet near the welding cord; (b) in the heat-affected zone (HAZ) of the longitudinal weld; (c) in the fusion zone (FZ) of the longitudinal weld.

To find out if the large amount of oxides presented in figure 3 is the result of the intercrystalline corrosion developed in the compressed air cylinder wall from the corrosive interior environment, a lot of metallographic samples for purity, from the outer surface of the compressed air cylinder, an area almost unaffected by corrosion were prepared (figure 4). The results are clear. The steel unexposed to corrosion shows a specific grade of pure materials (Pmax = 1.5). The oxides

IOP Conf. Series: Materials Science and Engineering **485** (2019) 012024 doi:10.1088/1757-899X/485/1/012024

represented in figures 1, 2, 3 are oxides due to the corrosion processes developed in the wall thickness from the interior of the cylinders formed during working process.

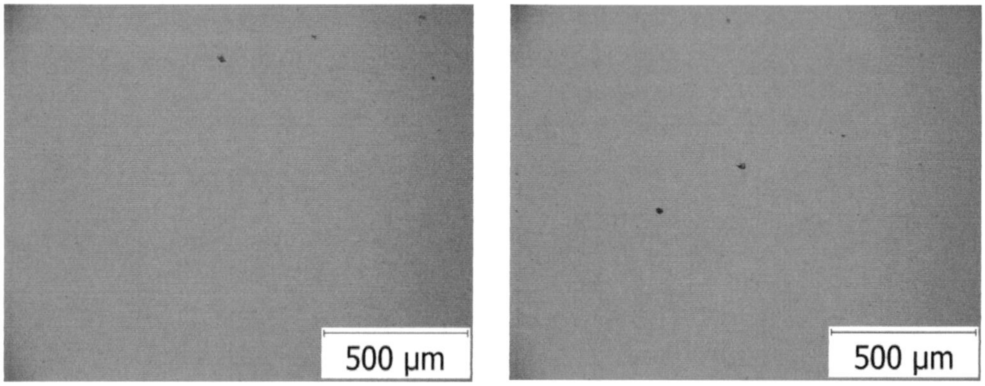

(a) (b)

Figure 4. Analysis of non-metallic inclusions of steel on the sample from the outer surface of the compressed air cylinder (100X).

2.4. Microstructural analysis of metallographic etched samples

In order to highlight the nature of the structural constituents, their size and distribution, metallographic samples grinded, polished to mirror gloss and etched with Nital metallographic reagent (3% nitric acid in ethyl alcohol) were performed. The microstructural analysis on the transverse surfaces through the compressed air cylinder wall was carried out. Samples from the body of the damaged compressed air cylinder and the welding area were also taken (coded P_1).

3. Results and discussions

Figure 5 shows the microstructures in the compressed air cylinder wall on the sample P1 taken perpendicular to the welded joint. The microstructures present the structural aspects in the steel sheet material (base material-BM), the heat affected zone (HAZ) and the fusion zone (FZ).

The microstructures of figures 5a, 5b, 5c and 5d, made in the HAZ/BM area, highlight the non-uniform corrosion of the compressed air cylinder wall. Thus, near the welded joint, the steel wall measures 1239-1324 μm. Approaching the welded area, the HAZ area caused an acceleration of corrosion, and the wall thinned considerably to just 1023 μm. The microstructures are made with an Olympus metallographic microscope equipped with a digital imaging camera, with the possibility of automatically assign the microstructure magnification and with automatic software for linear measurements.

The corrosion process, so complex and non-uniform, is explained, on the one hand by the fact that a rusty surface always exhibits a larger area of corrosion compared with the same glossy surface and, on the other hand, corrosion occurs both on the irregularities picks and in their depth, where a multitude of microcracks that can develop rapidly and can contribute to the degradation of the steel are formed. Fine grains embedding a very large grain boundary area increasing accelerated intercrystalline corrosion (grain or subgrade grain boundaries are in fact dislocation walls, areas with a huge amount of linear imperfections that conserve high free energy, high instability, a high corrosion susceptibility, so a low corrosion resistance).

The presence of fine tertiary cementitious precipitates -Fe_3C_{III} - many in number on the grain boundary can lowers the corrosion resistance of steel by the occurrence of galvanic couplings between the carbides and the neighboring areas between the grains. The heat - affected zone (HAZ) is of particular importance for the welded joint quality of the as it connects the seam to the welded piece. It's an area with fine grains, as fine as approaching to the seam area.

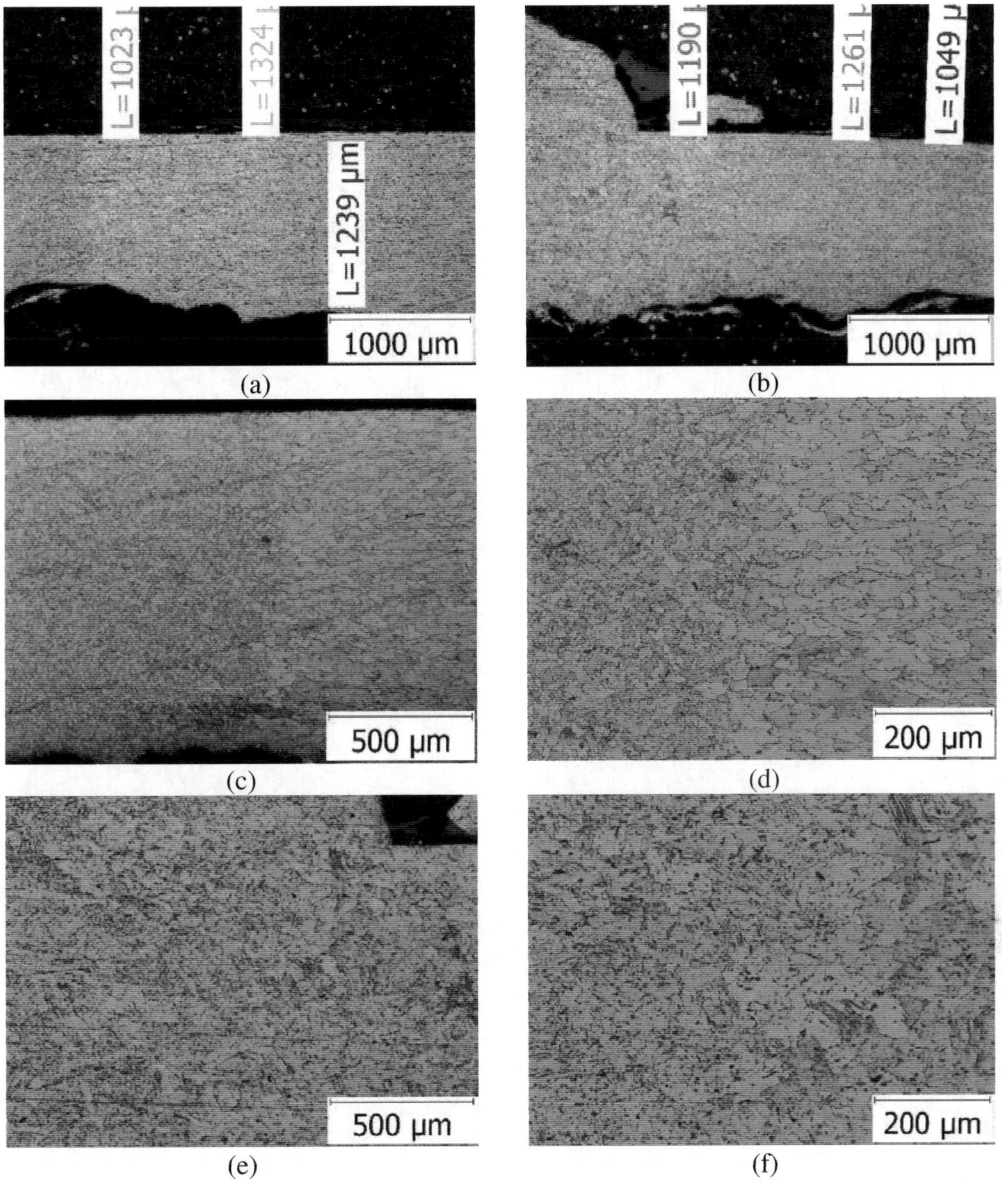

Figure 5. Microstructures of sample 1 (P_1) in cross section through the compressed air cylinder wall: (a) in HAZ/BM area (50X), (b) in FZ/HAZ area (50X), (c) and (d) in HAZ/BM area at (100X) and (200X), (e) and (f) in FZ area at (100X) and (200X).

Figures 5e, 5f, highlight a Widmastätten ferrite-pearlite/pearlite coarse structure with acicular separation in dendritic grains due to fast cooling. This structure has low ductility and tenacity. By removing from the seam area, the heating temperature decreases, the grains diminish and the Widmanstätten character decreases gradually. The columnar, lacy and dendritic grains, specific of FZ can increase accelerated corrosion in the stitch thickness, as evidenced by the polished and unetched section of figures 6a and 6b. In this heterogeneous, tense structure, the direction of propagation of catastrophic rupture through HAZ is highlighted. In the welding cord, in the FZ area, the microstructures of figure 6 shows the Widmanstätten coarse ferrite-pearlite structure with columnar, lace-shaped separations within the grains. The presence of non-metallic inclusions (welding slag), porosity, shape and distribution of heterogeneous structural constituents can increase an accelerated local corrosion.

IOP Conf. Series: Materials Science and Engineering **485** (2019) 012024 doi:10.1088/1757-899X/485/1/012024

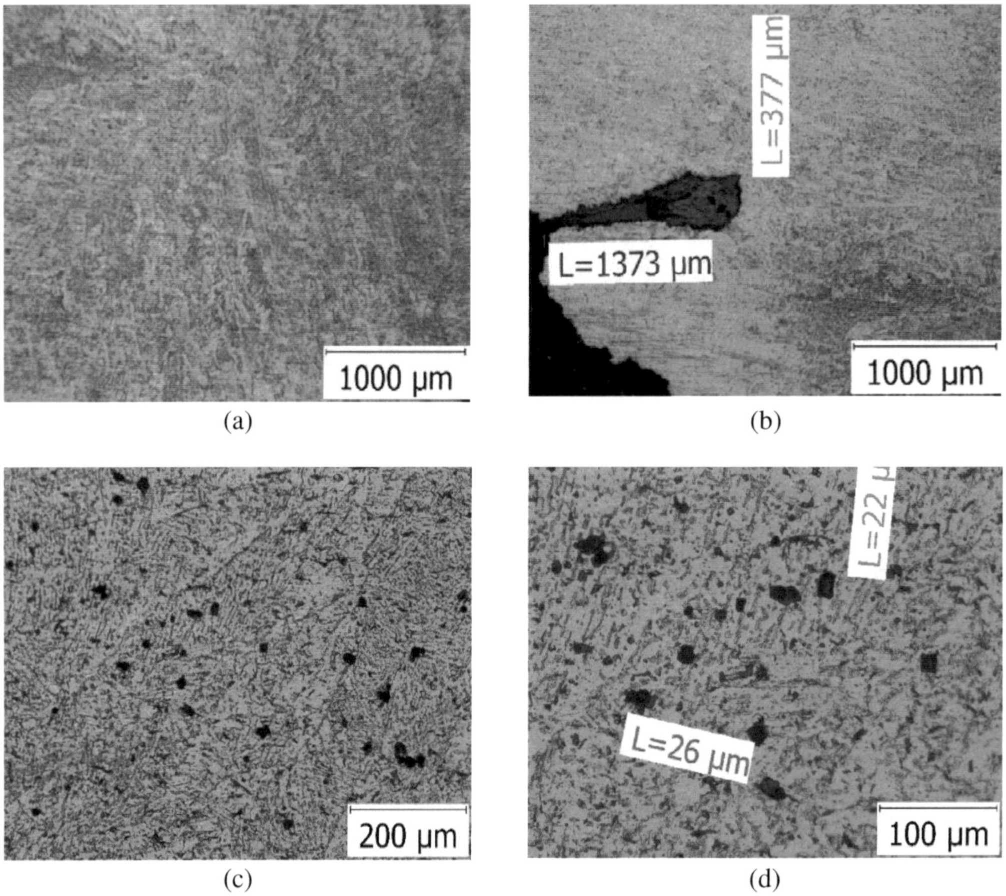

Figure 6. Microstructures of sample 1 (P1) in cross section of the compressed air cylinder wall: (a) and (b) in the FZ area (50X), (c) and (d) in the FZ area at (200X) and (400X).

The pitting (punctiform) corrosion as indicated in figures 6c, 6d, is manifested more pronounced. The rupture was propagated, as shown, in the most dense area in local [1, 8- 11] corrosion microvolts.

Intercrystalline corrosion associated with pitting corrosion developed between the dendrite branches leads to intense metal degradation in wall depth and, in extreme cases, causes the perforation of the metal wall, obviously, increasing the development of explosive rupture [12-16].

4. Conclusions

The steel sheet from which the compressed air cylinder was made corresponds chemically and structurally for this purpose.

The damage by explosion of compressed air cylinder was caused by excessive corrosion inside the cylinder. The corrosion was occurred generalized throughout the cylinder, but catastrophic (intergranular, pitting) only in the welding cord and heat-affected zone (HAZ).

The area from which explosive rupture was initiated was the heat-affected zone (HAZ) in the inner area of the weld, because it had a significantly lower thickness than the thickness of the welding cord. Excessive corrosion of the welding cord (FZ area) and the heat-affected zone (HAZ) was increased by the tensions and structural conditions different from those specific to the steel sheet of the base material, as well as by the manufacturer's location of the condensate drain hole near the longitudinal welding cord (the welding cord has been in most of the time in contact with the condensate water coming from the city traffic air which is the more polluted).

IOP Conf. Series: Materials Science and Engineering **485** (2019) 012024 doi:10.1088/1757-899X/485/1/012024

References

[1] Bosich J. 1977 Prevention de la corrosion P.Y.C.

[2] Uhlig, H.H., Revie R. W. 1985 Corrosion and Corrosion Control, 3rd ed., Wiley, New York, p. 347.

[3] Braham C. 1986 Corrosion sous contraintes susceptibilité d'un acier inoxydable austenitique. Influence de l'ecruisage et des contraintes residuelles, Materiaux et Technique.

[4] Constantinescu A. 1976 Detectarea şi măsurarea coroziunii, Editura Tehnică Bucureşti .

[5] Dragulanescu E., D.Iordache, G.Oprea, I.Gherghescu 1998 Considerations sur les relations de fragmentation des copeaux utilisables en conception des outils de coupe *UPB Scientific Bulletin, Series D* vol.60 **3-4** pp 91-94

[6] Gherghescu I. Le role et la place des traitements thermiques dans la science des materiaux 1999 *UPB Scientific Bulletin, Series B: Chemistry and Materials Science* 61 **1-2** pp 279-284 ISSN 1454-2331

[7] I. Gherghescu, V.Constantin 1998 Correlation du comportement en corrosion d'un alliage a memoire de forme Ni50Ti48Nb2 avec l'elaboration *UPB Scientific Bulletin Series B: Chemistry and Materials Science* Vol.60 **3-4** pp105-110

[8] Jérôm M, Galland J and Perrot P 1993 Évolution d'un processus de corrosion généralisée. Analyse critique et comparative de différents méthodes électrochimiques, Matériaux et Techniques, pp 10-11.

[9] Manemc J 1997 Caracterisation des états de surface en métallurgie et plus particulierement dans le domaine des aciers spéciaux, *Matériaux et Techniques*

[10] M. Sohaciu, S. Ciucă, D. Savastru, G. Coman, A. Predescu, A. Berbecaru, C. Cotruţ, E. Matei, I. A. Gherghescu, C. Predescu 2016 Influence of turbine blades fabrication conditions on their lifetime *Optoelectronics and Advanced Materials – Rapid Communications* 10 **3-4** pp. 257 – 261

[11] M. Ciurdaş, I. A. Gherghescu, S. Ciucă, A. D. Necşulescu, C. Cotruţ, R. E. Dumitrescu 2018 Heat Treatment Influence on the Corrosion Resistance of a Cu-Al-Fe-Mn Bronze, *Rev Chim-Bucharest* 69 **5** pp. 1055-1059

[12] I.Gherghescu, V.Constantin 1998 Comportement en corrosion d'un alliage a memoire de formeNi50Ti48Nb2 dans des milieux chimiques divers *UPB Scientific Bulletin Series B: Chemistry and Materials Science* 60 **3-4** pp 137-142

[13] Oniciu L. 1986 Coroziunea metalelor Editura Ştiinţifică şi Enciclopedică Bucureşti.

[14] Uhlig H. 1970 Corrosion et protection Dunod Paris.

[15] Zamfir S., Vidu R. and Brînzoi V. 1994 Coroziunea materialelor metalice E.D.P. Bucureşti

[16] Gurau G, Gurau C, Bujoreanu LG, Sampath V 2017 *IOP Conf. Ser. Mater. Sci. Eng* ICIR EUROINVENT (Iasi) Vol 209

IOP Conf. Series: Materials Science and Engineering **485** (2019) 012025 doi:10.1088/1757-899X/485/1/012025

The influence of vacuum degassing on improving the quality of ship's steels

F Potecaşu*, C Istrate and O Potecaşu
Faculty of Engineering, "Dunărea de Jos" University of Galati, Domnească Street, 47, RO-800008, Galati, Romania

*fpotec@ugal.ro

Abstract: In this paper, the authors chose to study the non-alloy steel grade S235JR, developed in the LD (Lintz-Donawitz) converter from Arcelor Mittal Galati plant, considering the quality and purity improving of the produced steel as a main objective. The correlation of mechanical properties with steel purity, as a result of the liquid steel processing to the continuous casting machine (CC) using three cases was followed. In the first case, the final chemical analysis of the liquid steel was taken at the end of the blowing operation without desulphurisation and degassing. In the 2nd case, the final chemical analysis of the liquid steel for CC was taken after the desulphurisation and chemical and thermal homogenization in the Ladle Furnace (LF) installation. In the 3rd case, the final chemical analysis of the liquid steel for CC was taken after the vacuum degassing, i.e after the treatment in the Ruhrstahl Heraeus (RH) installation.

1. Introduction

Steel is an important material in the dynamics of today's industrial development. The increased production is accompanied by a continued concerning regarding the production technologies improvement requirements regarding the metallurgical products quality. The naval board is a thick, hot rolled sheet with high corrosion and abrasion resistance, used in the shipbuilding industry for shipbuilding and various component parts. In recent years, safety and sustainability of ships and the marine environment protection have become of great importance. An increase in steel production for the shipbuilding sector by introducing and improving fast procedures regarding steel production and casting in Arcelor Mittal Galati Company was achieved. Thus, the Linz-Donawitz (LD) process by oxygen injection, leading to an increased productivity and cost reduction in manufacturing was introduced [1, 2]. This process has been improved and steel production in combustion insulated converters has been created, which generated metallurgical products with high quality. In order to meet the increasing requirements of steel quality, a series of technologies and techniques to treat the liquid steel are currently widely applied, aiming to transfer (or finish) the metalurgical processes outside of the aggregate [1- 5]. A particular feature of these steel treatment processes lies in the fact that, in most cases, an conjugate effects (deoxidation, desulphurisation, degassing) [6-11] are obtained. These effects lead to the following main technical and economical advantages: reducing fuel and electricity consumption as a result of reducing batch time in primary aggregates; improving the steel quality by reducing the content of non-metallic and gas inclusions, obtaining a high accuracy and homogeneous chemical composition; reducing the consumption of ferroalloys for alloying due to better assimilation of alloying elements; automatic management of technological processes [10-16].

2. Materials and methods

The chemical composition and mechanical properties for non-alloyed steel grade S235JR according to EN 10025-2 2004 / AC: 2005 are given in Table 1 and Table 2.

According to the classification system EN10025-2, it means: S - naval steels; 235 - minimum value of tensile strength for wall thickness ≤ 16 mm in N / mm^2; JR - quality class with minimum impact energy at 27J to +20° C; AR - the austenitic rolling path in which the recrystallization phase exists; S235JR, structural steel: $Rp_{0,2}$ = 235Mpa, 27J to 20°C (R) - steel standard 10025-2.

Table 1. Chemical composition for steel S235JR.

C max	Mn max	S max	P max	N max	Cu max
0,17 - 0,19	1,40 - 1,50	0,035 - 0,045	0,035 - 0,045	0,012 - 0,014	0,55 - 0,45

Table 2. Mechanical properties for steel S235JR according to EN 10025-2 2004 / AC:2005.

Tensile strength R_m [N/mm^2]	Minimum yield strength R_{eH} [N/mm^2]	Minimum elongation A% L	Minimum elongation A% T	Hardness HB	Minimum impact energy at 20^0 C KV[J]
360-510	235	26	24	104-154	27

The production of the S235JR naval steel was done in the LD converter. The mechanical properties correlation with steel purity, as a result of liquid steel processing to the continuous casting machine (CC) in three cases was followed.

In case 1, the final chemical analysis of the liquid steel for CC was taken at the end of the blowing operation, without desulphurisation and degassing.

Table 3. Final chemical composition of the liquid steel for CC without desulphurisation and degassing (charge A).

Series A	**Chemical composition, [%]**			
Step 1 (at the end of the blowing operation)	C=0.0201;	Mn=0.1341;	P=0.0108;	S=0.012;
Step 2 (the final chemical analysis in CC)	C=0.1328;	Si=0.1838;	Mn=0.6848;	P=0.0175;
	S=0.0067;	Al=0.0416;	Ti=0.0012;	V=0.0021;
	Ca=0.0005;	Cu=0.033;	Ni=0.0163;	Cr=0.0345;
	Mo=0.0034;	Nb=0.0013		

In the 2nd case, the final chemical analysis of the liquid steel for CC was taken after the desulphurisation, chemical and thermal homogenization in the Ladle Furnace LF.

In the 3rd case, the final chemical analysis of the liquid steel for CC was taken after vacuum degassing, i.e after the treatment in the Ruhrstahl Heraeus (RH) installation. For the analysis, three differently charges of liquid steel processing of (A, B, C) were selected.

In tables 3, 4, 5 the final chemical composition for each charge was presented. In the first step of each series only four elements were determined: carbon, manganese, phosphorus and sulphur.

IOP Conf. Series: Materials Science and Engineering **485** (2019) 012025 doi:10.1088/1757-899X/485/1/012025

Table 4. Final chemical composition of the liquid steel from CC after desulphurization, chemical and thermal homogenization in Ladle Furnace installations (charge B).

Series B	**Chemical composition, [%]**			
Step 1 (at the end of the blowing operation)	C=0.0244;	Mn=0.1974;	P=0.0217;	S=0.017;
Step 2 (the final chemical analysis for CC after desulphurisation, chemical and thermal homogenization)	C=0.1221; S=0.0081, Cu=0.012; Nb=0.002;	Si=0.2116; Al=0.0435; Ni=0.0079; B=0.0003	Mn=0.6606; Ti=0.0021; Cr=0.0199;	P=0.0161; V=0.002; Mo=0.0007;

Table 5. Final chemical composition of the liquid steel from CC after vacuum degassing and treatment in Ruhrstahl Heraeus (RH) (charge C).

Series C	**Chemical composition, [%]**			
Step 1 (at the end of the blowing operation)	C=0.0282;	Mn=0.068;	P=0.0046;	S=0.0103;
Step 2 (the final chemical analysis from CC after vacuum degassing)	C=0.119; S=0.003; Ni=0.008; Nb=0.002,	Mn=0.7158; Al=0.021; V=0.002; B=0.0001;	Si=0,203; Cu=0.011; Mo=0.001; N_2=0.0069;	P=0,010; Cr=0.022; Ti=0.002; H_2=0.0002

3. Results and discussions

3.1 Mechanical tests

Table 6 presents the mechanical properties values of S235JR non-alloy steel samples obtained for different liquid processing cases (A, B, C series). The mechanical tests were performed on samples corresponding to the different liquid processes for three charges. Tensile strength (R_m), yield strength (R_{eH}) and elongation (A) were determined. The results obtained and their histograms are shown in figures 1, 2, 3.

All the analyzed samples are in correlation with the S235JR non-alloy steel grade prescriptions, but the highest tensile strength values were obtained for the C-series with the lowest grade of non-metallic inclusions, corresponding to the finest grain, with the largest real grains.

Table 6. Mechanical properties values for series A (without desulphurisation and degassing), series B (with desulphurization, chemical and thermal homogenization in Ladle Furnace installation), series C (with vacuum degassing).

		R_m, [MPa]	**R_{eH}, [MPa]**	**A, [%]**
Series A	Sample 1	425	304	34
	Sample 2	430	318	34
Series B	Sample 1	414	275	30
	Sample 2	438	292	33
	Sample 3	438	292	33
Series C	Sample 1	438	308	30
	Sample 2	441	296	30
	Sample 3	469	366	31

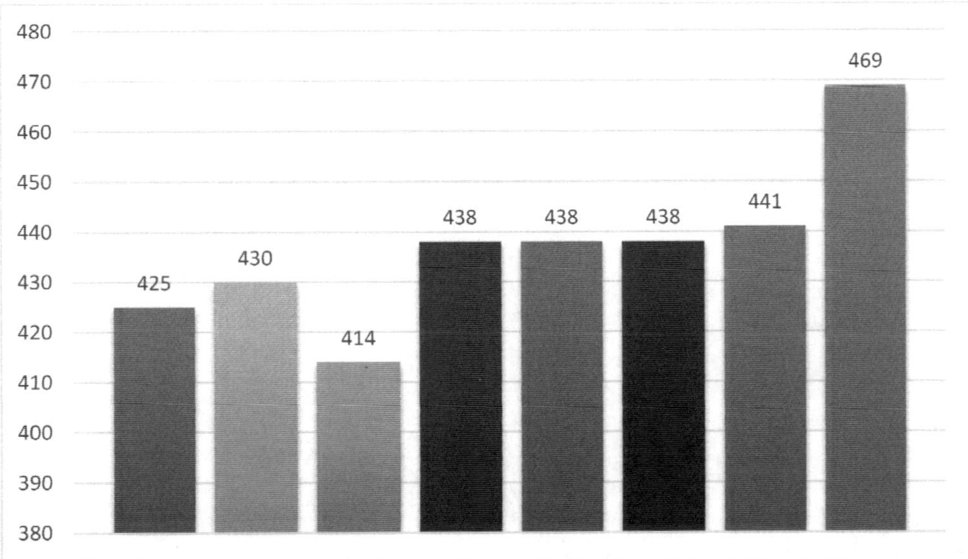

Figure1. Tensile strength variation Rm [MPa] for all samples from A series (without desulphurisation and degassing), B (with desulphurization, chemical and thermal homogenization in Ladle Furnace installations), C (with vacuum degassing).
Series A - (Rm=425, 430 [MPa])
Series B - (Rm=414, 458, 438[MPa])
Series C - (Rm=438, 441, 469 [MPa])

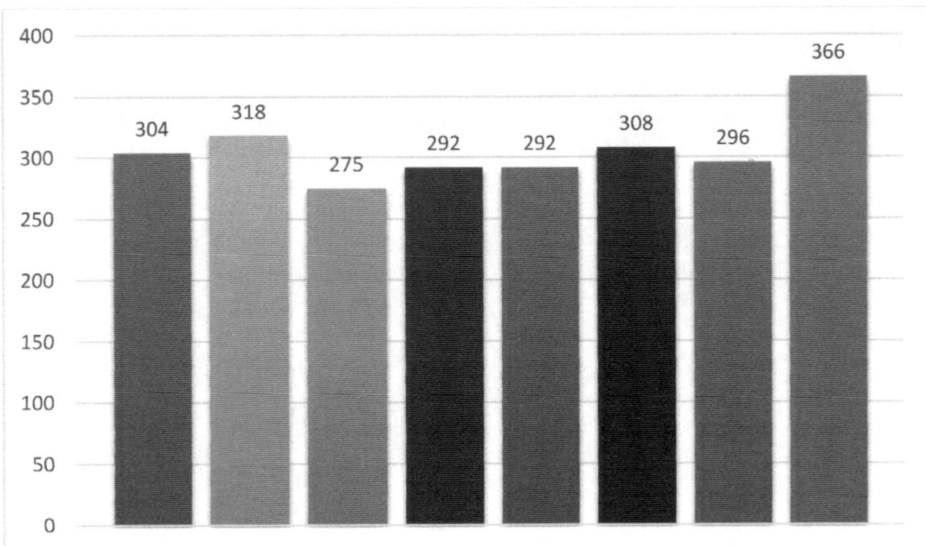

Figure 2. Yield strength variation ReH [MPa] for all samples from A series (without desulphurisation and degassing), B (with desulphurization, chemical and thermal homogenization in Ladle Furnace installations), C (with vacuum degassing).
Series A – (ReH=304, 318 [MPa])
Series B – (ReH =275, 292, 292 [MPa])
Series C – (ReH =308, 296, 366 [MPa])

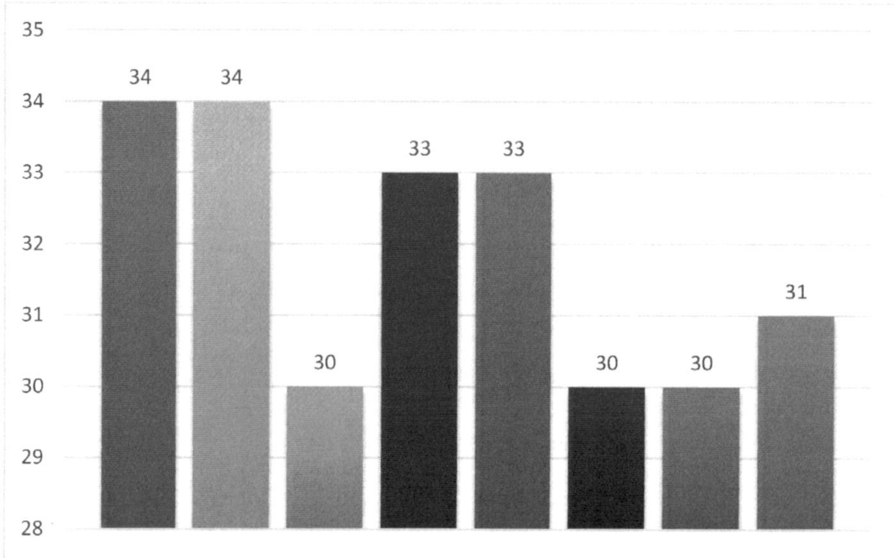

Figure 3. Elongation A [%] variation for all samples from A series (without desulphurisation and degassing), B (with desulphurization, chemical and thermal homogenization in Ladle Furnace installations), C (with vacuum degassing).
Series A – (A =34 , 34 [%])
Series B - (A =30, 33, 33 [%])
Series C – (A =30, 30, 31 [%])

3.2. Microscopic analysis of non-metallic inclusions

Qualitative analysis of non-metallic inclusions according to STAS 5949-80, on polished and unetched samples was performed. The qualitative analysis of non-metallic inclusions in the light field, provides information on: contour, shape, distribution, fragility, influence of etchants. In the dark field, transparency (luminous inclusions), opacity (dark inclusions), and the color of the inclusions are highlighted.

The quantitative determination of non-metallic inclusions in steels is made using the comparison method with the standard charts and indicates the steel impurity degree. The polished and unetched samples are analyzed under a light microscope at a magnification of 100x.

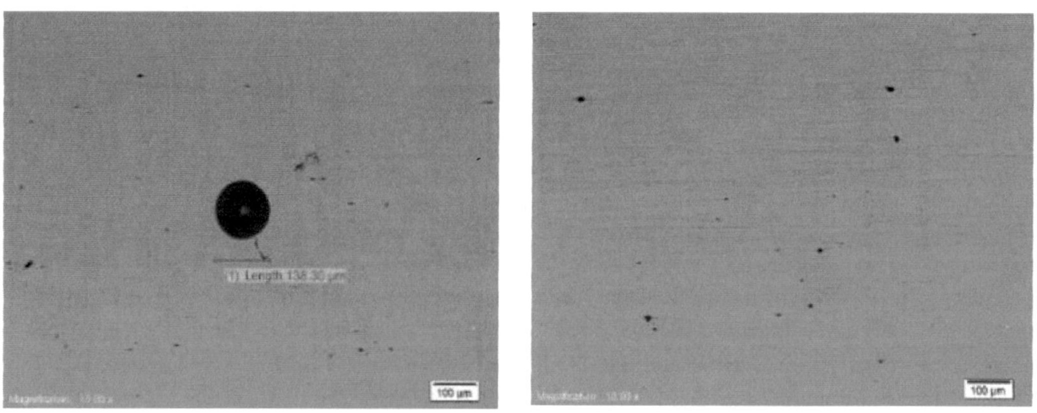

(a) (b)

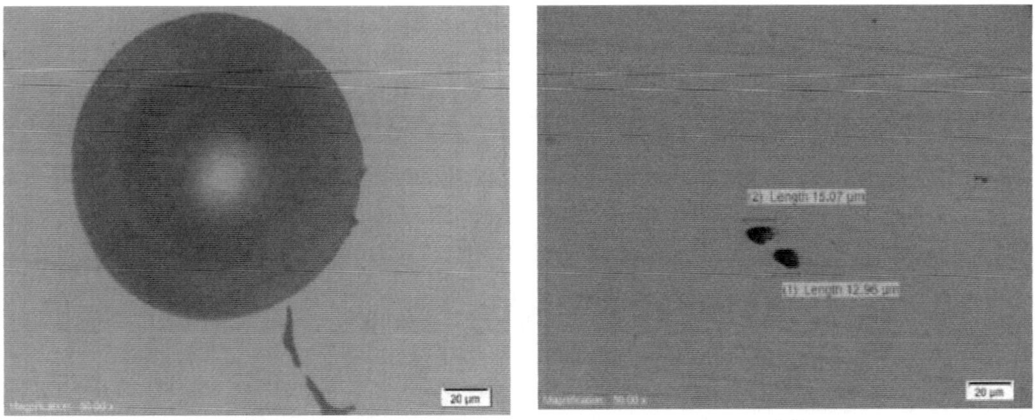

(c) (d)

Figure 4. The appearance of non-metallic inclusions on the unetched surface of the A series (without desulphurisation and degassing): (a) and (b) 100X; (c) and (d) 500X.

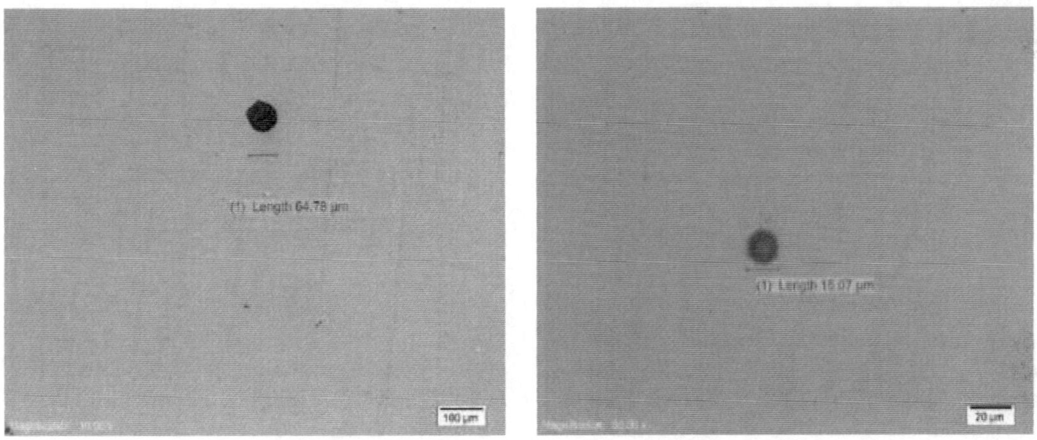

(a) (b)

(c) (d)

Figure 5. The appearance of non-metallic inclusions on the unetched surface of the B series (with advanced desulphurization, chemical and thermal homogenization): (a) and (b) 100X; (c) and (d) 500X.

The inclusions types that are determined by different standards charts are: OL + OP - linear and point oxides; S-sulphides; SF + SP - fragile and plastic silicates and SN - non-deformable silicates; NT + NA - titanium and aluminum nitrides. The analysis consists in examining the sample at a microscope on the whole surface prepared and comparing the most impure visual fields with the standard charts with levels from 1 to 5.

In figures 4, 5, 6 are presented the microstructures of unetched samples for non-metallic inclusions rate for the charges corresponding to the series A, B, C. The inclusion level for the A series (without desulphurisation and degassing) was P_A = 3.5. For the B-series (with advanced desulphurization, chemical and thermal homogenization), the inclusion level was P_B = 2.5. For C series (with vacuum degassing), the inclusion level achieved was P_C = 1.0.

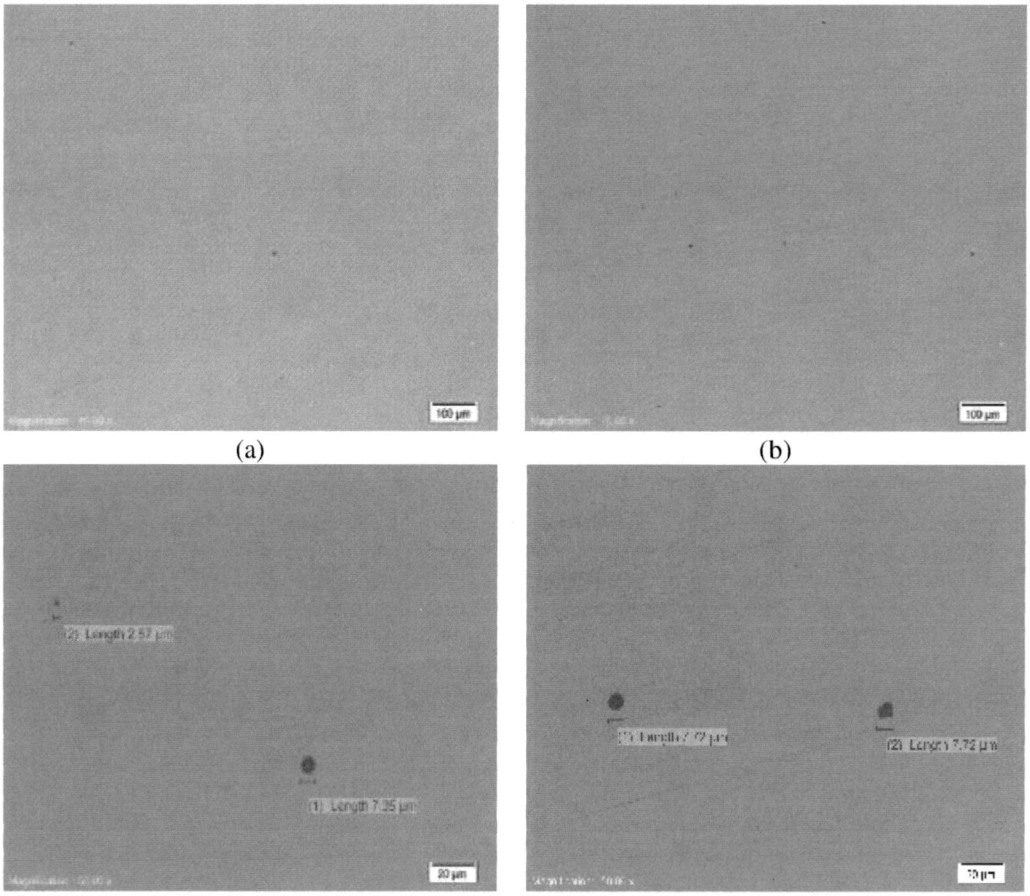

(c) (d)

Figure 6. The appearance of non-metallic inclusions on the unetched surface of the C-series (with vacuum degassing): (a) and (b) 100X, (c) and (d) 500X.

The objective of this research was to demonstrate that improving the steel quality sheets by production of charges with increasingly low inclusions (i.e with increasing purity) had as effect the finishing and homogenize the grains, tensile strength improvement, elongation remaining at the values required by the current standard.

The research has shown that the target requirements can be achieved through a rigorous technology and process-controlled processing. Increased quality means fewer scrapes and relatively lower production costs.

IOP Conf. Series: Materials Science and Engineering **485** (2019) 012025 doi:10.1088/1757-899X/485/1/012025

4. Conclusions

In order to improve the steel quality and purity, an important place is occupied by the functional and technological adjustment of RH-type vacuum steel treatment plants equiped.

Research has shown that treatment in RH has led to: an increased in purity due to degassing; reducing the H_2 content; reducing of O_2 content by vacuum de-oxidation; decreasing content of non-metallic inclusions; the possibility to produce steels with narrow limits of the chemical composition; vacuum steel carburizing, reducing the liquid metal viscosity; reducing the N_2 content.

References

[1] Butnariu I and Geantă V. 1993 Tehnologii speciale de elaborare şi rafinare a oţelului ed. Universitatea Politehnica Bucureşti.
[2] Choudhury A 1990 *Vacuum Metallurgy* ASM International
[3] Dragomir I 1996 *Câteva aspecte ale tratamentului oţelului în vid* Metalurgia **3**
[4] Hahn F and Haasteren H 1993 *Proc. of Int. Symp. on Low-Carbon Steels for the 90's* ed. The Minerals, Metals and Materials Society Pittsburg
[5] Inoue, S.; Furuno, Y.; Usui, T. & Miyahara, S. 1992 Acceleration of Decarburization in RH Vacuum Degassing Process *ISIJ International* **32** pp.120-125.
[6] Lee M, Simpson I and Jahanshahai S 2001 *Proc. of Oxygen in Steelmaking: Towards Cleaner Steels* London
[7] Kinsman G, Hazedean G and Davies M 1989 Physico-chemical factors affecting the Vacuum Deoxidation of Steel *J. of the Iron and Steel Institute* **207** 11
[8] Kumar S, Nerurkar H and Ballal N 1993 *Proc. of Int. Symp. on Low-Carbon Steels for the 90's* The Minerals Metals and Materials Society Pittsburg
[9] Kumar S and Reddy R 1993 *Proc. of Int. Symp. on Low-Carbon Steels for the 90's*, ed. The Minerals, Metals and Materials Society Pittsburg
[10] Tripşa I and Pumnea C 1984 *Retopirea şi rafinarea oţelurilor* ed. Tehnică Bucureşti
[11] Tripşa I and Pumnea C 1981 *Dezoxidarea oţelurilor* ed. Tehnică Bucureşti
[12] Shima H, Hoshijima Y, Fukuda K and Onuki K 1995 *7th Int. Conf. on Injection Metallurgy* Lulea Sweden
[13] Millman M 1999 Secondary steelmaking developments in British Steel *Ironmaking and Steelmaking* **26** 3
[14] Stephen F 2009 Improvement of the desulfurization process by slag composition control in the ladle furnace *Minerals and Metallurgical Engineering* p 66
[15] Andersson M, Mselly M and Nzott M 1999 *ISIJ International* **39** pp 1140-1149
[16] Gurau C, Gurau G, Alexandru P 2013 *METALURGIA INTERNATIONAL* Vol. 18 37-41

IOP Conf. Series: Materials Science and Engineering **485** (2019) 012026 doi:10.1088/1757-899X/485/1/012026

Magnetoelastic properties in polycrystalline Fe-Pd based ferromagnetic shape memory alloys

M Sofronie[1], B Popescu[1], AD Crisan[1], AR Lupu[2,3], F Tolea[1], M Valeanu[1]

[1] National Institute of Materials Physics, POB MG-7, 77125 Bucharest-Magurele, Romania

[2]Victor Babes National Institute of Pathology, Bucharest, Romania

[3]"Cantacuzino" National Medico – Military Institute for Research and Development, Bucharest, Romania

felicia.tolea@infim.ro

Abstract. Two series of Mn substituted Fe-Pd ferromagnetic shape memory ribbons were studied by X-ray diffraction, calorimetric and thermo-induced strain measurements. Linking the structural, magnetic and thermal expansion analysis on pristine ribbons and those subjected to different subsequent thermal treatments, highlights the role of substitution and of heat treatment of the martensitic transformation (MT). The melt–spinning technique removes the precipitation of undesirable b.c.t. irreversible phase and the addition of Mn, as a third alloying element, promotes the stabilization of the f.c.t. martensite. The linear thermal expansion measurements with and without static magnetic field provide information on the easy magnetization axis in the martensitic structure. Preliminary *in vitro* cytotoxicity tests generally indicate no toxicity or a tendency towards cell proliferation induced by the tested materials.

1. Introduction

Specific to the shape memory alloys (SMA) is the martensitic transformation (MT), a thermo-elastic, reversible structural phase transition between a high symmetry phase (austenite) and a lower one (martensite). On cooling, the high temperature austenite phase undergoes a diffusionless transformation in which atoms shift cooperatively reducing the symmetry and forming the low temperature martensite phase. Ferromagnetic SMA (FSMA) are materials with the MT temperature lower than magnetic order-disorder transition temperatures. The Fe-Pd based ferromagnetic shape memory alloys are promising materials for applications, due to better mechanical properties and a higher magnetic order-disorder transition temperature (~760 K) than Ni-Mn-Ga FSMA. Moreover, the biocompatibility is a significant benefit of Fe-Pd (30 at. %) compared to Ni_2MnGa based Heusler alloy allowing to use this material for biomedical applications, such as stents or pumps [1]. In this system was found that MT is characterized by various possible martensite phases with different degrees of distortions: face-centered tetragonal (f.c.t.) and body-centered tetragonal (b.c.t.). By cooling, a thermo-elastic, reversible transformation from face-centered cubic (f.c.c.) austenite to f.c.t. martensite takes place. When the cooling process continues the tetragonal distortion is enhanced until a new irreversible transformation (IMT) f.c.t.-b.c.t. may be evidenced. The martensitic transformation (MT) from f.c.c. to f.c.t. in disordered Fe-Pd alloys is observed in a narrow composition range, between 29 and 32 at. % Pd [2]. In order to avoid the formation of the undesirable b.c.t. martensite, so to stabilize the f.c.t. martensite, as well as to manipulate the transformation temperatures, the addition of a third alloying element became a challenge. The literature reports a significant increase of transformation temperature (333 K) and a decrease of the non-reversible transformation temperature for Fe-Pd-Mn ternary bulk system [3]. Rapid solidification, by using a melt-spinning technique, is an effective processing route to obtain ribbons with a non-equilibrium structure, in which the high temperature f.c.c. structure can be frozen as single phase [4]. The present paper emphasizes the influence of Mn addition and thermal treatments in $Fe_{69}Pd_{30}Mn_1$ and $Fe_{67}Pd_{30}Mn_3$ melt-spun ribbons on the MT evolution, magneto-elastic properties and biocompatibility, from the perspective of using these materials in applications.

2. Experimental

Alloy ingots of $Fe_{69}Pd_{30}Mn_1$ and $Fe_{67}Pd_{30}Mn_3$ were prepared under Ar protective atmosphere, by arc-melting of high purity elements and the procedure was repeated several times in order to ensure homogeneity. The procedure was following by melt-spun into ribbons using melt-spinning technique (3 mm wideness, 10-20 mm lengthiness and 25-30 μm thickness). The pristine ribbons (denoted PMn1 and PMn3) were thermal treated in vacuum quartz ampoules at 950°C for different duration, the samples being denoted ST (short time - 10min at 950°C) and LT (long time-30min at 950°C). Each thermal treatment (TT) was followed by a direct quenching of the sample in iced water. The structural investigations were done by X-ray diffraction (XRD) using a Bruker D8 Advantage diffractometer in the Bragg–Brentano geometry, with Cu Kα radiation, at room temperature (RT). The phase transformation temperatures were determined by using a differential scanning calorimeter (DSC) model 204 F1 Phoenix (Netzsch), with a scanning rate of 10 K/min. The chemical composition and the surface morphology of the ribbons were examined by Energy Dispersive X-Ray Spectroscopy (EDX) and Scanning Electron Microscopy (SEM) in a Zeiss Evo 50 XVP microscope. Magnetic measurements were performed by MPMS-SQUID-QD magnetometer in RSO mode, with magnetic field along ribbon direction. The linear thermal expansion (LTE) measurements have been realized by means of two strain gauges, one was glued along the ribbon length and the other serving as reference. The "Vishay Micro-Measurements Model P3 strain indicator and recorder" and the magnetic platform Cryogenic Ltd. were used for fulfilling these determinations. In-plane strains, with the field applied parallel to the ribbon plane were record as a function of temperature. Osteoblast-like cells (MG63) were seeded in 24-well culture plates, in volumes of 1 ml of Dulbecco's Modified Eagle Medium/Nutrient Mixture F-12 (DMEM-F12) complete culture medium (10% fetal bovine serum-FBS), 1% L-glutamine and 1% antibiotics), at a density of 10^5 cells/cm^2. After cell adherence and growth (24h), the culture medium was discarded and replaced with fresh medium (1 ml/well) containing floating ferromagnetic ribbons with total surface area of 10, 10/2 and 10/4 mm^2 respectively. The cells were treated for 24h. Following treatment, the culture medium was removed and replaced with MTT [3-(4,5-Dimethylthiazol-2-yl)-2,5 diphenyltetrazolium-bromide] solution (300 ul/well of 1 mg/ml MTT in phosphate buffered saline (PBS). After 2 hours of incubation, the MTT solution was discarded and 300 ul of dimethyl sulfoxide (DMSO) were added to each well in order to dissolve the produced formazan crystals. The absorbance of the formazan solution was determined at 540 nm with a Thermo Multiskan EX spectrophotometer. Cell viability was quantified and expressed as percent versus control.

3. Results and discussions

DSC scans have been performed in wide temperature range, between 90 and 400 K and it is important to note that MT was evidenced, without IMT for all investigated alloys, as a result of the melt spinning technique. Figure.1 shows DSC records during the cooling/heating sequences which have allowed highlighting MT and identifying its characteristic parameters: martensite start (Ms) and finish (Mf), austenite start (As) and finish, (Af) and the transformation enthalpy (H), calculated as average between the direct and reverse transformation enthalpy. DSC results reveal sharp peaks, the transformation enthalpy is enhanced and Ms drastically increase (~ 40 K) with increasing of Mn substitution [5]. The martensite start temperature (Ms) and the transformation enthalpy (associated with the DSC peak area) values are depicted in Figure.2. The structural transformation on the PM3 sample could not be evidenced by this method probably due to its low enthalpy. Can be seen a surprisingly lower MT temperature for

IOP Conf. Series: Materials Science and Engineering **485** (2019) 012026 doi:10.1088/1757-899X/485/1/012026

this sample, determined by thermo-magnetic measurements (marked with dashed line). By reaching the atomic order after the thermal treatments, the effect of Mn substitution is to slightly enhanced of the MT temperatures (~5 K) for Mn 1 samples and drastically increasing (~60 K) for Mn3 samples.

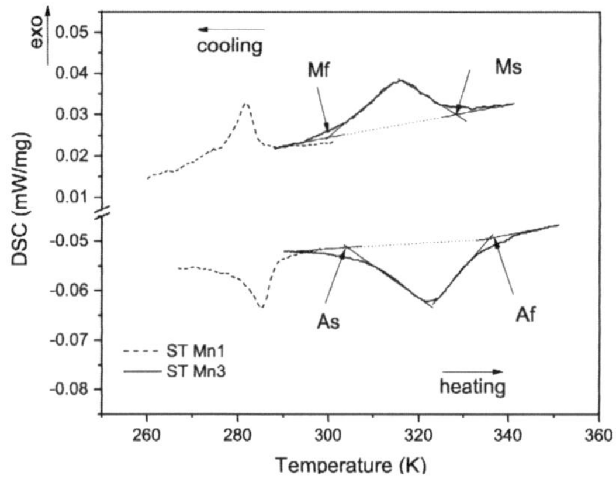

Figure.1. DSC scans after cooling/heating sequence for ST Mn1 and ST Mn3.

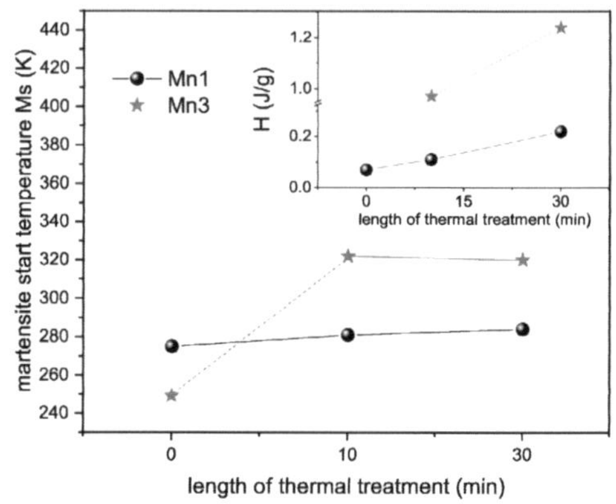

Figure.2. Ms for Mn1 and Mn3 samples (dashed line indicates Ms for P Mn3, determined by thermo-magnetic measurements); Inset: the transformation enthalpy (H).

The transformation enthalpy has a monotonous increase with TTs for Mn1 samples. It has to be noticed the rise of the transformation enthalpy, up to 1.24 J/g, (for LT Mn3), close to that registered in Fe-Pd singlecrystal (30at.% Pd), but 2.5 times higher than one for polycrystalline alloys [2]. The X-ray diffraction patterns (not shown) at room temperature indicate that pristine samples (P Mn1 and P Mn3) exhibit a single highly textured f.c.c. phase, the high temperature stable phase retained in the ribbons due to the rapid cooling, specific for the melt spinning technique. The TTs at high temperature (950°C) induced the f.c.c. microstructure ordering and elimination of the quenched-in stress tensions arising from preparation technique for Mn1 samples. A mixture of f.c.c./f.c.t. phases and rather small content of another phase rich in Pd and Mn (@) was found for TTs Mn3 samples. The SEM images (Figure.3. a-b) indicate small grains and typical twin variant band morphology characteristic to the martensitic phase for Mn3 samples. In addition, the presence of precipitates can be distinguish in the composite matrix, small (0.5-1μm) and rather numerous, which are mainly located at the grain boundaries.

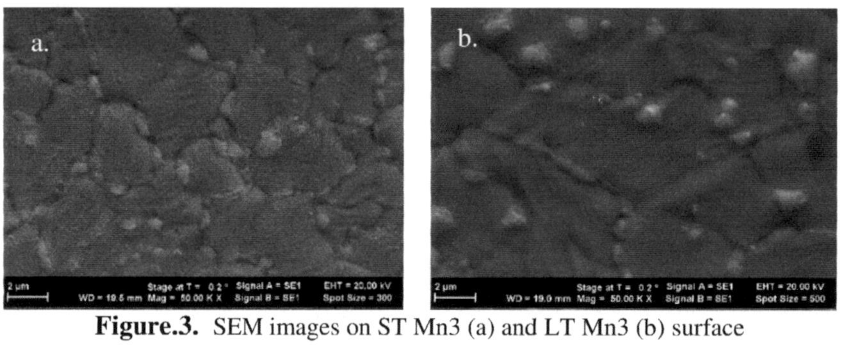

Figure.3. SEM images on ST Mn3 (a) and LT Mn3 (b) surface

The presence of precipitates is the results of the particularly low solubility of Mn in the Fe-Pd system [3]. EDX analysis indicates a decreasing of Pd content in the ST Mn3 and LT Mn3 matrix, supported by the existence of precipitates, which have the chemical compositions close to $Fe_5Pd_6Mn_2$. For Fe-Pd shape memory alloys, a decrease in the Pd content results in an increase of the MT temperatures [6], in agreement with calorimetric results (Figure.2.).

IOP Conf. Series: Materials Science and Engineering **485** (2019) 012026 doi:10.1088/1757-899X/485/1/012026

The thermo-magnetic measurements performed at higher temperature (above to 300 K) show the influence of the matrix composition changes on the order-disorder magnetic transition for ST Mn3 and LT Mn3 samples (Figure.4). On the heating curve, the magnetization variation in 300-400 K temperature range is the results of the magneto-crystalline anisotropy change when the structure recovered the austenitic state, after reversible MT. In addition, the abrupt decreases of the magnetization curve observed in 500-600K temperature range, indicates the magnetic order-disorder transition (T_C- Curie temperature). This temperature was determined by derivate of magnetization versus temperature (Inset Figure.4.). During the thermomagnetic measurements, the temperature is increased and then lowered slowly, a process that induces the decomposition of the metastable f.c.c. phase in $L1_0$–f.c.t. phase and a Fe-rich phase with b.c.c. structure. For ST Mn3 sample the decomposition of the f.c.c. phase is not complete; so, on the cooling branch besides to the magnetization jump at 770K, the Curie temperature for the $L1_0$ phase, a small amount of the austenite with f.c.c. structure having T_C= 615 K may be evidenced.

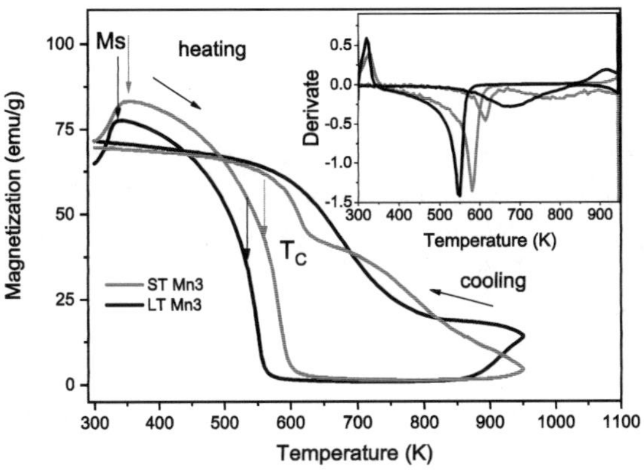

Figure.4. Low field thermo-magnetic measurements (100 Oe) at high temperatures (above to 300 K) of ST Mn3 and LT Mn3 ribbons (inset: Derivate of magnetization versus temperature)

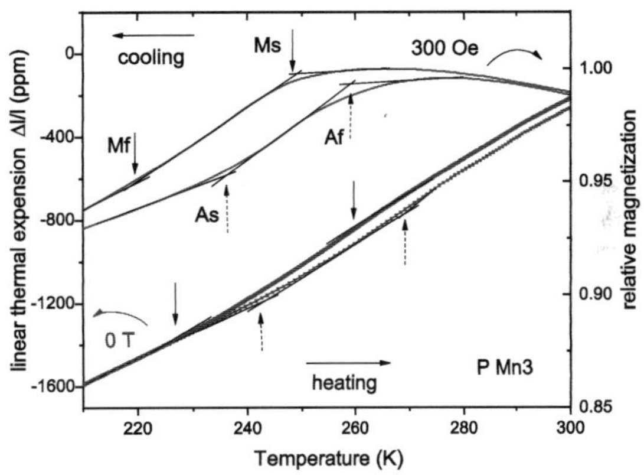

Figure.5. LTE curve during ZFC (zero field cooling) process versus relative magnetization (300 Oe) for ST Mn3 and LT Mn3 samples. The arrows indicate Ms, Mf (solid line) and As, Af (dash line) temperatures

The LT Mn3 sample has a complete decomposition of f.c.c. phase in the stable $L1_0$ phase (T_C = 680 K) and a second Fe-rich one with T_C > 900K on cooling branch of magnetization curve and the absence of any other phases. This evolution of thermo-magnetic curves is strongly influence by the process of nucleation and subsequent growth of the precipitates that are influence by the thermal treatment duration. Therefore, increasing the length to 30 min induces a homogenization of multi-phase structure and a decrease of T_C, with reduction of Fe concentration in the sample matrix.

The linear thermal expansion (LTE) results provide information about the relative length-change (Δl/l) of samples in cooling/heating sequence, with and without applied magnetic field. The change in length is proportional with temperature for a solid material and can be expressed as $\Delta l/ l_i = \alpha \Delta T$, ($\Delta l = l_f - l_i$ and $\Delta T = T_f - Ti$) where l_i and l_f represent, the initial and final lengths with the temperature change from T_i to T_f and α is the coefficient of LTE. First, the LTE curves were measured by cooling and heating between 320 and 200 K without magnetic field followed by the same cooling/heating sequence with magnetic field (1T) applied in austenitic phase, parallel to the ribbons length. On cooling and passing through the MT in ZFC (zero field cooling) mode, the ribbons undergo a continuous contraction as a result of the tendency of the variants to accommodate the strain, without preferred direction of growth, in order to minimize the elastic energy and maintain the shape of the ribbons [7]. It's important to notice that, the ZFC curve shows the spontaneous contraction and a weak thermal hysteresis characteristic to MT in 225-270 K temperature range for P Mn3. Therefore, the MT temperatures were estimate, with values in close agreement with those obtained by the thermomagnetic measurements at low magnetic field (300 Oe) (Figure.5). A slightly discrepancy between these temperatures (~10K) obtained from thermal expansion and thermomagnetic data arises from instrumental error, especially for the thermomagnetic ones (it is impossible to place the thermocouple on the sample surface). Thus, the linear thermal expansion measurements in ZFC mode represent a powerful tool to evidence MT characteristic temperatures in SMA. The MT has been identified by LTE measurements (Figure.6. and Figure.7.) and the values are in close agreement with those obtained by calorimetric measurements (Figure.2.). The arrows indicate Ms (solid line) and Mf (dash line) temperatures. Due to the high temperature limitation of the LTE measurement system (320K), the martensite start (Ms) temperature for ST Mn3 and LT Mn3 samples cannot be detected (Figure.7.).

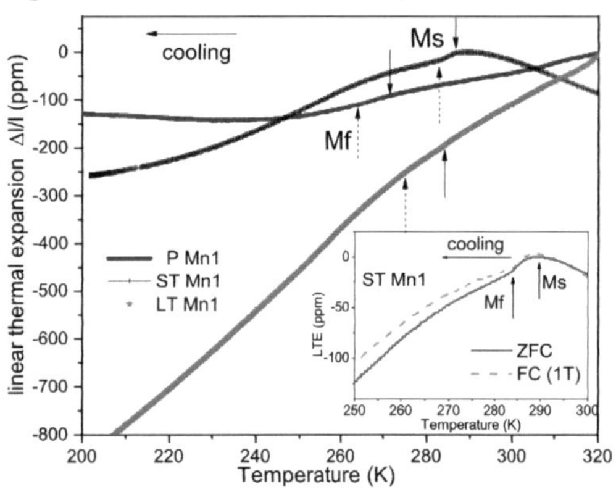

Figure.6. LTE curves during cooling (ZFC) for Mn1 samples. Inset: LTE curves during the cooling process, with (FH) and without (ZFH) magnetic field on ST Mn1.

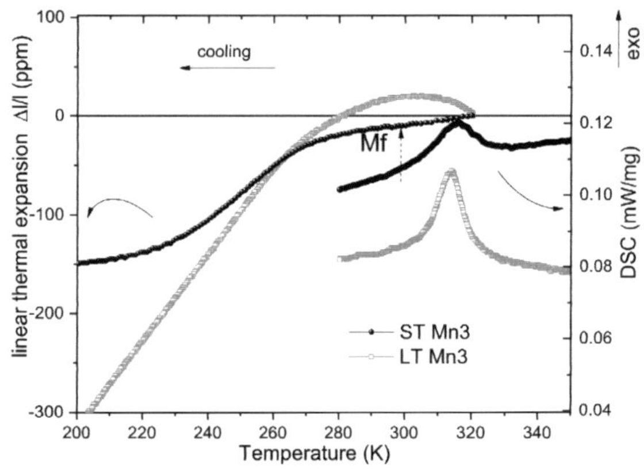

Figure.7. LTE curves during cooling (ZFC) versus DSC for ST Mn3 and LT Mn3 samples.

A recent study indicates that after annealing in $Fe_{66.8}Pd_{30.7}Mn_{2.5}$ bulk material, the dislocations and martensite variants increase their mobility. Hence, after TT, the decrease of quenched-in-defects, leads to decrease in the amount of obstacles which favour the twin boundaries movement. The LTE values at 200K provide the next comparison: $\Delta l/l_{LT\ Mn1} > \Delta l/l_{ST\ Mn1} > \Delta l/l_{P\ Mn1}$ (Figure.6.) and $\Delta l/l_{LT\ Mn3} > \Delta l/l_{ST\ Mn3}$ (Figure.7.). This relationship points to an increase of the active phase fraction after subsequent TT accordingly increasing the amount of the austenitic phase able to undergo MT assumption sustained by a similarity in the enthalpy values (see Inset Figure.2.). The smallest values of thermal expansion achieved in Mn3 TTs ribbons as compared to Mn1 TTs, can be a consequence of the presence of small precipitates. The LTE behaviour on cooling/heating sequences under magnetic field (1T) (Inset Figure.6.) reflects the competition between the preferential orientation under magnetic field and the spontaneous self-accommodation of martensitic variants in each grain in the polycrystalline textured samples. When applied magnetic field in austenite state, decrease the temperature and passing through MT, the martensite variants nucleate and grow along the field direction, in order to minimize the Zeeman energy. This nucleation mechanism is influence by the grain size or the existence of precipitates, cracks within the grain or at the grain boundaries [8]. The variants with the easy magnetic axis along the applied field direction grow at the expense of others, giving rise to the reduction of the contraction in the field direction. In the Inset of Figure.6., the decrease of LTE values with applied magnetic field, bellow Mf, confirms the previous report that the long axis *a* is the easy axis of magnetization for the martensitic state in Fe-Pd system [9].

The preliminary cell viability results (Figure.8.) generally indicate no toxicity or a tendency towards cell proliferation induced by the tested materials. However, in the case of ST Mn3 sample, significant cytotoxic effects seem to occur particularly at the highest surface (concentration) tested. Taking into account the significant increase of cell viability (i.e. cell number) for ST Mn3_S ½, we consider that the viability decrease corresponding to ST Mn3_S1 sample is due to density dependent inhibition of cell growth in culture rather than its cytotoxic effects.

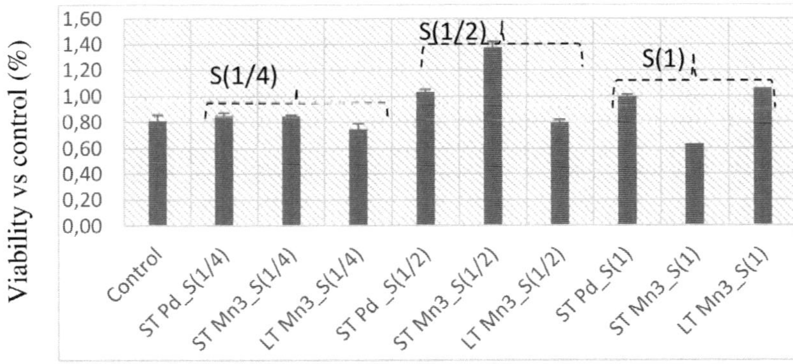

Figure.8. Cell viability for samples with different surface areas: S1 (10 mm^2), S1/2 (10/2 mm^2), S1/4 (10/4 mm^2)

Considering the tumor nature of the used cells, it is premature to conclude whether the obtained preliminary results indicate lack of cytotoxicity or possible tumor's cells proliferation. To assess the biocompatibility of the studied materials, further complex tests on multiple, normal and tumor, cell lines are required.

4. Conclusions

The $Fe_{69}Pd_{30}Mn_1$ and $Fe_{67}Pd_{30}Mn_3$ pristine ribbons have single f.c.c. phase with high atomic disorder, without irreversible martensitic transformation, as characteristic features induced by the processing route. Subsequent TTs promote an increasing of the MT temperatures as well as changes of the transformation enthalpy. The segregation and growth of precipitates after TTs for $Fe_{67}Pd_{30}Mn_3$, on the account of the main transformable phase is considered to contribute essentially to the suddenly increase of the MT temperature (via the depletion of the main phase in Mn and Pd content) and the smallest values of the linear thermal expansion. The magnetoelastic behaviour reflects the competition between the preferential orientation under magnetic field and the spontaneous self-accommodation of martensitic variants in each grain in the polycrystalline textured samples and indicates that the long axis *a* of the ribbons is the easy axis of magnetization for the martensitic state. The preliminary cell viability results indicate a dependent inhibition of cell growth in culture for $Fe_{67}Pd_{30}Mn_3$ ribbons, rather than a cytotoxic effect.

Acknowledgment
The authors thank PhD. M. Enculescu for assistance with the SEM facility. This work was supported by Grant of Romanian Ministry of Research and Innovation, CCCDI-UEFISCDI, project number PN-III-P1-1.2-PCCDI-2017-0062/contract no.58/component project no.2, within PNCDI III.

References

[1] Ma Y Zink M Mayr SG 2010 Appl. Phys. Lett. 96 213703.
[2] Sofronie M Tolea F Kuncser V Valeanu M Filoti G 2015 IEEE Trans. Mag. 51 2500404.
[3] Sánchez-Alarcos V Recarte V Pérez-Landazábal J I González MA Rodriguez- Velamazan JA 2009 Acta Materialia 57 4224.
[4] Vokoun D Hu C T 2002 J. Alloys Compd. 346 147.
[5] Tolea F Tolea M Sofronie M Valeanu M 2015 Solid State Comunications 213 37
[6] Oshima R 1981Scripta Metall 15 829.
[7] Albertini F Morellon L Algarabel P A ,M R Ibarra, LPareti, Z. Arnold, G. Calestani, 2001, J. Appl. Phys. 89 (10), 5614.
[8] Sofronie M Tolea F Kuncser V Valeanu M 2010 J. Appl Phys. 107 113905.
[9] Cui J Shield TW James R D 2004 Acta Mater. 52 35.

Graphene Post-Processing

B Tincu[1, 2, *], A Avram[1], V Tucureanu[1, 3], A Matei[1], C Marculescu[1], T Burinaru[1, 4], F Comanescu[1], M Popescu[1] and M Avram[1]

[1]National Institute for Research and Development in Microtehnologies (IMT-Bucharest), Bucharest, Romania

[2]University Politehnica of Bucharest, Faculty of Applied Chemistry and Materials Science, Bucharest, Romania

[3]Transilvania University of Brasov, Department of Materials Science, Brasov, Romania

[4]University Of Agronomic Sciences and Veterinary Medicine of Bucharest, Faculty of Veterinary Medicine, Anatomic Pathology Department, Bucharest, Romania

*bianca.tincu@imt.ro

Abstract. Graphene is a two-dimensional wonder material, used for electronic and medical applications. To date, various synthesis method of graphene layers is proposed, such as: SiC epitaxial growth, Chemical Vapor Deposition (CVD) and mechanical exfoliation. Graphene single layer can be rapidly produced by CVD. Although the transfer process on different substrates has been researched, a post-processing step after graphene growth for understanding influence of radical of carbon unreacted in graphene properties has not been studied. Deposition of graphene on metal transition substrate involves many reactions, and is not clear what happens with the unreacted methane. The scanning electron microscopy is the best method to visualize this unreacted species as a thin film which covers the graphene layer. The post-processing of graphene after CVD is a crucial step for the performance of graphene-based devices. The graphene film is characterized morphologically and structurally before and after the post-processing step, with the scope of removing the unreacted film and investigating the influence of this step on the graphene properties. To identify the specific vibration of graphene layer before and after post-processing step, Raman spectroscopy has been used. Here, we investigate the quality of CVD graphene before and after removing unreacted hydrocarbon, to better understand the importance of post-processing process for device applications, before the graphene transfer.

1. Introduction

Graphene, two-dimensional alternative of graphite, is one atom layer of carbon atoms [1–3] arranged in a hexagonal lattice. Graphene is the best conductor of electricity and a great material with remarkable properties (mechanical [4–6], electrical [7, 8]) and one of the most studied area in the recent years [9], [10]. Andre Geim and Konstantin Novoselov discovered graphene in 2004 and they won the Nobel Prize for Physics in 2010 [11].

The history of graphene dates back from 1840, since Schafhaeutl report an experiment about the oxidation and exfoliation of graphite with a variety of interpolating [12–15]. The next step, in 1859 was marked by graphene used in attempts to a British chemist, Brodie to characterize the molecular weight

IOP Conf. Series: Materials Science and Engineering **485** (2019) 012027 doi:10.1088/1757-899X/485/1/012027

of the graphite. In 1962, Boehm et al. has studied the characterization by transmission electron spectroscopy of some thin lamellas obtained by oxidation of graphite and reducing the graphene layers of graphite and for this proposed term of graphene, but in 1997 the term of graphene will be formalized by IUPAC. Graphene has been obtained by the method of chemical vapor deposition (CVD) [16–21] since 2006. Currently, graphene growth by CVD is the most studied method that can produce graphene in significant quantities, and on a high quality. The process is based on the carbon saturation of a transition material regarding the exposure to a carbon-hydrogen gas mixture at high temperature. The various hydrocarbons such as methane, ethylene, acetylene or benzene are decomposed on the surfaces of some transition metal substrates as: Co [23], Ni [24–26], Cu [27]. Copper is an excellent candidate for the preparation of large area of single layer graphene with a uniform thickness (95%) due to the solubility of carbon in copper [28]. Most of the studies directed the post-processing step for the graphene transfer on a different substrate. After a structural analysis of the samples obtained in our laboratory by Nanofab 1000, we observed a film formed on the graphene layer grown on Cu substrate, and as we want to remove it this study aims on the aspect of the post-processing step as cleaning samples after CVD growth of graphene. Graphene has particularly properties and the edges of the graphene are more reactive atoms that have a unique energy that is not present in the bulk. Similar to the carbon nanotubes, this can be either zigzag or armchair. The introduction of the defects is an important step for many applications. The structure of the surface on which graphene deposition occurs is also a field of study for understanding the connection between Cu substrate and CVD graphene growth [29]. The literature does not report a study on this issue and it is not clear what happens with the unreacted methane, after the graphene CVD process.

The introduction of this post-processing step brings many advantages in the next uses of this product. Through this study we have seen getting defects after post-processing, which leads to increased chemical reactivity. Most of the studies bring into discussion obtaining graphene free defects, useful aspect and well defined in electronic applications, but exactly these defects [30] give the remarkable properties of graphene, well illustrating in chemical applications [31]. These defects are the results of post-processing step due to rearrangements of the graphene structure and the possibility of forming at the edge to zigzag and armchair arrangement. But, it is important to note that this experiment is not focused on the analysis of defects, it is aimed at the removal of the unreacted hydrocarbon.

2. Experimental

2.1 Materials

For the experiments of the graphene growth at high temperature, as a catalyst was used the copper foil with a thickness of 35 μm (Graphene Platform Corporation, Japan: Cooper Foil Special for CVD Single Layer Graphene, specification of Copper Foil, Copper purity: over 99.95%, thickness: 0.035 mm). Methane (CH4-99.9%) and Hydrogen (H2-99.9%) were used as precursor gases for graphene growth, connected to the Nanofab system. Acetic acid, deionized water and isopropyl alcohol (Sigma Aldrich) were used for the pre-cleaning step of copper foil. Acetone and isopropyl alcohol were used for post-cleaning step for samples with the graphene single layer on copper foil.

2.2 Pre-processing step

For the processes of graphene growth, the copper foil was used as a substrate with a thickness of 35 μm (Graphene Platform Corporation, Japan), delivered in the form of sheets with the size of 210 mm x 300 mm. For the CVD experiments the copper sheets have been cut using a catheter for pieces of 20 mm x 20 mm. Before processing, each substrate has been cleaned up in acetic acid for 10 minutes, for pickling surface. This step was followed by a rise in deionized water for 10 minutes, then maintained in isopropyl alcohol until use. The substrate has dried with nitrogen before placing in equipment by flushing.

2.3 Graphene synthesis by thermal chemical vapor deposition

For the graphene growth has been used PlasmaPro100 equipment, design Nanofab 1000 (Oxford

Instruments, UK) dedicated for growth processes of carbon materials. In order to obtain graphene are used hydrocarbon-based reactants. In this experimental paper we used the methane. Due to the strong C-H bond of methane molecule (440 kJ / mol), its thermal decomposition occurs at temperatures from more than 1200 degrees Celsius. The graphene growth process by CVD method has been of one hour in presence of methane and hydrogen followed by cooling the reactor and then the sample is transferred in Load Lock. The graphene growth by CVD method includes five steps: vacuum, heating, thermal treatment, growth and cooling down.

2.4 Post-processing step

Due to the presence of pellicular film on graphene growth on copper, observed by scanning electron microscopy, we tried different cleaning processes, with various steps. In this experiment, the samples are exposed to a post-cleaning step: in acetone and isopropyl acid. For the removal of any residue it was used the heat treatment at 400°C degrees in an inert atmosphere.

The best results have been observed in samples where the post-processing involves bringing to the boiling point of acetone and maintaining the samples for 5 minutes, rinsed in isopropyl acid and dried on the plate. Samples have been exposed to the thermal analysis.

3. **Results**

Raman spectroscopy has been used to identify the specific vibrational modes of the graphene layer. We used a high-resolution spectrometer LabRam HR 800. The LabRAM HR 800 Raman spectrometer in a setup with the HeNe Laser at 633 nm were used for spectra acquisition. Scanning electron microscopy (SEM) was used to investigate layers of graphene from the early stages to continuous films. For the SEM investigation was used a Line Raith microscope.

3.1 Raman Spectroscopy

We have investigated CVD graphene on copper samples before and after post-processing by taking into account the main characteristic Raman bands of graphene which are [32–34]: D band- defects band, its intensity is proportional with the density of defects in the investigated area, its position being around 1332 cm^{-1},G band- the first optical phonon corresponding to graphene, its position being around 1582 cm^{-1}, 2D band- the overtone of D band. This band appears even in the absence of D band.

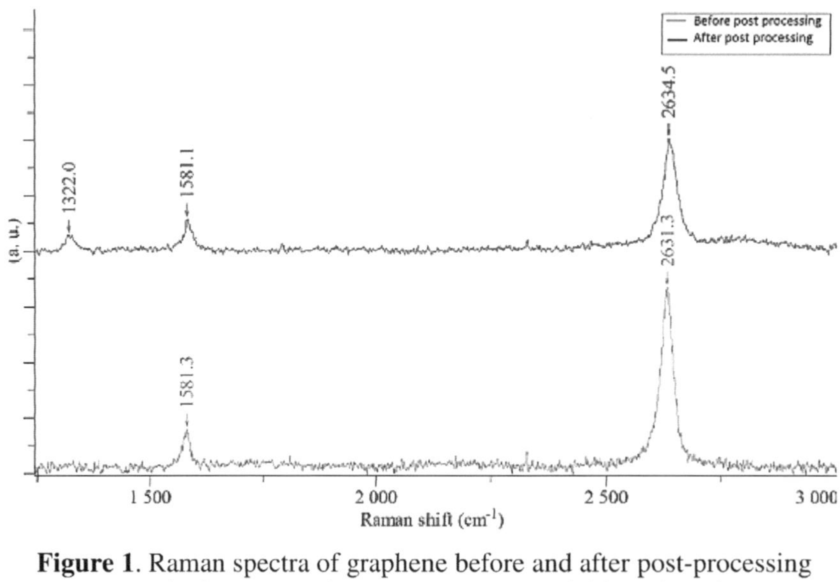

Figure 1. Raman spectra of graphene before and after post-processing

In Figure 1 are comparatively presented Raman spectra acquisitioned on CVD graphene grown on copper substrate before and after post-processing. Both spectra are revealing the characteristic Raman bands of single layer Graphene: D, G and 2D bands.

In Figure 2 are comparatively presented the defect and quality ratios extracted from CVD graphene Raman spectra before (figure 2a) and after (figure 2b) post-processing. After post-processing, the density of defects has increased which is confirmed by the defect's ratio. In the investigated areas the quality ratio (I2D/IG) has values larger than 2 being more uniform after post-processing.

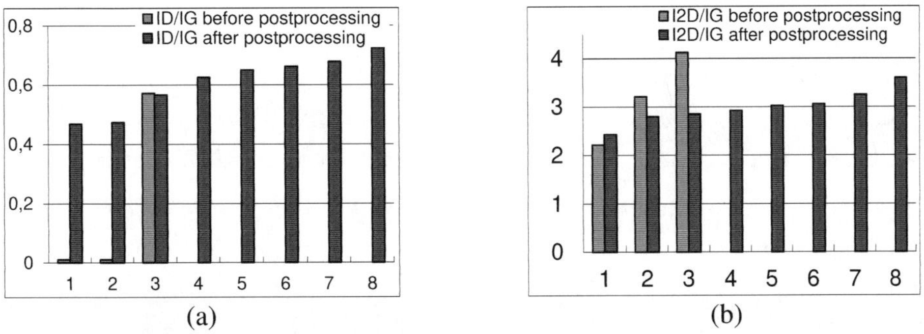

Figure 2. Defect (a) ID/IG) and quality ratio (b) I2D/IG) extracted from the Raman spectra of single layer CVD graphene samples before and after post-processing

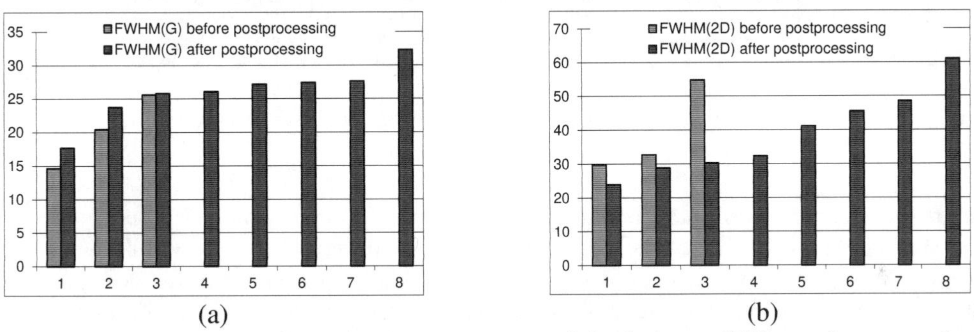

Figure 3. Parameters extracted from the Raman spectra of single layer CVD graphene samples before and after post-processing: FWHM corresponding to: a) G; b) 2D Raman bands

In Figure 3 are comparatively presented FWHM values corresponding to G and 2D bands before and after post-processing. G band FWHM values are higher than 17 cm^{-1} which indicates a large strain in graphene especially after post-processing. 2D band FWHM is higher than 25 cm^{-1} before and after post-processing which suggests a high degree of crystalline disorder. After post-processing graphene D and 2D bands position are blue shifted, D band position being in range 1313 – 1333 cm^{-1}, 2D band position being in range 2613 – 2648 cm^{-1}.

3.2 Surface morphology of graphene on cooper foil

The morphology of single layer graphene grown on Cu foil can be observed using SEM techniques, due to the different energy levels of electrons back-scattered from the materials, which results in images with different shades of grey, depending on the number of graphene layers present. SEM images show the graphene layer to be non-uniform, but graphene domains are clearly visible on copper substrate. The importance of the post-processing step is clearly observed during SEM investigations, as can be seen in Figure 4a and figure 4c where the surface morphology of the graphene is obscured by unreacted hydrocarbon by-products. This post-processing step, not only removes deposition by-products, but also introduces defects in the graphene layers, which are very important for the reactivity of graphene structure. During the graphene growth, CH_4 concentration is stable, thermally decomposed on the surface of the copper where the hydrogen sites are previously formed. Controlling the debit of precursor gases is an advantage for graphene CVD synthesis, but the reaction of methane and hydrogen is not controlled and secondary products can appear on the surface.

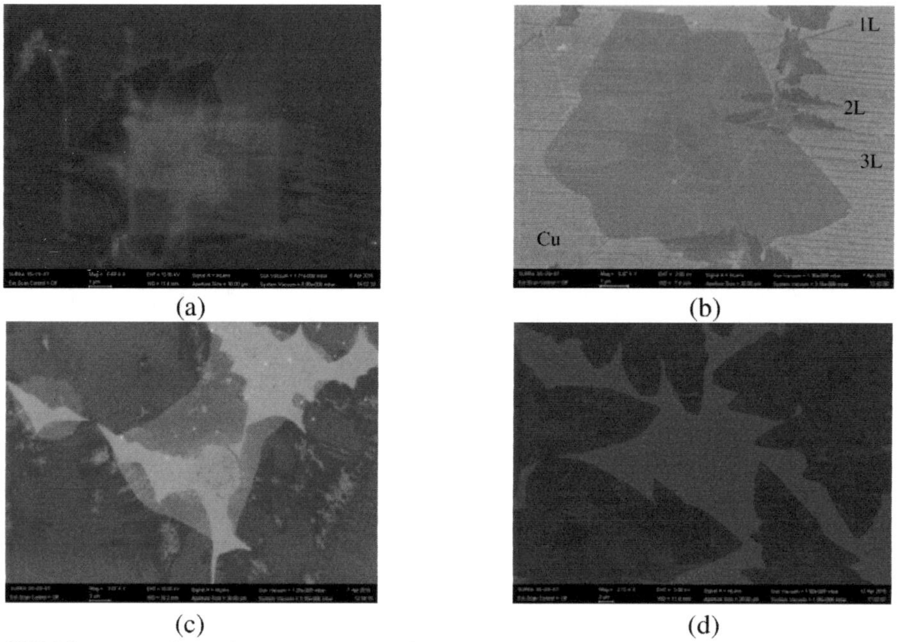

Figure 4. SEM images of graphene on copper foil: (a), (c) before post-processing, (b), (d) after post-processing

After the growth step, an additional thermal annealing at 400°C helps remove these unreacted groups and causes stress induces the defects in graphene structure. In Figure 4b and figure 4d, we can observe the graphene morphology after the post processing step, revealing the shape of the domains, as well as additional multi-layered features. In figure 4b multiple layers of graphene: single-, double- and tri-layer (1L, 2L and 3L) grown around the nucleation center can be observed.

4. Conclusions

We investigated the importance of post-processing step for CVD graphene layer with Raman and SEM methods. It has been found that the post-processing step adds to the graphene structure defects. The defects are very useful for chemical applications. The step of introducing defects in the graphene structure is very important when in the applications the structure is needed as such.

The post-processing step has been optimized as a result of the characterizations accomplished during the experiments of CVD graphene synthesis. Our results are beneficial for the uses of graphene in different applications, by the control of defects.

Acknowledgments

This work was supported by a grant of the Ministry of National Education and Scientific Research, STAR 639/2017 and PN-III-P1-1.2-PCCDI-2017-0214 (Project No. 3PCCDI/2018).

References

[1] Geim A K and Novoselov K S 2007 *Nature Materials* **6** 183-191

[2] Bolotin K I, Sikes K I, Jiang Z, Klima M, Fudenberg G, Hone J, Kim P, Stormer H L 2008 *Solid State Commun.* **146** 351–355

[3] Dean C R, Young A F, Meric I, Lee C, Wang L, Sorgenfrei S, Watanable K, Taniguchi T, Kim P, Shepard K L and Hone J 2010 *Nat. Nanotechnol.* **5** 722–726

[4] Cranford S W and Buehler M J 2011 *Carbon N. Y.* **49** 4111–4121

[5] Zhang Y Y and Gu Y T, 2013 *Comput. Mater. Sci.* **71** 197–200
[6] Georgantzinos S K, Giannopoulos G I and Anifantis N K 2016 *J. Appl. Phys.* **120** 014305
[7] Vicarelli L, Heerema S J, Dekker C and Zandbergen H W 2015 *ACS Nano* **9** 3428–3435
[8] Berger D and Ratsch C 2016 *Phys. Rev. B - Condens. Matter Mater. Phys.* **93** 1–9
[9] Zhu Yanwu, Murali Shanthi, Cai Weiwei, Li Xuesong, Suk Won Ji, Potts Jeffrey R and Ruoff Rodney S. 2010 *Adv. Mater.* **22** 3906–3924
[10] Syama S and Mohanan P V 2016 *Int. J. Biol. Macromol.* **86** 546–555
[11] Narula U and Tan C 2016 *Appl. Sci.* **6** 204
[12] Yan L, Zheng Y B, Zhao F, Li S, Gao X, Xu B, Weiss P S and Zhao Y 2012 *Chem Soc Rev* **41** 97–114
[13] Obeng Y and Srinivasan P 2011 *Electrochem. Soc. Interface* **20** 47–52
[14] Dreyer D R, Ruoff R S and Bielawski C W 2010 *Angew. Chemie - Int. Ed.* **49** 9336–9345
[15] Heller E J, Yang Y, Kocia L, Chen W, Fang S, Borunda M, Kaxiras E 2016 *ACS Nano* **10** 2803 2818
[16] Brian Derby 2010 *Engineering Materials and Processes* New York
[17] Li X. *et al.* 2009 *Science* **324** 1312
[18] Reina A, Jia X, Ho J, Nezich D, Son H, Bulovic V, Dresselhaus M S and Kong J 2009 *Nano Letters* **9** 30-35
[19] Wan X, Chen K and Xu J 2014 *Small* **10** 4443–4454
[20] Qi J L, Zheng W T, Zheng X H, Wang X, and Tian H W 2011 *Appl. Surf. Sci.* **257**, 6531– 6534
[21] Mahmood Aliofkhazraei, Nasar Ali, William I. Milne, Cengiz S. Ozkan, Stanislaw Mitura, Juana L. Gervasoni, 2016 *Graphene Science Handbook* New York
[22] Shapira P, Gok A and Salehi F 2016 *J. Nanoparticle Res.* **18** 269
[23] Ago H, Ito Y, Mizuta N, Yoshida K, Hu K, Orofeo C M, Tsuji M, Ikeda K and Mizuno S 2010 *ACS Nano* **4** 7407–7414
[24] Liu W, Chung C H, Miao C Q, Wang Y J, Li B Y, Ruan L Y, Patel K, Park Y J, Woo J, Xie Y H 2010 Thin solid films **518** S128–S132
[25] Reina A *et al.* 2009 *Nano Lett.* **9** 30–35
[26] Zhang Y, Zhang Y, Gomez L, Ishikawa F N, Madaria A, Ryu K, Wang C, Badmaev A and Zhou C 2010 *J. Phys. Chem. Lett* **1** 3101–3107
[27] Mattevi C, Hokwon K and Chhowalla M 2011 *J. Mater. Chem.* **21** 3324–3334
[28] López G A and Mittemeijer E J 2004 *Scr. Mater.* **51** 1–5
[29] Tian J, Hu B, Wei Z, J Y, Luo Z, Xia M, Pan Q, Liu Y 2014 *Appl. Surf. Sci.* **300** 73–79
[30] Eftekhari A and Garcia H 2017 *Mater. Today Chem.* **4** 1–16
[31] Ţucureanu V, Matei A, and Avram A M 2016 *Crit. Rev. Anal. Chem.* **46** 502–520
[32] Malard L M, Pimenta M A, Dresselhaus G and Dresselhaus M S 2009 *Phys. Rep.* **473** 51–87
[33] Costa S D, Righi A, Fantini C, Hao Y, Magnuson C, Colombo L, Ruoff R S, Pimenta M A 2012 *Solid State Commun.* **152** 1317–1320
[34] Ferrari A C 2007 *Solid State Commun.* **143** 47–57

IOP Conf. Series: Materials Science and Engineering **485** (2019) 012028 doi:10.1088/1757-899X/485/1/012028

Effect of Cr, C, B and Mo substitutions on the structure and magnetic properties of Zr-Co Rare-Earth-free magnetic alloy

F. Tolea, P. Palade, M. Sofronie, B. Popescu and V. Kuncser

National Institute of Materials Physics, POB MG-7, 77125 Bucharest-Magurele, Romania

felicia.tolea@infim.ro

Abstract. Acquiring new permanent magnets without Rare-Earths is a new challenge in the field of magnetic materials and Co-rich Co-Zr alloys seem to be a solution to be pursued. The microalloying effects of different elements on the structure and stabilization of hard magnetic Zr_2Co_{11} phase of Zr-Co alloy have been analysed in the past and it was shown that the Zr_2Co_{11} non-cubic high anisotropy structure can be stabilized by the partial substitution of Co in the ZrCo alloy with elements with high atomic radius. In the present work, we made substitutions of up to 5% alloying elements such as Mo, C, B, Cr in order to highlight how the element with the small / large atomic radius stabilize the nano-hard magnetic phase. The structural and magnetic properties of the new $Zr_{18}Co_{77}Cr_3X_2$ (X = C, B, Mo) prepared bulk alloys have been analysed and related to those of the original $Zr_{13}Co_{87}$.

1. Introduction

The exponential development of science and technology, and the need for magnetic energy, make strong magnets indispensable. At the base of many magnetic materials are rare earths (RE). However, the current geopolitical and geostrategic context, as well as the RE high costs, make it imperative to find magnetic materials without rare earths for strong permanent magnets. Since, fundamentally, this is not impossible researchers from all over the world are looking for alternatives.

To obtain a permanent magnet with high energy product $(BH)_{max}$ one has to start from a material with a high Curie temperature, a large saturation magnetization and an appreciable magnetic anisotropy. The first two requirements are fulfilled by the 3d- elements that are additionally cheap and affordable. The drawback of the 3d elements is their low magnetic anisotropy due to the weak spin-orbit coupling. However, there are ways to improve magnetocrystalline anisotropy like doping with elements to change the structure from cubic to tetragonal (e.g. Fe-N), by combining 3d elements with 5d and searching for new compounds or re-examination of less studied alloys [1]. Very good results have been obtained, for instance, in FePt alloys showing tetragonal $L1_0$-ordered structure and a $(BH)_{max}$=54 MGOe [2, 3], but the high cost of Pt limits its use for commercial applications. Mn-based alloys, like MnBi, MnAl, MnGa, are widely investigated due to their high constant of uniaxial anisotropy (K1) and Curie temperature (Tc), but up to now the published results are modest, the highest value for $(BH)_{max}$ being only 7.7 MGOe [4 - 9].

Recently, Co-rich transition metal $Co_{100-x}TM_x$ alloys are reported as good candidates for the replacement of rare earth based magnetic materials. One of them is Co-Zr alloy, in which phase diagram is the metastable $Co_{11}Zr_2$ phase formed by a peritectic reaction $L + Zr_6Co_{23} \rightarrow Zr_2Co_{11}$ at high temperatures [10]. Recent experimental results have proved that the rhombohedral Zr_2Co_{11} has hard magnetic properties [11]. The Zr_2Co_{11} is polymorphic, having the pseudohexagonal, orthorombic or rhombohedral crystalline structures, with closely competitive energies [12]. Applying a classical melting procedure of the component elements, the alloy is obtained in the cubic structure $Co_{23}Zr_6$, with soft magnetic properties. But, it was shown that the Zr_2Co_{11} non-cubic high anisotropy structure can be stabilized in ZrCo alloy by the partial substitution of Co with elements with high atomic radius [13].

Therefore we made substitutions of up to 5% of alloying elements like Mo, C, B, Cr in order to highlight the role of the atomic radius of the substitution elements in the nucleation and stabilization of hard magnetic phase. The structural and magnetic properties of new $Zr_{18}Co_{77}Cr_3X_2$ (X = C, B, Mo) prepared as bulk alloys have been analysed and related to those of $Zr_{13}Co_{87}$. Starting from the hard Zr_2Co_{11} phase with uniaxial anisotropy and high Curie temperature (500°C) but with low saturation magnetization at room temperature, the aim of this study is to achieve a potentially performing material for permanent magnets.

2. Experimental details

Bulk alloys with nominal compositions $Zr_{13}Co_{87}$, $Zr_{18}Co_{77}Cr_3C_2$, $Zr_{18}Co_{77}Cr_3B_2$, $Zr_{18}Co_{77}Cr_3Mo_2$, denoted in the following ZrCo, CrC, CrB and respectively CrMo, were prepared by arc melting in argon atmosphere from high purity (99,99%) constituent elements. In order to obtain a better homogeneity, the samples were re-melted four times.

The Diffractometer Bruker D8 Advance with Cu Kα radiation was used for the crystalline structure investigation. Magnetic measurements were made with a Quantum Design ; - Physical Properties Measurement System (PPMS) working in vibrating sample magnetometry (VSM) mode.

3. Results and discussions

3.1. Structural characterization

The type and size of the unit cell of the polymorph Zr_2Co_{11} phase are still under debate. In bulk samples multiple phases and small grain size, that enlarge the diffraction lines, make difficult to determine the atomic crystalline structures of this compound using standard X-ray techniques. However, the development and optimization of this material for practical applications is conditioned by the clarification of its crystalline structure [12]. The XRD patterns recorded for ZrCo, CrC and CrB reveal the coexistence of cubic Zr_6Co_{23}, hexagonal Co – rich soft magnetic phases and Zr_2Co_{11} in orthorhombic symetry. The crystallite size, calculated using the Scherrer formula, varies between 7 nm and 20 nm for ZrCo, CrC, CrB alloys. Diffraction maxima belonging to the rhombohedral phase appear in the diffractogram specific to the CrMo sample. Rhombohedral structure of hard magnetic phase Zr_2Co_{11} is often obtained for stoichiometry $ZrCo_{5.1}$ [5].

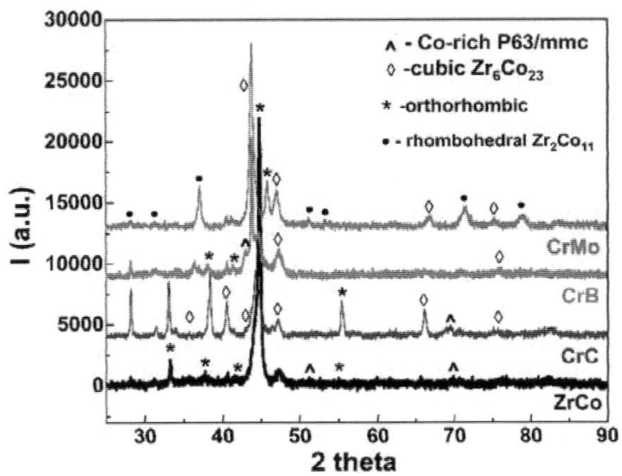

Figure 1. XRD pattern recorded at room temperature for all investigated samples.

Based on Rietveld refinements using MAUD code [14] a small amount of $ZrCo_{5.1}$ hard magnetic phase with rhombohedral structure (about 5%) was detected in CrMo sample, besides the main soft magnetic phase $Co_{2.23}Zr_{0.77}$ with cubic structure (Fd-3m group, fitted lattice constant a=0.6857(5) nm and crystal size of 23(2) nm) and the minority Co-rich phase with hexagonal structure (P63/mmc group) with smaller crystal size of 10(1) nm. These values of the crystallite size agree well with that obtained by using Scherrer method.

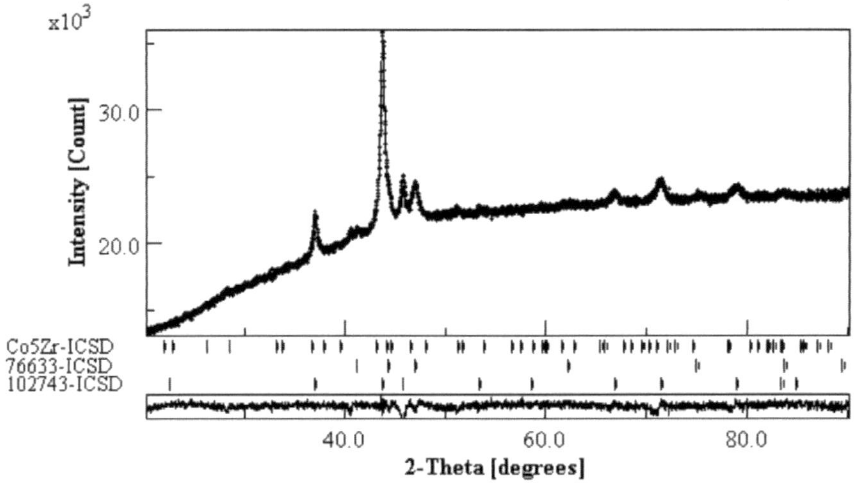

Figure 2. Rietveld analysis of the XRD data for the sample denoted CrMo

The presence in the early stage of the rhombohedral phase is the evidence that CrMo alloy is a good precursor for a hard magnetic material without rare earths.

3.2. Magnetic characterization

In Figure 3.a are presented the temperature dependences of magnetization measured in 2T for the under discussion alloys. It is to note the very slow magnetization increase with the temperature reduction especially for the reference alloy, suggesting a Curie temperature much higher than room temperature.

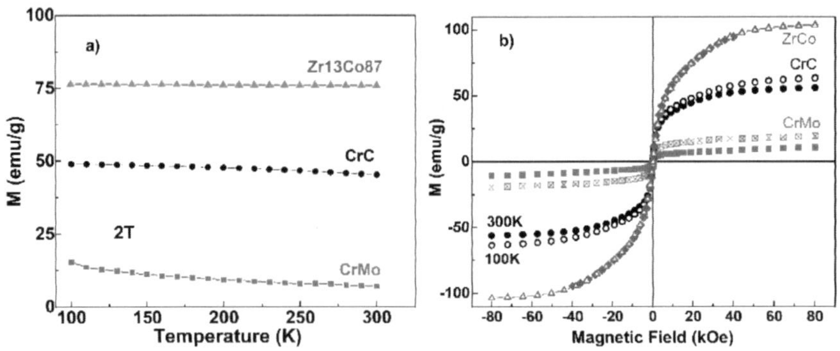

Figure 3. (a) The overall behavior of the Field Cooling (FC) curves measured in 2T applied magnetic field; (b) Hysteresis loops collected at 100K (empty symbols) and 300K (full symbols) in the field range from –80 kOe to 80 kOe for ZrCo, CrC and CrMo samples.

Hysteresis loops collected at 100K and 300K in the field range from -80 to 80 kOe on ZrCo, CrC and CrMo samples depicted in Fig.3b, show the high saturation field, meanning saturation of magnetization difficult to reach. This trend suggests a randomly oriented magnetic anisotropy mechanism (RMA). The nanometric dimensions of crystallites and the presence of a disordered spin structure with large local anisotropy generate such random magnetic anisotropy behavior. Disordered systems with RMA show distribution of the anisotropy energies, which involve magnetic moments with different easy magnetization axes in various sites, due to local crystal fields. In addition, it is worth noting the decrease of magnetization in the CrC substitution alloy, which is even more pronounced in the CrMo substitution alloy. This behaviour suggests a non parallel coupling between the cobalt magnetic moments with those of the substitution atoms (Cr and Mo). Theoretical and experimental studies have revealed that RMA suppresses the ferromagnetic coupling of magnetic moments resulting a speromagnetic structure of spins. The large anisotropy randomly oriented owning to the random atomic arrangement strongly influences magnetic properties. If we analyze the hysteresis curves, in small fields is present the specific behavior of magnetic domains. At higher fields, the speromagnetic behavior is obvious: the magnetic moments disposed within a cone begin to orient in the direction of the applied field. It is worth mentioning here, that in the RMA, the speromagnetic behavior is defined by a 4π solid angle, meaning randomly oriented spins. Besides this local behavior, we also have an exchange interaction that is different from pure Co, because we do not only have Co atoms. Therefore, the high magnetization in the ZrCo sample is due to the strong exchange interaction through the direct spatial overlap of the 3d shells in the alloy, but is less than pure Co, due to Zr-induced 3d-4d hybridization. The hybridization differs depending on the substitution element, and in the other samples is added the hybridization due to substitutions of Cr, C, B and Mo.

From hysteresis loops it is evident the decrease of magnetization at saturation with the increase of atomic radius of substitution element. On the contrary, the coercive field increases with the increase of the radius of the substitution element. Room temperature hysteresis loops evidenced lower magnetization than those recorded at 100K. The magnetization diminishing can be justified by increased magnetic fluctuation when measuring temperature approaches Currie point. Additionally, low values of magnetization at saturation obtained for sample with substitutions can be the effect of the presence of oxides. According to RMA model, the bulk alloys exhibit a spin-glass-like behaviour, when the dimensions of the crystallites are located in the nanoscale region [13, 15, 16]. The nanometric crystallite size is also the justification for the small coercive fields in some alloys. According to the J. Löffler and W. Wagner's [17] regarding the relation between the crystallite size and magnetic properties, if the crystallite size is comparable to the effective domain-wall width of nanostructured bulk materials, the magnetization of neighboring grains may be aligned. The magnetocrystalline anisotropy constant is then averaged over several crystallite inducing a decrease of the the coercive field with decreasing the crystallite size. As highlighted in insets of Fig. 4, the coercive field values are small, depending on the substitution elements, ranging from 20 Oe for ZrCo to 200 Oe for the CrMo sample. The coercive field increased by an order of magnitude compared to the others in the CrMo sample. The X-ray diffraction on the same sample reveals the appearance of the rhombohedral phase. So, we have proved that substitution with Mo, the element with the largest atomic radius of the elements chosen by us, is responsible for the occurrence of the rhombohedral phase, which give hard magnetic magnetic properties.

This behaviour, together with the weak variation of magnetization vs temperature and the high value of the magnetic field required to achieve saturation, suggests that the RMA model explains the magnetic behavior of these alloys.

4. Conclusion

The presence of Random Magnetic Anisotropy and the behaviour of magnetization in high fields (saturation is difficult to reach) show that the material is a good precursor in order to obtain Permanent Magnets and can be used as a substitute for magnetic materials based on Rare Earths, after proper processing and thermal treatments.

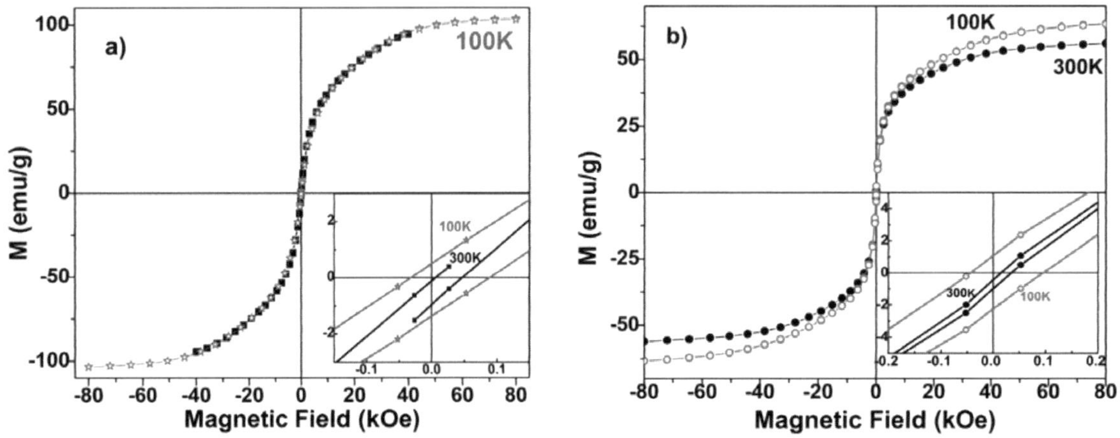

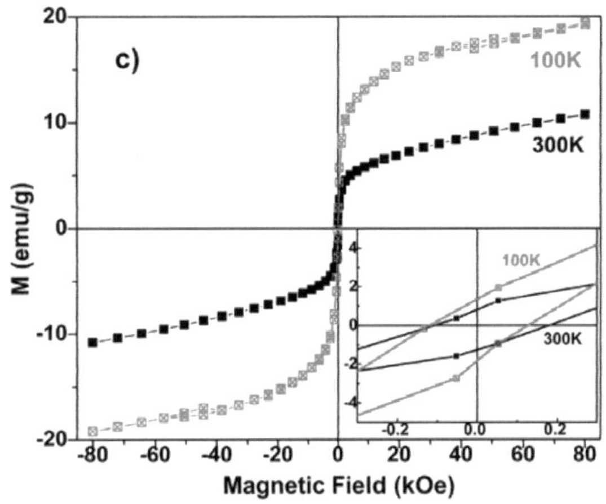

Figure 4. (a) Hysteresis loops for ZrCo; the same for (b) CrC and (c) CrMo samples. The detail about coercitivity are shown in inset bottom right. In CrMo sample is obvious that the saturation of magnetizations is difficult to achieve.

Acknowledgments
This work was supported by PN III nr. 47 PCCDI/2018, Project 3.

References

[1] Kuz'min M. D, Skokov K. P, Jian H, Radulov I and Gutfleisch O 2014 *J. Phys.: Condens. Matter* **26** 064205
[2] Crisan O, Crisan A. D, Mercioniu I, Nicula R, Vasiliu F 2016 *J. of Mag. and Mag. Mat.*, **401** 711
[3] Crisan O, Crisan A.D, Randrianantoandro N, Nicula R, Burkel E, 2007 *J. All. Comp.* **440** L3-L7
[4] Coey J. M. D 2012 *Scr. Mater.* **67** 524
[5] Balamurugan B, Das B, Zhang W. Y, Skomski R and Sellmyer D. J 2014 *J. Phys.: Condens. Matter* **26** 064204
[6] Balke B, Fecher G H, Winterlik J and Felser C 2007 *Appl.Phys. Lett.* **90** 152504
[7] Park J H, Hong Y K, Bae S, Lee J J, Jalli J, Abo G S, Neveu N,Kim S G, Choi C J and Lee J G 2010 *J. Appl. Phys.* **107** 09A731
[8] Yang J B, Kamaraju K, Yelon W B, James W J, Cai Q and Bollero A 2001 *Appl. Phys. Lett.* **79** 1846
[9] Rao N V R, Gabay A M and Hadjipanayis G C 2013 *J. Phys.D: Appl. Phys.* **46** 062001
[10] Palit M, Chelvane Arout J, Basumatary H, Babu Aravindha D and Kamat S.V 2015 *J. Alloys and Comp.* **644** 7-12

[11] Zhang W.Y, Li X. Z, Valloppilly S, Skomski R, Shield J.E and Sellmyer D. J 2013 *J. Phys. D* **46** 135004
[12] Zhao X, Nguyen M. C, Zhang W. Y, Wang C. Z, Kramer M. J, Sellmyer D. J, Ke L. Q, Antropo, A. P and Ho K. M 2014 *Phys. Rev. Lett.* **112** 045502
[13] Xu S, Xu D, Hou Z, Zhang J, Su F, Sun L and Wang W 2012 *Cond. Matt. Phys.* **2** 197
[14] Lutterotti L 2010 *Nucl. Instrum. Meth. B* **268** 334
[15] Luo W, Nagel S. R, Rosenbaum T.F and Rosensweig R. E 1991 *Phys. Rew. Lett.* **19** 2721
[16] Hattori Y, Fukamichi K, Suzuki K, Aruga-Katori H and Goto T 1995 *J. Phys.: Condens. Matter* **7** 4193
[17] Löffler J. F and Wagner W 1998 *Phys. Rew. B* **57** 2915

Institute of Physics
Dirac House, Temple Back
Bristol BS1 6BE UK

ISSN: 1757-8981
ISBN 978-1-5108-8482-3

9 781510 884823

CURRAN ASSOCIATES INC.
proceedings.com